MATHEMATICS RESEARCH DEVELOPMENTS

KALMAN FILTERING

MATHEMATICS RESEARCH DEVELOPMENTS

Additional books in this series can be found on Nova's website under the Series tab.

Additional E-books in this series can be found on Nova's website under the E-book tab.

ENGINEERING TOOLS, TECHNIQUES AND TABLES

Additional books in this series can be found on Nova's website under the Series tab.

Additional E-books in this series can be found on Nova's website under the E-book tab.

MATHEMATICS RESEARCH DEVELOPMENTS

KALMAN FILTERING

JOAQUÍN M. GOMEZ
EDITOR

Nova Science Publishers, Inc.
New York

Copyright © 2011 by Nova Science Publishers, Inc.

All rights reserved. No part of this book may be reproduced, stored in a retrieval system or transmitted in any form or by any means: electronic, electrostatic, magnetic, tape, mechanical photocopying, recording or otherwise without the written permission of the Publisher.

For permission to use material from this book please contact us:
Telephone 631-231-7269; Fax 631-231-8175
Web Site: http://www.novapublishers.com

NOTICE TO THE READER

The Publisher has taken reasonable care in the preparation of this book, but makes no expressed or implied warranty of any kind and assumes no responsibility for any errors or omissions. No liability is assumed for incidental or consequential damages in connection with or arising out of information contained in this book. The Publisher shall not be liable for any special, consequential, or exemplary damages resulting, in whole or in part, from the readers' use of, or reliance upon, this material. Any parts of this book based on government reports are so indicated and copyright is claimed for those parts to the extent applicable to compilations of such works.

Independent verification should be sought for any data, advice or recommendations contained in this book. In addition, no responsibility is assumed by the publisher for any injury and/or damage to persons or property arising from any methods, products, instructions, ideas or otherwise contained in this publication.

This publication is designed to provide accurate and authoritative information with regard to the subject matter covered herein. It is sold with the clear understanding that the Publisher is not engaged in rendering legal or any other professional services. If legal or any other expert assistance is required, the services of a competent person should be sought. FROM A DECLARATION OF PARTICIPANTS JOINTLY ADOPTED BY A COMMITTEE OF THE AMERICAN BAR ASSOCIATION AND A COMMITTEE OF PUBLISHERS.

Additional color graphics may be available in the e-book version of this book.

LIBRARY OF CONGRESS CATALOGING-IN-PUBLICATION DATA

Kalman filtering / [edited by] Joaqumn M. Gomez.
 p. cm.
 Includes index.
 ISBN 978-1-61761-462-0 (hardcover)
 1. Kalman filtering. I. Gomez, Joaqumn M., 1960-
 QA402.3.K276 2010
 629.8'312--dc22
 2010031195

Published by Nova Science Publishers, Inc. † New York

CONTENTS

Preface vii

Chapter 1 Optimization of Kalman Filtering Performance in Received Signal Strength Based Mobile Positioning 1
J. G. Markoulidakis

Chapter 2 Application of Kalman Filtering in Power Systems: Harmonic Distortion and Voltage Events 43
Julio Barros, Enrique Pérez, Ramón I. Diego and Matilde de Apráiz

Chapter 3 Statistical State-Space Modeling via Kalman Filtration 77
Marek Brabec

Chapter 4 Forecasting the Weekly US Time-Varying Beta: Comparison between Garch Models and Kalman Filter Method 111
Taufiq Choudhry and Hao Wu

Chapter 5 Ensemble Forecasting through Evolutionary Computing and Data Assimilation: Application to Environmental Sciences 135
M. Kashif Gill, Mark Wigmosta, Andre Coleman and Lance Vail

Chapter 6 Attitude Determination Using Kalman Filtering for Low Earth Orbit Microsatellites 161
Si Mohammed and Mohammed Arezki

Chapter 7 Design of Extended Recursive Wiener Fixed-Point Smoother and Filter in Continuous-Time Stochastic Systems 197
Seiichi Nakamori

Chapter 8 Kalman Filtering Approach to Blind Separation of Independent Source Components 225
Andreas Galka and Tohru Ozaki

Chapter 9	Using a Restricted Kalman Filtering Approach for the Estimation of a Dynamic Exchange-Rate Pass-Through *Rafael Martins de Souza, Luiz Felipe Pires Maciel and Adrien Pizzinga*	255
Chapter 10	Quantized Kalman Filtering of Linear Stochastic Systems *Keyou You and Lihua Xie*	269
Chapter 11	Kalman Filter to Estimate Dynamic and Important Patterns of Interaction between Multiple Variables *Harya Widiputra, Russel Pears and Nikola Kasabov*	289
Chapter 12	Kalman Filters Family in Geoscience and Beyond *Olivier Pannekoucke and Christophe Baehr*	321
Index		377

PREFACE

The purpose of the Kalman filter is to use measurements that are observed over time that contain noise (random variations) and other inaccuracies, and produce values that tend to be closer to the true values of the measurements and their associated calculated values. The Kalman filter has many applications in technology, and is an essential part of the development of space and military technology. This book presents topical research data in the study of Kalman filtering, including Kalman filtering in the detection and analysis of voltage dips, short interruptions and overvoltages in voltage supply; statistical state-space modeling using Kalman filtration; and attitude estimators based on Kalman filtering for application on low Earth orbit microsatellites.

Chapter 1 - The current paper considers the problem of Mobile Terminal (MT) positioning based on a time series of Received Signal Strength (RSS) measurements and provides statistical estimators of the MT- Base Station (BS) distance which substantially improve the MT positioning accuracy. The set of MT-BS distance estimators defined for optimizing RSS based positioning developed in is used as a basis. The problem that is analyzed refers to the adaptation of these estimators to the presence of Kalman filtering in the MT position calculation process. Three different Kalman filtering options can be applied at different stages of the MT position calculation process: (a) RSS Kalman filtering, (b) MT-BS distance Kalman filtering and (c) MT position coordinates Kalman filtering. To identify the optimal MT-BS distance estimators at the presence of Kalman filtering a method exploiting the characteristics of the steady state Kalman filter performance is developed. This novel method allows for the definition of refined MT-BS distance estimators matching the nature of the Kalman filter process. The results indicate that the resulting estimators provide good efficiency for all Kalman Filter options.

Chapter 2 - Kalman filtering is a digital signal processing tool that has been extensively used in many electric power system applications. Voltage and current phasors, power system frequency, voltage flicker, high-impedance faults, harmonic distortion, voltage dips, voltage unbalance, high-frequency transients and other power system magnitudes can be successfully computed using Kalman filters. This chapter provides a general overview of the application of Kalman filters in electric power systems, focusing on their use for the analysis of two of the most important electric power quality disturbances in modern day power systems: harmonic distortion and voltage events. Section 2 gives a short introduction to Kalman filtering and a review of the main applications of Kalman filtering in electric power systems. Section 3 describes the practical application of Kalman filtering in the estimation of magnitude and phase angle of harmonics and interharmonics in voltage and current waveforms and section 4 explains how Kalman filters can be used for detection and analysis of voltage dips, short interruptions and overvoltages in voltage supply.

Chapter 3 - In this paper, the authors will start with a brief review of the theory underlying the Kalman filter (KF) applications in statistical modeling based on the state-space approach. In particular, they will stress the prediction error decomposition (achievable through the KF application to the analyzed time series) as a highly effective way of computing the likelihood function, useful when maximum likelihood estimate (MLE) of certain structural parameters is attempted. One-step prediction errors (evaluated at the MLE of structural parameters) are useful for other purposes as well, including the model diagnostics. Similarly, KF (evaluated at the MLE of structural parameters) is used to estimate state variables.

Next, the authors will illustrate how the state-space modeling and KF can be useful for solving practical problems from several interesting real-life applications.

Firstly, the state-space approach and subsequent (extended) KF estimation will be shown as a valuable tool for estimation of time-varying parameters (radon entry rate and air exchange rate) describing radon concentrations in houses. The model here will be built on two underlying differential equations summarizing the radon and tracer dynamics. As such, the methodology can be viewed as a flexible way to approach functional data analysis.

Secondly, the authors will show how the Kalman filtration can be useful for estimation of underlying (de-noised) growth curve of children. As demonstrated in Brabec 2004, KF is useful here not only for direct filtering of an individual growth curve, but also for effective information retrieval from relatively complex (semi-longitudinal) data containing irregular measurements for many individual children.

In addition to the previous examples oriented to univariate time series modeling (but with multivariate state-space), the authors will consider also multivariate approach useful for individualized natural gas consumption modeling. There, it is useful to account for gradual evolution of individual long-term consumption averages, or to allow for (individual scaling factor)*(shape of consumption dynamics in daily resolution) interaction which is not present in the current generation of gas consumption models.

Chapter 4 - This paper investigates the forecasting ability of four different GARCH models and the Kalman filter method. The four GARCH models applied are the bivariate GARCH, BEKK GARCH, GARCH-GJR and the GARCH-X model. The paper also compares the forecasting ability of a non-GARCH model in the Kalman method. Forecast errors based on twenty US companies' weekly stock return forecasts (based on estimated time-varying beta) are employed to evaluate the out-of-sample forecasting ability of both the GARCH models and the Kalman method. The results when measuring forecast errors overwhelmingly support the Kalman filter approach. Among the GARCH models, the BEKK model appears to provide somewhat more accurate forecasts than the other bivariate GARCH models. However, the predominance of BEKK over the other GARCH models is not significant. Jel Classification: G1, G15

Chapter 5 - A distributed modeling system for short-term to seasonal streamflow forecasts with the ability to utilize daily remotely-sensed snow cover products and real-time streamflow and meteorology measurements is presented herein. The modeling framework employs the state-of-the-art data assimilation and evolutionary computing strategies to accurately forecast environmental variables i.e., streamflow. Spatial variability in watershed characteristics and meteorology is represented using a raster-based computational grid. Snow accumulation and melt, simplified soil water movement, and evapotranspiration are simulated in each computational unit. The model is run at a daily time-step with surface runoff and subsurface flow aggregated at the watershed scale. The model is parameterized using a multi-objective evolutionary computing scheme using Swarm Intelligence. This approach allows the model to be updated with spatial snow water equivalent from National Operational Hydrologic Remote Sensing Center's (NOHRSC) Snow Data Assimilation (SNODAS) and observed streamflow using an ensemble Kalman-based data assimilation strategy that accounts for uncertainty in weather forecasts, model parameters, and observations used for updating. The daily model inflow forecasts for the Dworshak Reservoir in north-central Idaho are compared to observations. The April-July volumetric forecasts issued by the U.S. Army Corps of Engineers (USACE) for Water Years 2000 – 2007 are also compared with model forecasts. October 1 and March 1 volumetric forecasts are comparable to those issued by the USACE's regression based method. An improvement in March 1 forecasts is shown by pruning the initial ensemble set based on their similarity with the observed meteorology. The short-term (one-, three-, and seven-day) forecasts using Kalman-assimilation of streamflow show excellent agreement with observations. The scheme shows great potential for the use of data assimilation in modeling streamflow and other environmental variables.

Chapter 6 - As part of the attitude determination and control system on board microsatellites, the kalman filter is implemented to generate an estimate attitude upon sensor observation.

This chapter presents an overview and analysis of some attitude estimators based on Kalman filtering for application on low Earth orbit microsatellites. Various steps of the design and in orbit implementation of the estimators are described. These estimators are used during nominal attitude of the microsatellite. One type of attitude estimator is based on small Euler angles. For this mode two estimator versions have been used, the full estimator and its reduced version. This estimator is useful for on board computation where energy resources are very limited. Another type of attitude estimator is based on the quaternion parameter. The advantage of the quaternion parameter is the none singularity. For this mode two estimator versions have been used, the six state and the seven state estimators.

Numerical analysis and in orbit test results clearly indicate that the reduced version is better than the full version for specifically attitude determination and control system configuration of a microsatellite. Also, the results prove that the six state estimator is better than the seven state estimator in terms of computational demand.

Chapter 7 - This paper, at first, designs the extended recursive Wiener fixed-point smoother and filter in continuous-time wide-sense stationary stochastic systems. It is assumed that the signal is observed with the nonlinear mechanism of the signal and with the additional white observation noise. The estimators use the information of the system matrix F for the state vector $x(t)$, the observation vector C for the state vector, the variance $K(t,t) = K(0)$ of the state vector, the nonlinear observation function and the variance R of the white observation noise. F, C and $K(0)$ are usually obtained from the auto-covariance function of the signal.

Secondly, by using the covariance information of the signal and the observation noises, the extended fixed-point smoother and filter for white plus colored observation noise are proposed.

Numerical simulation examples are shown to demonstrate the validity of the proposed estimation algorithms.

Chapter 8 - The problem of extracting a set of independent components from given multivariate time series data under the assumption of linear instantaneous mixing can be addressed within a Kalman filtering framework. For this purpose, the authors introduce a new class of state space models, the Independent Components State Space Model (IC-SSM). The resulting algorithm has several attractive features: It takes temporal correlations within the data into account; it allows for the presence of observation noise; it can deal with both gaussian and non-gaussian source distributions; it provides a representation for the main frequencies present in the data; and it succeeds in distinguishing between dependencies which are introduced by the mixing step, and dependencies which represent coincidental dependencies resulting from finite time series length. IC-SSM is compared with five well-known algorithms for Independent Component Analysis (ICA). Through simulations the authors show that these algorithms in most cases produce estimates of the source components for which the residual mutual information is too small, as compared to the correct value. Two sets of simulations are carried out, linear stochastic low-order processes driven by gaussian and Laplacian noise, and nonlinear deterministic processes displaying sub-gaussian distributions. Finally, an example of the analysis of a real biomedical time series is presented.

Chapter 9 - In this paper the authors propose linear state space models to estimate the time-varying pass-through of Brazilian price indexes against the US Dollar/Real exchange rate from 1999 to 2007. The methodological framework encompasses the restricted Kalman Filtering under a reduced modeling approach, under which it becomes possible to check whether some economic hypotheses are supported by the data. The paper has three main targets. The first is to decide whether models of null (or of full) pass-through are acceptable to the price indexes investigated here. The second is to carryout likelihood ratio tests for the significance of some economic exogenous variables, which are termed determinants in this paper and are theoretically associated with the pass-through. The third is to analyze the behavior of the Kalman Filter estimates of the pass-through from the best models.

Chapter 10 - In recent years, networked systems such as wireless sensor networks (WSNs) have gained popularity in the research community due to their broad potentialmilitary and civilian applications. WSNs are generally composed of a large number of low-quality sensors equipped with limited communication capabilities and limited energy. However, a collective of these nodes can form a powerful network for information gathering and processing. Due to limited communication capacity and also for the sake of energy saving, the number of bits of information to be transmitted between nodes is quite restrictive. Therefore, signals such as sensor measurements are to be severely quantized before their transmissions. This introduces interesting and challenging problems such as what information is to be transmitted and how many bits are needed to represent the information in order to achieve a given performance.

Chapter 11 - Estimating state of nature processes has been a very challenging task for human being. These processes include not only those which occur in the biological field, i.e. interaction between genes; or those which emerge in the ecological field, i.e. how humidity is related to the level of sun ray, wind speed and rain; but also processes which exist in the global financial area, i.e. how a stock market in a specific country influences or is being influenced by the other stock markets in different countries directly or indirectly, and how the macro economic factors in a single country are related to each other from time to time. Being able to model past and present states of these processes would lead us to the possibility of understanding how various variables or things in this world interact, which in the end would offer us the knowledge to estimate future states of the systems in which these processes take place.

Chapter 12 - Being able to predict the weather is one of the greatest challenges of mankind. This success relies on the Kalman filter equations, and its various generalization or approximations. The aims of the chapter is to see why Kalman equations are needed and also to provide various generalization and approximation of information dynamics.

In: Kalman Filtering
Editor: Joaquín M. Gomez

ISBN: 978-1-61761-462-0
© 2011 Nova Science Publishers, Inc.

Chapter 1

OPTIMIZATION OF KALMAN FILTERING PERFORMANCE IN RECEIVED SIGNAL STRENGTH BASED MOBILE POSITIONING

J. G. Markoulidakis[*]

Vodafone-Panafon, Technology Strategic Planning,
R&D Dept., Tzavella 1-3, Halandri, Greece

ABSTRACT

The current paper considers the problem of Mobile Terminal (MT) positioning based on a time series of Received Signal Strength (RSS) measurements and provides statistical estimators of the MT- Base Station (BS) distance which substantially improve the MT positioning accuracy. The set of MT-BS distance estimators defined for optimizing RSS based positioning developed in [1] is used as a basis. The problem that is analyzed refers to the adaptation of these estimators to the presence of Kalman filtering in the MT position calculation process. Three different Kalman filtering options can be applied at different stages of the MT position calculation process [2]: (a) RSS Kalman filtering, (b) MT-BS distance Kalman filtering and (c) MT position coordinates Kalman filtering. To identify the optimal MT-BS distance estimators at the presence of Kalman filtering a method exploiting the characteristics of the steady state Kalman filter performance is developed. This novel method allows for the definition of refined MT-BS distance estimators matching the nature of the Kalman filter process. The results indicate that the resulting estimators provide good efficiency for all Kalman Filter options.

1. INTRODUCTION

Mobile Terminal (MT) positioning is the key enabler of the so called Location Based Services (LBS) in mobile communications. A wide variety of MT positioning techniques has

---

[*] Corresponding author: Tel. +30 210 6703143, Email: Yannis.Markoulidakis@vodafone.com

been proposed in the literature while some of these techniques are currently in commercial operation. Each MT positioning technique is exhibiting different characteristics referring to the accuracy, the deployment cost, the support of legacy MTs, etc. To deal with the issue of required network investment and the modernization of MTs the roadmaps of all positioning technologies start with low cost and low accuracy techniques (e.g., mobile network Cell Identity based) and evolve in the long term towards the Assisted Global Positioning System (A-GPS) the best performing positioning technique in terms of accuracy and reliability. However, the relatively slow commercial take up of LBS combined with the associated cost indicate the fact that there will be a relatively long period of time for which legacy 2G or 2G/3G MTs will not be equipped with GPS receivers on a worldwide basis. On the other hand, even after the introduction of A-GPS, the limited availability of satellite signal in indoor and certain outdoor environments (urban canyon effect) as well as other issues like time-to-first-fix or battery consumption, indicate the need for hybrid techniques which combine A-GPS with a cellular based positioning. This fact indicates that there is still space for enhanced network based positioning techniques which combine adequate accuracy with low deployment costs.

One of the low cost MT positioning techniques is RSS based positioning. The MT performs regular RSS measurements from the primary and the neighboring cells both when idle or during a call. These measurements can then be exploited for MT positioning based on two main steps: (a) MT-BS distance estimation based on path-loss empirical models like Hata [3] or Cost 231 [4] and (b) trilateration of the estimated distances between the MT and three different BSs. In this paper we analyze the time-series RSS based MT positioning which corresponds to two main concepts: (a) Statistical Terminal Assisted Mobile Positioning (STAMP) [5] where the MT while in idle mode stores historical RSS measurements from the primary and the neighboring cells and exploits this time series of measurements at the setup phase of an LBS session for better positioning accuracy and (b) MT tracking [6] where after the setup of a tracking application a periodic monitoring of RSS and Timing Advance parameters can lead to MT positioning and tracking. The main issue with RSS based positioning is the relatively high error which occurs in the estimation of MT-BS distance because of the stochastic nature of the radio propagation environment. In techniques that time-series RSS measurements are considered, Kalman filtering (typical approach in navigation also applied in [6]) appears to be a very effective method in order to reduce the error of the individual RSS measurements and improve the MT positioning accuracy.

As shown in [1] due to the nature of the path loss propagation models is such that the estimation of the MT-BS distance is a biased estimator. Moreover, the application of trilateration influences the way that RSS stochastic nature affects the MT positioning accuracy. In the analysis presented in [1] a set of four different MT-BS distance estimators are defined so as to optimize the resulting MT positioning accuracy. Moreover, the estimators developed in [1] have been applied to time-series based positioning at the presence of Kalman filtering leading to better accuracy. This raises the question whether the estimators developed in [1] are the optimum ones at the presence of Kalman filtering.

In [2] another aspect of the MT position calculation process at the presence of a time-series of measurements is considered. In particular, three different options regarding the stage at which Kalman filtering is applied have been defined: (a) RSS Kalman filtering, (b) MT-BS distance Kalman filtering and (c) MT position coordinates Kalman filtering. The analysis of [2] indicated that the RSS Kalman filtering leads to the best accuracy followed by MT-BS distance Kalman filtering and MT coordinates Kalman filtering. This fact introduces an additional complexity to the problem raised in the previous paragraph: what is the set of MT-BS distance estimators which optimizes the performance of MT positioning process at the presence of any of the three Kalman filtering options defined in [2].

The current paper explores the ability to apply the MT-BS estimators defined in [1] so as to further improve the MT positioning accuracy that the Kalman filtering options provide. The first step of the analysis considers the performance of the three alternative Kalman filtering options defined in [2] for the various MT-BS distance estimators defined in [1]. The results indicate that depending on the Kalman filtering option a different MT-BS estimator leads to an optimum positioning accuracy. Moreover, no improvement is achieved for RSS Kalman Filtering while the accuracy of the other two options (MT-BS distance Kalman Filtering and MT Position Coordinates Kalman Filtering) is being improved from the employment of the MT-BS distance estimators defined in [1]. The explanation of this fact relies on the way that the error is propagated in the non-linear MT position estimation process. The paper then taking into account the specific characteristics of the MT position calculation process as well as the steady state Kalman filter performance and develops a set of refined estimators. The results of this analysis indicate that the employment of the refined MT-BS distance estimators improves the accuracy achieved in all three Kalman filtering options. Moreover, the optimized MT positioning accuracy is in the same range for all three Kalman filtering options a fact which allows for higher flexibility in adopting the proposed MT positioning technique in a real application.

The paper is organized as follows: Section 2 provides a short overview of the MT positioning techniques. Section 3 provides an overview of the MT-BS distance estimators defined for the optimization of RSS positioning based on a single RSS measurement sample. Section 4 provides and overview of the STAMP concept. Section 5 provides an overview of the Kalman filtering options. Section 6 provides the analysis of the MT-BS distance estimators that optimize the performance of the three Kalman filter options. Section 7 provides the conclusions and future research challenges.

2. OVERVIEW OF MOBILE TERMINAL POSITIONING TECHNIQUES

The MT positioning techniques, developed so far, are characterized by a trade off between the offered accuracy and the deployment costs. In principle, the higher the accuracy of an MT positioning technique, the higher the deployment cost and the higher the value of the commercial services that can be offered. The existing MT positioning techniques can be divided into three main categories: (a) network based, (b) handset based and (c) hybrid technologies [7]. The main methods, which have been proposed so far, are the following:

Cell Global Identity (CGI) and CGI plus Timing Advance (CGI+TA): CGI is the serving cell and TA is a parameter that allows for an estimation of the MT-BS distance (in increments of 550m in GSM [8], [17]). The CGI alone or the CGI plus the TA parameter can be exploited leading to accuracy that varies with the cell size from 100m to 1100m. No terminal modification is required.

Enhanced CGI (ECGI): Similar to CGI+TA exploiting in addition the RSS of the neighbouring cells which are used to estimate the distance between the MT and at least three BSs. Then using methods like tri-lateration, the terminal position is identified with an accuracy of 250m-1000m. No terminal modification is required. An alternative to this method is to use pattern matching techniques [15] so as to compare the RSS measurements of the terminal with the database of predicted RSS levels. In this case the measured RSS from all neighbouring cells is taken into account (6 at maximum).

CGI + Common Pilot Channel Signal Strength for 3G (CGI+CPCSS): Similar technique to the ECGI for 3G systems. Using CPCSS of neighbouring cells [9] the distance between the terminal and a set of base stations is estimated [10]. The accuracy of this method in urban areas can be between 100m and 400m driven also by the 3G cell size in such areas.

Uplink-Time Difference Of Arrival (U-TDOA): The network measures the Time Of Arrival (TOA) of a known signal sent from the MT and received at four or more LMUs (Location Measurement Units) [11]. Based on the known positions of the LMUs, the terminal position can be calculated via hyperbolic trilateration. No terminal modification is required.

Angle Of Arrival (AOA): AOA requires directional antennas or antenna arrays at the BS. The network measures the AOA from at least two BSs and the terminal position is determined through triangulation with an average accuracy of ~300m. Line-of-sight is essential for accurate terminal positioning in this technique.

Enhanced - Observed Time Difference (E-OTD): The terminal measures the Observed Time Difference (OTD) between arrivals of bursts for pairs of BSs. As BSs are not synchronized, the network must also measure the so-called relative time difference. At least three distinct pairs of geographically separated BSs are needed to provide an accuracy of 50m-300m.

GPS (Global Positioning System): A time-based method where the GPS receiver equipped terminal calculates its position based on the time-difference observed in the received GPS satellites' signals. A number of three (for 2D positioning) or four satellites (for 3D positioning) are required in this case. GPS has very high accuracy outdoors (5m-50m), but is complicated in indoor environment or certain urban areas as it requires line of sight with a number of GPS satellites.

Assisted GPS (A-GPS): In A-GPS the mobile network assists the GPS enabled terminal with GPS related information (e.g., frequencies of visible satellites, timing information, differential GPS corrections, etc.) This improves the "Time to First Fix" and the terminal battery consumption. A-GPS can be either terminal or network based and the accuracy it achieves is 5m-50m.

Hybrid Techniques: The nature of the MT positioning techniques is such that hybrid techniques have been proposed aiming at exploiting at a complementary way the advantages of the various techniques. A typical example is a low cost / low accuracy network based technique which provides MT position at all environments (e.g. indoor) and at a relatively short time period combined with A-GPS which faces some limitations in indoor environments and still the time to first fix may be an issue.

WiFi based techniques: with the increasing presence of wifi access points especially in urban environments and the trend of wifi-enabled MTs an alternative (or complementary to the above) technique has been also proposed in the literature. According to such techniques the wifi Identity or even the RSSI can be exploited for locating the terminal.

3. MOBILE TERMINAL – BASE STATION DISTANCE ESTIMATORS IN RSS BASED POSITIONING

The current section provides an overview of the research results presented in [1]. Figure 1 depicts the MT position calculation process for RSS based positioning (i.e., MT positioning based only on RSS measurements or ECGI technique where TA is also considered). In this case a single sample of RSS measurements collected at the setup phase of the LBS session is used so as to calculate the MT position. As it can be seen from this figure the MT position calculation process right after the measurement collection phase comprises of two main steps: (a) estimation of the MT- BS distance for three different BSs and (b) trilateration which leads to the estimation of the MT coordinates (2D positioning is assumed in this case).

Apparently, the error in terminal positioning for RSS based techniques depends both on the error of the RSS measurements as well as the way this error is propagated through the involved calculation steps of the process depicted in Figure 1.

3.1. Error Associated to the RSS Measurements

In mobile communications it is well known that the signal strength received from a BS at a certain location is temporal and consists of two main factors [12], [13]: (a) the slow fading or local mean which is related to path loss and terrain nature and (b) the short-term or fast fading which is caused by multipath effects. The slow fading factor is characterized by lognormal behavior while the relevant pdf (probability density function) for short-term fading is Rayleigh or Rician, depending on whether there is line of sight or not [12].

According to the standard MT operation RSS is periodically measured based on averaging over a time period (in GSM this is 480msec while in dedicated mode and 2-5sec while in idle mode [14]). According to [12] short-term fading can be smoothed-out for a moving MT by averaging over a time period during which the MT covers a distance of 40 to 80 wavelengths. Therefore, we can assume that in the MT collected RSS measurements the short-term fading has, to a certain degree, been smoothed-out. The remaining slow fading component expressed in decibels can be modeled as a Gaussian random variable with a standard deviation in the order of 8db-12db as estimated through field measurements [12]. To eliminate the impact of slow fading the signal should be averaged for longer time periods corresponding to a distance covered by the MT that is hundred times longer than the wavelength employed (in the order of 100m-200m for 900MHz). In such a case the averaged signal provides the so called median component which corresponds to the RSS which is a direct function of the path loss occurring due to the distance of the MT from the BS. Radio propagation loss models (e.g., Hata [3], Cost231 [4]) can be exploited in this case so as to estimate the distance between the MT and a BS.

3.2. RSS Based Estimation of the Distance between the Terminal and the Base Station

Path loss prediction models provide the RSS as a function of the distance of the MT from the BS. Such models include Hata [3] and Cost231 [4] models:

$$\text{Hata}: R_x(i) = -K(i) - [69,5 + 26,16\log(f(i)) - 13,82\log(h_{bs}(i)) + [44,9 - 6,55\log(h_{bs}(i))]\cdot \log(d(i)) - c(h_{mt})]$$
$$\text{Cost 231}: R_x(i) = -K(i) - [46,33 + 33,9\log(f(i)) - 13,82\log(h_{bs}) + [44,9 - 6,55\log(h_{bs})]\cdot \log(d(i)) - a(h_{mt}) + c] \quad (1)$$

Where $R_x(i)$ the RSS from BS(i) at the MT location in dbm, $K(i)$ is the BS(i) transmitting power in dbm, $f(i)$ the transmitter frequency in MHz, $h_{bs}(i)$ the BS(i) height in m, h_{mt} the terminal height in m, $d(i)$ the MT-BS(i) distance in km, $c(h_{mt})$, $\alpha(h_{mt})$ are functions depending on the environment (e.g., urban, suburban) and c is a model specific constant which also depends on the radio propagation environment. The RSS based estimation of the distance between the MT and the BS relies on the reverse calculations of propagation path loss prediction models:

$$\hat{d}_o(i) = 10^{-\left[\frac{\hat{R}_x(i)+A(i)}{B(i)}\right]}$$

Hata: $A(i) = K(i) + 69,5 + 26,16\log(f(i)) - 13,82\log(h_{bs}(i)) - c(h_{mt})$ (2)

Cost231: $A(i) = K(i) + 46,33 + 33,9\log(f(i)) - 13,82\log(h_{bs}(i)) - a(h_{mt}) + c$

$B(i) = 44,9 - 6,55\log(h_{bs}(i))$

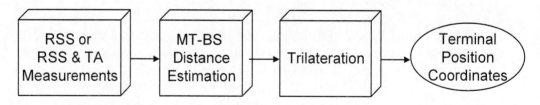

Figure 1. The basic processing steps for RSS based Terminal Positioning.

Where $\hat{d}_o(i)$ is the resulting estimation of the distance between the MT and BS i, $\hat{R}_x(i)$ is the measured RSS at the MT position. In the case that the path-loss prediction model is properly calibrated for a specific geographical region then we can assume that RSS can be modeled as a Gaussian distributed random variable. Based on Eq. (2) it can be concluded that the estimated MT-BS distance will follow a Lognormal distribution:

$$\hat{R}_x(i) \sim N(R_x(i), \sigma_{Rx}^2) \Rightarrow \hat{d}(i) \sim LN(\mu_d(i), \sigma_d^2(i))$$

$$\mu_d(i) = \ln[d(i)] = \ln\left[10^{\frac{R_x(i)+A(i)}{B(i)}}\right], \quad \sigma_d^2(i) = \left(\frac{\ln(10)}{B(i)}\right)^2 \sigma_{Rx}^2 = \beta(i) \cdot \sigma_{Rx}^2 \quad (3)$$

Where σ_{Rx}^2 is the variance of the RSS Gaussian distribution, $\mu_d(i)$ and $\sigma_d^2(i)$ is the mean value and the variance of the $\hat{d}_o(i)$ lognormal distribution respectively.

In practice the calibration of an empirical path loss propagation model is a process which requires extensive field measurements. Moreover, propagation predictions models have evolved based e.g. on ray tracing and accurate geographical layout data. However, the advanced propagation prediction models cannot be used in MT positioning because in this case the exact propagation path is unknown during the positioning process.

The distribution provided in Eq. (3) assumes that the error of the rest of the parameters involved in Eq. (1) is negligible compared to the RSS error. Moreover, the above equations consider a constant antenna gain for the entire sector covered by each cell (in practice antenna gain is higher at the cell main direction). From Eq. (2) the mean value, the variance and the Mean Squared Error (MSE) of the MT-BS distance can be estimated:

$$E(\hat{d}_o(i)) = d(i) \cdot e^{\beta(i)\frac{\sigma_{Rx}^2}{2}}$$

$$V(\hat{d}_o(i)) = d(i)^2 \cdot e^{\beta(i)\cdot\sigma_{Rx}^2} \cdot \left[e^{\beta(i)\cdot\sigma_{Rx}^2} - 1\right] \quad (4)$$

$$MSE(\hat{d}_o(i)) = d(i)^2 \cdot \left[e^{2\cdot\beta(i)\cdot\sigma_{Rx}^2} - 2\cdot e^{\beta(i)\frac{\sigma_{Rx}^2}{2}} + 1\right]$$

From Eq. (4) it is obvious that $E(\hat{d}_o(i)) \neq d(i)$ which means that the MT-BS distance estimator provided in Eq. (2) is biased. This fact creates the need to investigate the issue of optimizing the RSS based MT-BS distance estimator.

3.2.1. Unbiased Distance Estimator

At a given the MT-BS distance d(i), the unbiased distance estimator is provided by the following equation [1]:

$$\hat{d}_1(i) = c_1(i) \cdot \hat{d}_o(i) = e^{-\beta(i)\frac{\sigma_{Rx}^2}{2}} \cdot 10^{\frac{\hat{R}_x(i)+A(i)}{B(i)}} \quad (5)$$

Indeed the mean value of the distance estimation becomes:

$$E(\hat{d}_1(i)) = d(i) \tag{6}$$

The variance and the MSE in this case are equal (property of unbiased estimators):

$$V(\hat{d}_1(i)) = MSE(\hat{d}_1(i)) = d(i)^2 \cdot \left[e^{\beta(i) \cdot \sigma_{Rx}^2} - 1\right] \tag{7}$$

3.2.2. Minimum MSE Estimator of MT-BS Distance

The minimum MSE MT-BS distance estimator can be achieved by the introduction of a factor $c_2(i)$ and the minimization of the following function:

$$MSE(\hat{d}_2(i)) = d^2(i) \cdot \left[c_2^2(i) \cdot e^{\beta(i) \cdot 2\sigma_{Rx}^2} - 2c_2(i) \cdot e^{\beta(i) \cdot \frac{\sigma_{Rx}^2}{2}} + 1\right] \tag{8}$$

It is easy to show that the MSE is optimized for the following value of $c_2(i)$ [1]:

$$c_2(i) = e^{-\beta(i) \cdot \frac{3\sigma_{Rx}^2}{2}} \quad \text{i.e., } \hat{d}_2(i) = e^{-\beta(i) \cdot \frac{3\sigma_{Rx}^2}{2}} \cdot 10^{-\frac{\hat{R}_x(i) + A(i)}{B(i)}} \tag{9}$$

In this case the MSE, the mean value and the variance of the MT-BS distance are expressed as follows:

$$\begin{aligned}
\min\{MSE(\hat{d}_2(i))\} &= d^2(i) \cdot \left[1 - e^{-\beta(i) \cdot \sigma_{Rx}^2}\right] \\
E(\hat{d}_2(i)) &= d(i) \cdot e^{-\beta(i) \cdot \sigma_{Rx}^2} \\
V(\hat{d}_2(i)) &= d^2(i) \cdot e^{-\beta(i) \cdot 2\sigma_{Rx}^2} \cdot \left[e^{\beta(i) \cdot \sigma_{Rx}^2} - 1\right]
\end{aligned} \tag{10}$$

3.2.3. MT-BS Distance Estimation Accuracy Evaluation

Figure 2 provides a comparison of the theoretical mean squared root error of the MT-BS distance for the following estimators: (a) basic estimator (j=0, Eq. (2)), (b) unbiased estimator (j=1, Eq. (5)) and (c) minimum mean squared error estimator (j=2, Eq. (9)). As it can be seen from the figure the two proposed estimators deliver a significant improvement in the MT-BS distance estimation error especially for high MT-BS distance values.

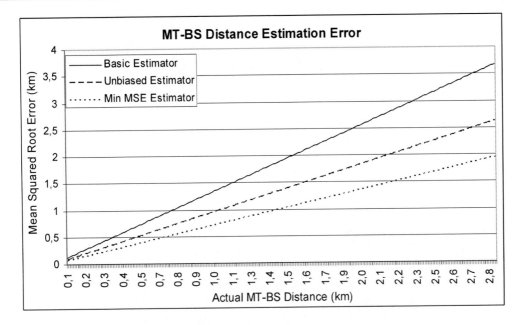

Figure 2. Accuracy of RSS based MT-BS distance estimation (h_{bs}=25m, σ_{Rx}=12db).

3.4. Trilateration Error Analysis

Trilateration is a geometrical method which based on the estimated distances between the MT and three different BSs it provides the MT position. Trilateration is expressed by the following equations:

$$(x-x_i)^2 + (y-y_i)^2 = d^2(i), \text{ for } i=1,2,3 \tag{11}$$

Where (x,y) are the MT position coordinates and (x_i,y_i) i=1,2,3 are the coordinates of the three BSs. By subtracting the equation corresponding to i=1 from the ones corresponding to i=2 and i=3 we get an equivalent set of equations:

$$(x-x_1)^2 + (y-y_1)^2 = d^2(1) \tag{12a}$$

$$-2x(x_1-x_i) - 2y(y_1-y_i) = d^2(1) - d^2(i) - x_1^2 + x_i^2 - y_1^2 + y_i^2, \; i=2,3 \tag{12b}$$

From the two equations of Eq. (12b) it is feasible to estimate the terminal position (x,y) in a 2D plane as follows:

$$x = \frac{2}{D}\left[-(y_2-y_3)\cdot d^2(1) + (y_1-y_3)\cdot d^2(2) - (y_1-y_2)\cdot d^2(3) - (y_1-y_2)\cdot(y_1^2-y_3^2+x_1^2-x_3^2) + (y_1-y_3)\cdot(y_1^2-y_2^2+x_1^2-x_2^2)\right]$$

$$y = \frac{2}{D}\left[(x_2-x_3)\cdot d^2(1) - (x_1-x_3)\cdot d^2(2) + (x_1-x_2)\cdot d^2(3) + (x_1-x_2)\cdot(y_1^2-y_3^2+x_1^2-x_3^2) - (x_1-x_3)\cdot(y_1^2-y_2^2+x_1^2-x_2^2)\right] \tag{13}$$

Eq. (13) for simplicity can also be written as:

$$z = \frac{1}{D}\left[\sum_{i=1}^{3} a_z(i) \cdot d^2(i) + \Omega_z\right], \quad i=1,2,3, \; z=x,y \tag{14}$$

Where Ω_z and $a_z(i)$, i=1,2,3, z=x,y can be directly derived from Eq. (13) and D is provided by the following equation:

$$D = 4 \cdot (x_1 - x_2) \cdot (y_1 - y_3) - 4 \cdot (x_1 - x_3) \cdot (y_1 - y_2) \tag{15}$$

As shown in [1] unless the three BSs are co-linear we have D≠0 which yields a unique solution to Eq. (12b). This property applies for any set of distances d(i) i=1,2,3 even at the presence of error MT-BS distance estimation. In the case that the MT-BS distance estimation is 100% accurate (i.e., no error exists) then the resulting coordinates provided in Eq. (14) are satisfying both Eq. (12a) and Eq. (12b) and correspond to the exact solution of the trilateration problem. On the other hand, in the presence of error Eq. (12a) will not be satisfied by the coordinates provided by Eq. (14) which in this case correspond to an estimation of the trilateration problem solution.

3.4.1. Trilateration Error for RSS Based Distance Estimation

As shown in [1], the MT position coordinates estimated through trilateration are provided by the following equation (i=1,2,3, j=0,1,2):

$$\hat{z}_j = \frac{1}{D}\left[\sum_{i=1}^{3} a_z(i) \cdot \hat{d}_j^2(i) + \Omega_z\right] = z + \frac{1}{D}\left[\sum_{i=1}^{3} a_z(i) \cdot \left[\hat{d}_j^2(i) - d^2(i)\right]\right], \; z=x,y \tag{16}$$

Where j is the index of the distance estimator (j=0 corresponds to basic distance estimator Eq. (2)). Assuming a given set of distances between the MT and three different BSs d(i) (i=1,2,3) the mean value and the variance of the estimated MT coordinates are the following (assuming that the random variables $\hat{d}_j(i)$, i=1,2,3 are independent):

$$E(\hat{z}_j) = z + \frac{1}{D}\left[\sum_{i=1}^{3} a_z(i) \cdot d^2(i) \cdot \left[c_j^2(i) \cdot e^{\beta(i) \cdot 2 \cdot \sigma_{Rx}^2} - 1\right]\right]$$

$$V(\hat{z}_j) = \frac{1}{D^2}\left[\sum_{i=1}^{3} a_z^2(i) \cdot d^4(i) \cdot c_j^4(i) \cdot e^{\beta(i) \cdot 4\sigma_{Rx}^2} \cdot \left[e^{\beta(i) \cdot 4\sigma_{Rx}^2} - 1\right]\right], \; z=x,y \text{ and } j=0,1,2 \tag{17}$$

And the MSE in the estimation of the terminal coordinates will be (z=x, y, i=1,2,3 and j=0,1,2):

Optimization of Kalman Filtering Performance in Received Signal Strength... 11

$$MSE(\hat{z}_j) = \frac{1}{D^2}\left[\sum_{i=1}^{3} a_z^2(i) \cdot d^4(i) \cdot \left[c_j^4(i) \cdot e^{8\beta(i)\cdot\sigma_{Rx}^2} - 2c_j^2(i) \cdot e^{2\beta(i)\cdot\sigma_{Rx}^2} + 1\right] + \sum_{i=1}^{3}\sum_{\substack{k=1\\k\neq i}}^{3} a_z(i) \cdot a_z(k) \cdot d^2(i) \cdot d^2(k) \cdot f_j(i,k)\right]$$

$$f_j(i,k) = \left[c_j^2(i) \cdot c_j^2(k) \cdot e^{2(\beta(i)+\beta(k))\cdot\sigma_{Rx}^2} - c_j^2(i) \cdot e^{2\beta(i)\cdot\sigma_{Rx}^2} - c_j^2(k) \cdot e^{2\beta(k)\cdot\sigma_{Rx}^2} + 1\right] \quad (18)$$

In the following we define another two MT-BS distance estimators based on the outcome of the trilateration process so as to provide unbiased estimation of the MT coordinates and a sub-optimal MSE.

3.4.2. Unbiased MT Coordinates Estimator

According to the analysis presented in [1], to achieve an unbiased estimation of the MT coordinates, the following MT-BS distance estimator is introduced:

$$\hat{d}_3(i) = c_3(i) \cdot \hat{d}_o(i), \quad \text{with} \quad c_3(i) = e^{-\beta(i)\cdot\sigma_{Rx}^2}, \; (i=1,2,3) \quad (19)$$

The mean value, the variance and the MSE of the MT coordinates estimation after trilateration becomes (i=1,2,3):

$$E(\hat{z}_3) = z, \qquad V(\hat{z}_3) = MSE(\hat{z}_3) = \frac{1}{D^2}\left[\sum_{i=1}^{3} a_z^2(i) \cdot d^4(i) \cdot \left[e^{4\beta(i)\cdot\sigma_{Rx}^2} - 1\right]\right], \; z=x, y \quad (20)$$

As expected the MSE equals the variance in this case as a property of an unbiased estimator.

3.4.3. Optimal and Suboptimal MT Coordinates MSE Estimators

According to the analysis presented in [1] the minimization of MT coordinates MSE leads to a solution which is a function of the actual MT-BS distances which are not known parameters. For that reason an alternative sub-optimal estimator is developed in [1]. This estimator is based on the observation that one way to reduce of the MSE of the MT coordinates estimation according to Eq. (16) is to minimize of the following quantities:

$$\min\left\{E\left[\left(\hat{d}_j^2(i) - d^2(i)\right)^2\right]\right\}, \; i=1,2,3 \quad (21)$$

This is achieved by the following estimator [1]:

$$\hat{d}_4(i) = c_4(i) \cdot \hat{d}_o(i), \text{ with: } c_4(i) = e^{-3\beta(i)\cdot\sigma_{Rx}^2}, \; i=1,2,3 \quad (22)$$

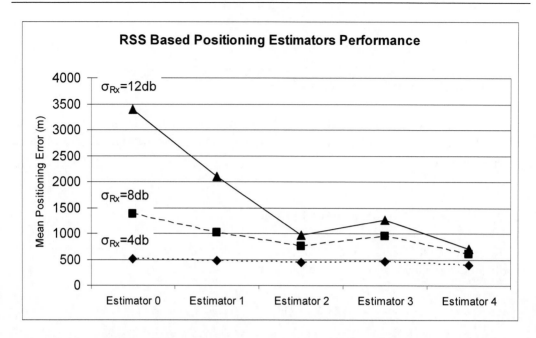

Figure 3. The performance of the proposed RSS estimators in RSS based positioning.

In this case the mean value and the variance of the estimated terminal coordinates will be (i=1,2,3):

$$E(\hat{z}_4) = z + \frac{1}{D}\sum_{i=1}^{3} a_z(i) \cdot d^2(i) \cdot \left[e^{-4\beta(i)\cdot\sigma_{Rx}^2} - 1\right]$$
$$V(\hat{z}_4) = \frac{1}{D^2}\sum_{i=1}^{3} a_z^2(i) \cdot d^4(i) \cdot e^{-8\beta(i)\cdot\sigma_{Rx}^2}\left[e^{4\beta(i)\cdot\sigma_{Rx}^2} - 1\right] \quad (23)$$

As it can be observed the mean value of estimator 4 is not equal to the actual value z and therefore this is a biased estimator. The resulting MSE will be (i=1,2,3):

$$MSE(\hat{z}_4) = \frac{1}{D^2}\left[\sum_{i=1}^{3} a_z^2(i) \cdot d^4(i) \cdot \left[1 - e^{-4\beta(i)\cdot\sigma_{Rx}^2}\right]\right] + \\ + \frac{1}{D^2}\sum_{\substack{k=1 \\ k\neq i}}^{3} a_z(i) \cdot a_z(k) \cdot d^2(i) \cdot d^2(k) \cdot \left[e^{-4[\beta(i)+\beta(k)]\sigma_{Rx}^2} - e^{-4\beta(i)\cdot\sigma_{Rx}^2} - e^{-4\beta(k)\cdot\sigma_{Rx}^2} + 1\right], z=x,y \quad (24)$$

3.5. Terminal Position Estimation Error

Based on the analysis of the error introduced in the estimation of the terminal coordinates through trilateration method we can now study the error in the estimation of the terminal position $\hat{e}_j$:

$$\hat{e}_j = \sqrt{(\hat{x}_j - x)^2 + (\hat{y}_j - y)^2} = \sqrt{\Delta\hat{x}_j^2 + \Delta\hat{y}_j^2}, \; j=0,1,2,3,4 \qquad (25)$$

Where $\Delta\hat{z}_j = (\hat{z}_j - z)$, z=x,y is the MT coordinate estimation error and index j indicates the applied MT-BS distance estimator. As shown in [1] the pdf of the MT position error can be approximated by the following Rice distribution:

$$\hat{e}_j \sim Rice(u_e(j), \sigma_e(j)) \qquad (26)$$

The mean value and the variance of the MT positioning error can be approximated as follows:

$$E(\hat{e}_j) = \sigma_e(j) \cdot \sqrt{\frac{\pi}{2}} \cdot L_{1/2}\left(-\frac{u_e^2(j)}{2\sigma_e^2(j)}\right)$$

$$V(\hat{e}_j) = 2\sigma_e^2(j) + u_e^2(j) - \frac{\pi\sigma_e^2(j)}{2} \cdot L_{1/2}^2\left(-\frac{u_e^2(j)}{2\sigma_e^2(j)}\right) \qquad (27)$$

Where $L_v(x)$ denotes a Laguerre polynomial. The parameter $u_e(j)$ is provided by the following equation:

$$u_e(j) = \sqrt{E(\Delta\hat{x}_j)^2 + E(\Delta\hat{y}_j)^2} \qquad (28)$$

Regarding parameter $\sigma_e^2(j)$ the following approximation can be used [1]:

$$\sigma_e^2(j) = \frac{1}{2}\left[V(\Delta\hat{x}_j) + V(\Delta\hat{y}_j)\right] \qquad (29)$$

As highlighted in [1] the application of the unbiased MT coordinates estimator (j=3) leads to a zero average value for both $\Delta\hat{x}_3$ and $\Delta\hat{y}_3$. In this case $u_e(3)$ becomes also zero and the MT positioning error distribution can be approximated by a Rayleigh distribution.

3.6. Performance Analysis of Terminal Positioning Accuracy

To analyze the performance of the various MT-BS distance estimators presented in this paper a simulation model has been developed considering a geographical area with a set of BSs each supporting three cells (each cell corresponds to a 120° sector) of equal radius. During the simulation the position of the MT is randomly selected within this area and the RSS levels from each of the BS are estimated from Hata model (Eq. (2)). Then a Gaussian white noise with variance σ_{Rx}^2 is being added to the resulting RSS and the MT coordinates estimation process depicted in Figure 1 is followed in order to estimate the MT position. The estimated MT position is then compared to the actual MT position to determine the positioning error. The simulation model corresponds to an ideal and apparently non-realistic case where all assumptions of the theoretical framework presented in this paper are met. Therefore the simulation model corresponds to a best case scenario in terms of the achieved MT positioning accuracy improvement.

Figure 3 provides the resulting mean MT positioning error for a variety of σ_{Rx} (4db, 8db and 12db) and for the various estimators presented in this paper (indexed by j as Estimators 0,1,2,3,4) for a cell radius of 1500m. The figure indicates that the suboptimal MT coordinates MSE estimator (j=4) provides the best accuracy. Moreover, the minimum MSE estimator of MT-BS distance (j=2) outperforms unbiased MT coordinates estimator (j=3). It is clear that significant improvement is achieved by the proposed estimators especially for high values of RSS standard deviation (σ_{Rx}). However, even the best performing estimator (j=4) leads to an accuracy in the order of ~500m.

As shown in [1] the accuracy of plain RSS based and plain ECGI even with the best performing MT-BS distance estimator have similar performance with techniques like CGI and CGI+TA in an urban environment.

4. OVERVIEW OF STATISTICAL TERMINAL ASSISTED MOBILE POSITIONING (STAMP)

As mentioned in the introduction, STAMP is a method that can improve the accuracy of existing MT positioning techniques provided that the required parameters can be measured by the terminal while in idle mode operation. According to STAMP concept the terminal stores a time-series of measurements which correspond to a set of historical terminal positions. This set of measurements are exploited at the LBS session setup so as to provide improved MT positioning accuracy. The current section provides an overview of the main principles of STAMP concept based on [5].

4.1. STAMP Operation during Mobile Terminal Idle Mode

While the MT is in idle mode, it measures and stores a set of parameters associated with a preselected positioning technique (e.g., ECGI, GPS, etc.) called the "*Idle Mode Location Technique*" (**IMLT**). Not all terminal positioning techniques can be selected as IMLTs as some parameters like TA become available only during an existing connection between the MT and the BS (i.e., not in the idle mode). In the current paper we focus on the RSS based positioning and for that reason we consider that the MT while in idle mode it measures and stores the RSS measurements from the primary and the neighbouring cells. It should be noted that according to the standard MT idle-mode operation [14] the terminal monitors the RSS levels so as to apply the cell selection and cell re-selection procedure and therefore the only additional requirement for the MT is to store a set of these measurements.

According to STAMP concept the MT periodically measures the target parameters according to the following software-based mechanism:

The STAMP List: The list stored at the terminal containing the N_L most recent samples of measurements (e.g., N_L =20). N_L is defined as the size of the STAMP list.

Sample Adoption Condition: The condition applied in order to either reject or store a sample measurement. For example, a corrupted measurement can be rejected by the terminal. As discussed in [5], this condition may also be applied to increase STAMP efficiency.

The STAMP Sampling Period (T): The time period between two measurements.

The Time-stamp: A reference on the time that a measurement was collected. The time-stamp is also stored in the STAMP list together with the relevant RSS measurement samples.

4.2. STAMP Operation during LBS Session Initiation

At the set-up phase of an LBS session (e.g., to locate the nearest restaurant) the following actions are taking place:

The Application Set-Up Location Technique (ASLT): At the LBS session set-up phase a pre-selected terminal positioning technique is applied (the ASLT). ASLT, which can be any terminal positioning technique (the same as the IMLT or not), allows for an initial estimation of the current terminal position. Thus STAMP provides the flexibility for the selection of the proper IMLT, ASLT pair (e.g., techniques with complementary characteristics). In the current paper ECGI will be considered to be the ASLT technique.

Calculation of Past Terminal Positions: the measurements stored in the STAMP list are retrieved and processed according to IMLT leading to the estimation of a set of N_L (N_L is the STAMP list size) past MT positions.

Estimation of Best Fit Terminal Trajectory: The two previous steps of STAMP have resulted in a set of N_L+1 MT estimated positions. Based on standard statistical methods it is possible to estimate the terminal trajectory and improve the terminal positioning accuracy compared to a stand-alone ASLT approach.

The most common statistical method applied in modern navigation systems is Kalman Filtering [6], [12], [18]. The Kalman filter model exploited in [6] is also applicable for STAMP. In [6] the so called MT tracking problem is addressed i.e., tracking of MT position assuming an active connection with the network (which allows for constant monitoring of RSS and TA parameters at a period of 480msec in GSM). In the case of RSS based STAMP only a set of N_L RSS measurements are available from the measurements the MT has performed while in idle mode. The Kalman filter model is described by the following state equations:

$$\mathbf{X}_k = \mathbf{\Phi} \cdot \mathbf{X}_{k-1} + \mathbf{\Gamma} \cdot \mathbf{W}_k \qquad (29)$$

$$\mathbf{Y}_k = \mathbf{M} \cdot \mathbf{X}_k + \mathbf{U}_k \qquad (30)$$

where k represents time instance t_k, and:

$$\mathbf{X_k} = \begin{bmatrix} x_1(t_k) \\ x_2(t_k) \\ v_1(t_k) \\ v_2(t_k) \end{bmatrix}, \mathbf{Y_k} = \begin{bmatrix} y_1(t_k) \\ y_2(t_k) \end{bmatrix} \quad (31)$$

$\mathbf{X_k}$ represents the combined vector of the terminal position [x_1, x_2] and velocity [v_1, v_2] and $\mathbf{Y_k}$ the estimated terminal position at time instance t_k.

$$\mathbf{\Phi} = \begin{bmatrix} 1 & 0 & \Delta t & 0 \\ 0 & 1 & 0 & \Delta t \\ 0 & 0 & 1 & 0 \\ 0 & 0 & 0 & 1 \end{bmatrix}, \mathbf{\Gamma} = \begin{bmatrix} 0 & 0 \\ 0 & 0 \\ \Delta t & 0 \\ 0 & \Delta t \end{bmatrix}, \mathbf{M} = \begin{bmatrix} 1 & 0 & 0 & 0 \\ 0 & 1 & 0 & 0 \end{bmatrix} \quad (32)$$

where the parameter Δt equals to the STAMP period T.

$$\mathbf{W_k} = \begin{bmatrix} w_1(t_k) \\ w_2(t_k) \end{bmatrix}, \mathbf{U_k} = \begin{bmatrix} u_1(t_k) \\ u_2(t_k) \end{bmatrix} \quad (33)$$

$\mathbf{W_k}$ and $\mathbf{U_k}$ are Gaussian distributed random vector parameters with zero mean and the following co-variance matrices:

$$\mathbf{Q} = \begin{bmatrix} \sigma_Q^2 & 0 \\ 0 & \sigma_Q^2 \end{bmatrix}, \mathbf{R} = \begin{bmatrix} \sigma_R^2 & 0 \\ 0 & \sigma_R^2 \end{bmatrix} \quad (34)$$

The parameter σ_Q^2 is estimated either by the proposed approach of [6] or based on a more generic estimation of the mobility conditions in the area and σ_R^2 estimated based on the set of IMLT terminal position estimations $\mathbf{Y_k}$.

4.3. STAMP Implementation Requirements

It is obvious that the employment of STAMP entails additional software functionality at both the MT and the network. Apart from the STAMP list storage space requirements, the list will constantly be updated with new samples. The latter implies a possible increase in the terminal power consumption. The MT storage requirements for ECGI (the received signal levels of the primary and neighbouring cells will be stored) will be ~2 kbytes for a STAMP list size of N_L=30. Such a storage capacity is considered negligible for state of the art MTs. The impact of STAMP on the terminal power consumption is closely related to the type of

IMLT. For instance, in the case of ECGI the parameters that need to be stored by STAMP are monitored by the terminal as part of its standard idle mode operation [14]. Consequently the additional power consumption imposed by STAMP is marginal. However, using another technique (e.g., GPS), this might not be the case. Apparently, battery power consumption may influence the selection of the IMLT.

At the LBS session set-up phase, STAMP introduces additional processing requirements for the calculation of the terminal position. The processing can take place either at the MT or at the network side depending on the type of IMLT and ASLT techniques. Therefore, compared to a stand-alone ASLT solution, STAMP introduces some delay in the calculation of the terminal position. The latter could be treated by appropriate computing techniques such as parallel computation of the past N_L+1 number of terminal positions making thus the impact on position calculation delay marginal.

It is obvious that certain communication protocol requirements arise in the case that the terminal has to upload the STAMP list to the network. The Open Mobile Alliance (OMA) "User Plane Location Protocol" [19] can be utilised for such a purpose. In particular STAMP concept for IMLT based on RSS measurements can be supported from the SUPL 2.0 protocol standard [20].

4.4. STAMP Performance Overview

4.4.1. STAMP Performance for moving vs. not-moving MTs

According to the analysis presented in [5] the employment of STAMP for moving MTs is more efficient in terms of the resulting MT positioning accuracy versus the case of not-moving MTs. In the case of not moving MTs a systematic error is present in the whole set of RSS measurements due to the specific propagation conditions of the MT static position. On the other hand in the case of moving MTs the propagation conditions constantly change leading to a better statistical behavior of the RSS error. As a result STAMP application in the case of non-moving MTs provides some improvement however, it does not alleviate the systematic error while in the case of moving MTs STAMP leads to substantial accuracy improvement.

4.4.2. Application of MT-BS distance estimators in STAMP

As mentioned in the introduction MT positioning techniques which rely on a time-series of measurements include STAMP concept [5] (i.e., RSS measurements are collected while the MT is in idle mode) and the MT tracking technique [6] (i.e., measurements of RSS and TA are collected while the MT is in an active session with the network). The terminal position estimation process which is applicable to both MT tracking and STAMP is depicted in Figure 4. As it can be seen from this figure an adaptive Kalman filter is applied to the estimated terminal position coordinates. The application of Kalman Filter relies on the assumption that the difference between the estimated and the actual MT position coordinates ($\Delta \hat{z}_j$, z=x,y) follow a zero mean normal distribution. As shown in the analysis of the previous section of the paper this condition can be approximated by using the MT-BS distance estimator which leads to unbiased estimation of the terminal coordinates (estimator indexed by j=3).

Apparently if this condition is not satisfied the Kalman Filter performance will be biased through a systematic error.

Figure 5 provides the performance of STAMP versus the RSS standard deviation (σ_{Rx}) and the various proposed MT-BS distance estimators (indexed as j=0,1,2,3,4). As it can be seen, the best performing estimator for both techniques is the Unbiased MT coordinates estimator (j=3). As shown in [1] MT tracking provides higher accuracy versus STAMP. This is due to the fact that in MT tracking there is an active session between the MT and the network which makes TA parameter available (TA is not available in idle mode) while RSS measurements are collected every 480ms versus a period of 2 to 5sec in idle mode (in GSM).

As shown in [5] STAMP performance for the basic MT-BS estimator (j=0) delivers significant improvement vs. plain RSS based positioning (i.e. for j=0). However, as shown in [1] in this case STAMP performance approaches the one of the CGI or CGI+TA method in an urban environment. However, the best performing MT-BS distance estimator for both STAMP and MT tracking (unbiased MT position coordinates estimator j=3) clearly outperforms CGI, CGI+TA and ECGI techniques.

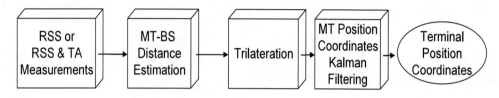

Figure 4. The MT positioning calculation process for MT tracking and STAMP.

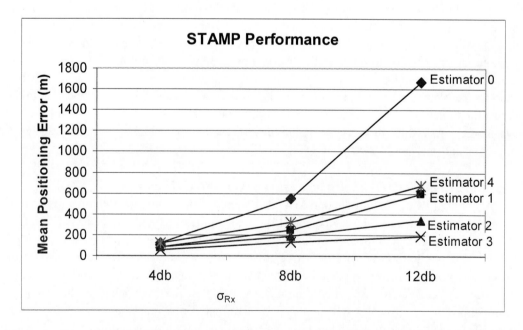

Figure 5. The performance of STAMP vs. the applied MT-BS estimator.

5. KALMAN FILTERING OPTIONS FOR STAMP

In [2] the application of Kalman filtering in MT position calculation process has been generalized by the introduction of three different options on the part of the process that the Kalman filter is introduced. Figure 6 presents the terminal position estimation process as well as the statistical filtering methods proposed in [2]: (a) RSS Kalman filtering (originally considered in [21]), (b) MT-BS distance Kalman filtering and (c) Terminal position Kalman filtering (common method used in the literature for mobile tracking [6], [12], [22], [23]).

The three Kalman Filter options of the process depicted in Figure 6 are described below:

Option A: RSS Kalman Filter. According to this option the RSS measurements are filtered through a Kalman filter so as to suppress the RSS error before this is propagated to the MT position calculation process. Pre-filtering may be applied for interpolation of corrupted or incomplete measurements and filtering out of extreme values. Of course a key issue which is dealt with later on in this paper is the fact that the set of monitored cells is changing over time especially for a moving MT.

Option B: MT-BS Distance Kalman Filter. In this option Kalman filtering is applied to the set of the resulting series of MT-BS distances. The scope of such an approach is to minimize the error in the MT-BS distance estimation before entering the trilateration process.

Option C: MT Coordinates Kalman Filter. In this option the common Kalman filtering approach followed in the literature (see [6]) is applied to process the set of MT positions for all the STAMP list measurements.

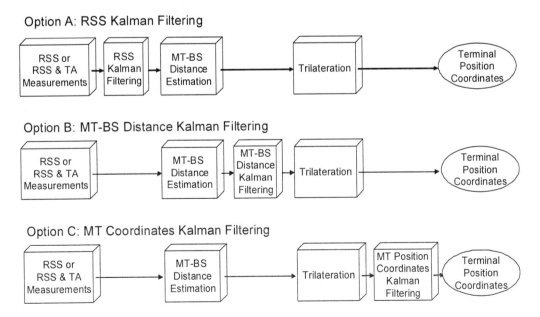

Figure 6. The terminal position estimation process in STAMP and MT tracking techniques.

It should be noted here that the application of multiple Kalman Filtering steps (e.g. combine some or all of the above three options) does not deliver additional improvement of the resulting accuracy. This is because of the fact that Kalman filtering delivers some gain provided that the input of the Kalman filter exhibits certain statistical characteristics including the assumption that the input corresponds to independent random variables. Moreover, the outcome of a Kalman filter provides a series of parameters which are not statistically independent any more. In this context the application of a second Kalman filter will not deliver any additional gain in error mitigation as its input will not correspond to independent random variables.

5.1. Option A: RSS Kalman Filter

In order to statistically filter the series of RSS of the STAMP list the Kalman filter used for trajectory tracking [6] is being properly modified. The proposed Kalman filter equations have the following form:

$$\begin{aligned} \mathbf{X}_k &= \Phi \cdot \mathbf{X}_{k-1} + \Gamma \cdot \mathbf{W}_k \\ \mathbf{Y}_k &= \mathbf{M} \cdot \mathbf{X}_k + \mathbf{U}_k \end{aligned} \tag{35}$$

where k represents time instance t_k and:

$$\mathbf{X}_k = \begin{bmatrix} R_x(BS_1, t_k) \\ R_x(BS_2, t_k) \\ R_x(BS_3, t_k) \\ v_{Rx}(BS_1, t_k) \\ v_{Rx}(BS_2, t_k) \\ v_{Rx}(BS_3, t_k) \end{bmatrix}, \quad \mathbf{Y}_k = \begin{bmatrix} \hat{R}_x(BS_1, t_k) \\ \hat{R}_x(BS_2, t_k) \\ \hat{R}_x(BS_3, t_k) \end{bmatrix} \tag{36}$$

$\mathbf{X}_k$ represents the combined vector of RSS values from three BSs ($R_x(BS_i, t_k)$, i=1,2,3) and the RSS rate of change.

$\mathbf{Y}_k$ represents the vector of the measured RSS levels $\hat{R}_x(BS_i, t_k)$, i=1,2,3.

Φ, Γ and $\mathbf{M}$ are the (6x6), (6x3) and (3x6) matrixes that describe the process as follows:

$$\Phi = \begin{bmatrix} 1 & 0 & 0 & \Delta t & 0 & 0 \\ 0 & 1 & 0 & 0 & \Delta t & 0 \\ 0 & 0 & 1 & 0 & 0 & \Delta t \\ 0 & 0 & 0 & 1 & 0 & 0 \\ 0 & 0 & 0 & 0 & 1 & 0 \\ 0 & 0 & 0 & 0 & 0 & 1 \end{bmatrix} \quad \Gamma = \begin{bmatrix} 0 & 0 & 0 \\ 0 & 0 & 0 \\ 0 & 0 & 0 \\ \Delta t & 0 & 0 \\ 0 & \Delta t & 0 \\ 0 & 0 & \Delta t \end{bmatrix} \quad \mathbf{M} = \begin{bmatrix} 1 & 0 & 0 & 0 & 0 & 0 \\ 0 & 1 & 0 & 0 & 0 & 0 \\ 0 & 0 & 1 & 0 & 0 & 0 \end{bmatrix} \tag{38}$$

Where, Δt equals to the measurements sampling period (e.g., according to the standard GSM MT operation Δt belongs in the range of 2sec-5sec). $\mathbf{W}_k$ is a Gaussian distributed random vector parameter with zero mean and a covariance matrix $\mathbf{Q}$ where $\sigma^2_{Qrx}(i)$ (i=1,2,3) represents the variance of the RSS change rate of BS(i):

$$\mathbf{W}_k = \begin{bmatrix} w_1(t_k) \\ w_2(t_k) \\ w_3(t_k) \end{bmatrix}, \mathbf{Q} = \begin{bmatrix} \sigma^2_{Qrx}(1) & 0 & 0 \\ 0 & \sigma^2_{Qrx}(2) & 0 \\ 0 & 0 & \sigma^2_{Qrx}(3) \end{bmatrix} \quad (39)$$

As shown in [2] based on a set of simplifying assumptions and based on path loss models like Hata [3] or Cost231 [4] we can approximate the variance of the RSS rate of change as follows:

$$\sigma^2_{Qrx}(i) \approx \left[\frac{[44,9 - 6,55 \cdot \log(h_{bs}(i))]}{d_{gm}(i) \cdot \ln(10)} \right]^2 \cdot \frac{\sigma^2_{Qmt}}{2} \quad (40)$$

Where σ^2_{Qmt} is the variance of the terminal velocity which is the one assumed in the Kalman filtering applied in the estimated positions [6]. Parameter $d_{gm}(i)$ represents the geometric mean of the distance between the MT and BS i:

$$d_{gm}(i) = \sqrt[N]{\prod_{k=1}^{N} d(i,t_k)} = 10^{-\left[\frac{\overline{R}_x(i)+A(i)}{44,9-6,55\log(h_{bs}(i))}\right]} \quad (41)$$

Where $\overline{R}_x(i)$ represents the average RSS level from BS(i).

$\mathbf{U}_k$ in Eq. (35) is also a zero mean Gaussian random vector parameter with a covariance matrix $\mathbf{R}$:

$$\mathbf{U}_k = \begin{bmatrix} u_1(t_k) \\ u_2(t_k) \\ u_3(t_k) \end{bmatrix}, \mathbf{R} = \begin{bmatrix} \sigma^2_{Rx} & 0 & 0 \\ 0 & \sigma^2_{Rx} & 0 \\ 0 & 0 & \sigma^2_{Rx} \end{bmatrix} \quad (42)$$

Where, σ_{Rx}^2 is the variance of RSS measurements. Based on [24] the standard deviation of the RSS is in the order of 8dbm therefore this input can be used for estimating $\mathbf{R}$.

As mentioned in [21], the main problem with the employment of the RSS Kalman filter is the fact that the filter ideally requires the RSS of the same set of cells to be filtered. However, in a real network the set of cells being monitored by the MT changes over time depending on the MT position, mobility and network topology. Since Kalman filter requires a certain set of steps in order to converge to an optimal solution, the issue of the dynamically changing set of cells may dilute the efficiency of RSS Kalman filtering. In [2] a solution to this problem has been proposed.

5.2. Option B: MT-BS Distance Kalman Filter

The second alternative method proposed in [2] is the statistical processing of the distances between the MT and the three BSs (see Figure 6). The Kalman filter provided in Eq. (35) is also considered in this case with appropriately adjusted parameters:

$$\mathbf{X_k} = \begin{bmatrix} d(1,t_k) \\ d(2,t_k) \\ d(3,t_k) \\ v_d(1,t_k) \\ v_d(2,t_k) \\ v_d(3,t_k) \end{bmatrix}, \quad \mathbf{Y_k} = \begin{bmatrix} \hat{d}(1,t_k) \\ \hat{d}(2,t_k) \\ \hat{d}(3,t_k) \end{bmatrix} \quad (43)$$

$\mathbf{X_k}$ in this model is the combined vector of MT-BS distance values from three BSs $d(i,t_k)$ and the distance rate of change $v_d(i,t_k)$ (i=1,2,3). $\mathbf{Y_k}$ represents the vector of the estimated MT-BS distances $\hat{d}(i,t_k)$, (i=1,2,3). Tables Φ, Γ, M are the ones provided in the RSS Kalman filtering model.

$\mathbf{W_k}$ is assumed to be a zero mean Gaussian random vector parameter with covariance matrix $\mathbf{Q}$:

$$\mathbf{Q} = \begin{bmatrix} \sigma_{Qd}^2(1) & 0 & 0 \\ 0 & \sigma_{Qd}^2(2) & 0 \\ 0 & 0 & \sigma_{Qd}^2(3) \end{bmatrix} \quad (44)$$

$\sigma_{Qd}^2(i)$ (i=1,2,3) represents the variance of the rate of change of the distance between the MT and BS(i).

As analysed in [2] we can estimate its variance as follows:

$$\sigma_{Qd}^2(i) \approx \frac{\sigma_{Qmt}^2}{2} \quad (45)$$

$\mathbf{U_k}$ is also assumed to be a zero mean Gaussian random vector parameter with a covariance matrix $\mathbf{R}$:

$$R = \begin{bmatrix} \sigma_{Rd}^2(1) & 0 & 0 \\ 0 & \sigma_{Rd}^2(2) & 0 \\ 0 & 0 & \sigma_{Rd}^2(3) \end{bmatrix} \quad (46)$$

Where, $\sigma_{Rd}^2(i)$ (i=1,2,3) is the variance of $\hat{d}(i)$ estimated distance.

Similarly with the case of RSS Kalman filter, the MT-BS distance Kalman filter faces the problem of dynamically changing set of monitored cells. The problem is tackled from the solution proposed in [2].

5.3. Option C: MT Position Kalman Filter

Option C is described in Section 4 where STAMP concept overview has been presented.

5.4. MT Positioning Accuracy vs. Filtering Option

In this section we provide an overview of the performance of the various Kalman filtering options based on the simulation model described in Section 3. The analysis presented in [2] indicated that RSS Kalman filtering (Option A) is the best performing option with MT Coordinates and MT-BS distance filtering options following. In the current analysis we consider also the possibility to apply the MT-BS distance estimators developed in [1] on the various Kalman filtering options aiming at further improvement of the MT positioning accuracy.

In Figures 7 and 8 we provide the performance of RSS, MT-BS distance and MT Coordinates Kalman filtering (Options A and B) respectively versus the variance of the RSS and versus the applied MT-BS distance estimator for a cell radius of 1500m. The performance of Option C (MT Coordinates Kalman filtering) has been provided in Figure 5. As it can be seen from these figures, depending on the applied Kalman filter option, the various MT-BS distance estimators lead to different positioning accuracy performance. In particular for RSS Kalman filtering (Option A) estimator 0 (plain estimator) provides the best performance, for MT-BS distance Kalman filtering (Option B) estimator 1 (unbiased MT-BS distance estimator) provides the best performance while for MT coordinates Kalman filtering (Option C) as presented in the section on STAMP concept overview, estimator 3 (unbiased MT coordinates estimator) provides the best performance.

The results presented in this section create an interesting research challenge: why the various Kalman filtering options have a different performance vs. the applied MT-BS distance estimators. This is the main driver of the novel contribution of the current research work which is presented in the next section.

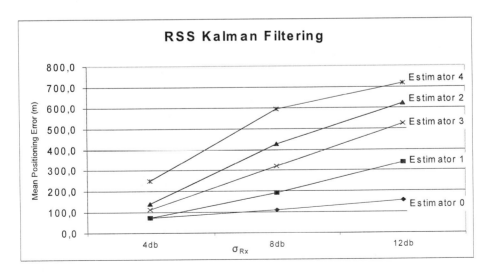

Figure 7. The performance of the RSS Kalman filtering vs. the applied MT-BS distance estimator.

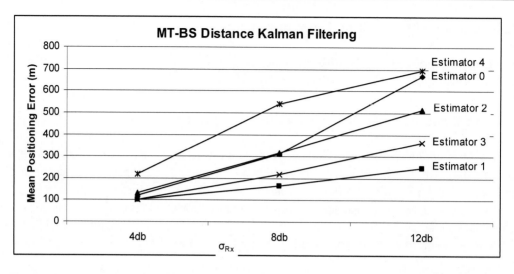

Figure 8. The performance of the MT-BS Distance Kalman filtering vs. the applied MT-BS distance estimator.

6. MT-BS Distance Estimators vs. Kalman Filtering Options

Based on the analysis presented in the previous paragraph it is evident that there is a need to understand the insights of the impact of Kalman filtering options vs. the applied MT-BS distance estimators. The analysis presented in this section addresses this problem. As a first step the steady state Kalman filter performance is analyzed and then a statistical equivalent model is built for each Kalman filtering option based on a set of assumptions on the impact of Kalman filtering on the statistical characteristics of its input and output. Based on this simplified model the MT-BS estimators are being revised so as to be adapted to the applied filtering option. The performance of the revised MT-BS estimators is then analyzed so as to assess the impact of this analysis on the achieved MT positioning accuracy.

6.1. Generalized Kalman Filter Model

Based on the analysis of the various Kalman filtering options it is evident that a generalized route tracking Kalman filter model is their common basis:

$$X_k = \Phi \cdot X_{k-1} + \Gamma \cdot W_k$$
$$Y_k = M \cdot X_k + U_k \qquad (47)$$

Where k represents time instance t_k and vectors $\mathbf{X_k}$ and $\mathbf{Y_k}$:

$$\mathbf{X_k} = \begin{bmatrix} x_k(1) \\ \dots \\ x_k(N) \\ v_k(1) \\ \dots \\ v_k(N) \end{bmatrix}, \mathbf{Y_k} = \begin{bmatrix} \hat{x}_k(1) \\ \dots \\ \hat{x}_k(2) \end{bmatrix} \quad (48)$$

The above vectors correspond to different parameters depending on the applied Kalman filter option: (a) RSS Kalman filtering i=1,2,3: $x_k(i)=R_x(i)$ the RSS from BS i, $v_k(i)$ is the rate of change of $R_x(i)$ and $\hat{x}_k(i) = \hat{R}_x(i)$ is the measured RSS, (b) MT-BS Distance Kalman Filtering i=1,2,3: $x_k(i)=d(i)$ the distance between the MT and BS i, $v_k(i)$ is the rate of change of d(i) and $\hat{x}_k(i) = \hat{d}(i)$ is the MT-BS distance estimated based on the measured RSS and (c) MT Coordinates Kalman Filtering i=1,2: $x_k(1)=x$, $x_k(2)=y$ the MT position coordinates, $v_k(1)= v_x(t_k)$, $v_k(2)= v_y(t_k)$ is the horizontal and vertical MT speed and $\hat{x}_k(1) = \hat{x}(t_k)$, $\hat{x}_k(2) = \hat{y}(t_k)$ is the MT coordinates estimated based on the measured RSS.

Moreover, Φ, Γ and M are the (2Nx2N), (2NxN) and (Nx2N) matrixes that describe the process as follows:

$$\Phi = \begin{bmatrix} \mathbf{I_N} & \Delta t \cdot \mathbf{I_N} \\ \mathbf{0_N} & \mathbf{I_N} \end{bmatrix}, \Gamma = \begin{bmatrix} \mathbf{0_N} \\ \Delta t \cdot \mathbf{I_N} \end{bmatrix}, \mathbf{M} = \begin{bmatrix} \mathbf{I_N} & \mathbf{0_N} \end{bmatrix} \quad (49)$$

Where I_N is the NxN unitary matrix and 0_N is the zero NxN matrix and Δt corresponds to the measurements sampling period. As mentioned above in RSS Kalman filtering N=3, in MT-BS Distance Kalman filtering N=3 and in MT Coordinates Kalman filtering N=2.

$\mathbf{W_k}$ is a zero mean Gaussian distributed random vector parameter which represents the process noise and a covariance matrix $\mathbf{Q}$ where $\sigma_Q^2(i)$ (i=1,...,N) represents the variance of $v_k(i)$:

$$\mathbf{W_k} = \begin{bmatrix} w_1(t_k) \\ \dots \\ w_N(t_k) \end{bmatrix}, \mathbf{Q} = \begin{bmatrix} \sigma_Q^2(1) & 0 & \dots & 0 \\ 0 & \sigma_Q^2(2) & \dots & 0 \\ \dots & \dots & \dots & \dots \\ 0 & 0 & \dots & \sigma_Q^2(N) \end{bmatrix} \quad (50)$$

U_k is also a zero mean Gaussian random vector parameter which represents the measurement noise with a covariance matrix **R**:

$$\mathbf{U_k} = \begin{bmatrix} u_1(t_k) \\ \cdots \\ u_N(t_k) \end{bmatrix}, \quad \mathbf{R} = \begin{bmatrix} \sigma_R^2(1) & 0 & \cdots & 0 \\ 0 & \sigma_R^2(2) & \cdots & 0 \\ \cdots & \cdots & \cdots & \cdots \\ 0 & 0 & \cdots & \sigma_R^2(N) \end{bmatrix} \quad (51)$$

Where, $\sigma_R^2(i)$ is the variance of $\hat{x}_k(i)$ (i=1,...,N) measurements.

6.2. Steady State Kalman Filter Performance

One of the key elements for understanding the role of Kalman filtering in the process of MT position estimation is the performance of the filtering process in steady state conditions. According to [25] in order to have a convergent steady state behavior the following conditions should hold:

(a) The pair {Φ, M} in the Kalman filter state-space Eq. (47), should be completely observable:

$$rank \begin{bmatrix} \boldsymbol{\Phi} \\ \mathbf{M} \cdot \boldsymbol{\Phi} \\ \cdots \\ \mathbf{M} \cdot \boldsymbol{\Phi}^{N-1} \end{bmatrix} = 2N \quad (52)$$

It is easy to show that the above condition is valid based on the fact that:

$$\mathbf{M} \cdot \boldsymbol{\Phi}^m = \begin{bmatrix} \mathbf{I_N} & \mathbf{O_N} \end{bmatrix} \cdot \begin{bmatrix} \mathbf{I_N} & m \cdot \Delta t \cdot \mathbf{I_N} \\ \mathbf{O_N} & \mathbf{I_N} \end{bmatrix} = \begin{bmatrix} \mathbf{I_N} & m \cdot \Delta t \cdot \mathbf{I_N} \end{bmatrix}, \, (m=1,\ldots,N-1) \quad (53)$$

(b) The pair {Φ, $G^{1/2}$} should be completely controllable:

$$rank \begin{bmatrix} \mathbf{G}^{1/2} & \boldsymbol{\Phi} \cdot \mathbf{G}^{1/2} & \cdots & \boldsymbol{\Phi}^{N-1} \cdot \mathbf{G}^{1/2} \end{bmatrix} = 2N \quad (54)$$

Where:

$$\mathbf{G} = \mathbf{G}^{1/2} \cdot \mathbf{G}^{T/2} = \boldsymbol{\Gamma} \cdot \mathbf{Q} \cdot \boldsymbol{\Gamma}^T = \Delta t^2 \cdot \begin{bmatrix} \mathbf{O_N} & \mathbf{O_N} \\ \mathbf{O_N} & \mathbf{Q} \end{bmatrix} \quad (55)$$

And therefore:

$$G^{1/2} = \Delta t \cdot \begin{bmatrix} O_N & O_N \\ O_N & Q^{1/2} \end{bmatrix}$$

$$G^{1/2} \cdot \Phi^m = \Delta t \cdot \begin{bmatrix} O_N & m \cdot \Delta t \cdot Q^{1/2} \\ O_N & Q^{1/2} \end{bmatrix} \quad (56)$$

Based on Eq. (56) it is easy to prove that Eq. (54) holds. As both Eq. (52) and Eq. (54) hold then there is a unique positive solution P in the following algebraic Riccati equation:

$$P = \Phi \cdot P \cdot \Phi^T + G - \Phi \cdot P \cdot M^T \cdot (M \cdot P \cdot M^T + R)^{-1} \cdot M \cdot P \cdot \Phi^T \quad (57)$$

The solution P of the algebraic Riccati equation determines the innovations covariance matrix:

$$S = M \cdot P \cdot M^T + R \quad (58)$$

And the Steady State Kalman gain K will be [25]:

$$K = P \cdot M^T \cdot S^{-1} \quad (59)$$

The analytical solution of the above Riccati equation requires the estimation of the eigenvalues of the following matrix [25]:

$$[I_N \quad -P] \cdot F \cdot \begin{bmatrix} P \\ I_N \end{bmatrix} = 0 \quad (60)$$

Where:

$$F = \begin{bmatrix} \Phi^{-1} & \Phi^{-1} \cdot G \\ -M^T \cdot R^{-1} \cdot M \cdot \Phi^{-1} & \Phi^T + M^T \cdot R^{-1} \cdot M \cdot \Phi^{-1} \cdot G \end{bmatrix} \quad (61)$$

$$F = \begin{bmatrix} I_N & -\Delta t \cdot I_N & O_N & -\Delta t^3 \cdot Q \\ O_N & I_N & O_N & \Delta t^2 \cdot Q \\ -R^{-1} & \Delta t \cdot R^{-1} & I_N & -\Delta t^3 \cdot R^{-1} \cdot Q \\ O_N & O_N & \Delta t \cdot I_N & I_N \end{bmatrix} \quad (62)$$

In this paper we have used an approximation for the solution of the above Riccati equation based on iteration.

In order to proceed with the analysis of the definition of revised estimators we have adopted a set of assumptions regarding the impact of Kalman filtering:

(a) In terms of statistical behavior, the pdf of the Kalman filter output can be approximated by the pdf of the input parameter.
(b) in the steady state the mean value of the Kalman filter output will be equal to the mean value of the Kalman filter input:

$$E(\hat{x}_{k|k}) = E(x_k) \qquad (63)$$

(c) in the steady state the variance of the Kalman filter output will be lower than the variance of the Kalman filter input representing the filter gain:

$$V(\hat{x}_{k|k}) = (\mathbf{I_N} - \mathbf{K} \cdot \mathbf{M}) \cdot V(x_k) \qquad (64)$$

Based on the state space Eq. (47), it is easy to show that for the Kalman filtering options considered in this paper Eq. (64) can be re-written as follows:

$$V(\hat{x}_{k|k}) = \begin{bmatrix} f(1) & 0 & \ldots & 0 \\ 0 & f(2) & \ldots & 0 \\ \ldots & \ldots & \ldots & \ldots \\ 0 & 0 & \ldots & f(N) \end{bmatrix} \cdot V(x_k) \qquad (65)$$

Where $0 < f(i) \leq 1$ represents the steady state Kalman Filter gain for parameter $x_k(i)$ (i=1,...,N).

6.3. Analysis of Option A: RSS Kalman Filter

Option A: The pdf of the Kalman Filter output

The measured RSS is considered to be following a Normal distribution as described in [1]:

$$\hat{R}_x(i) \sim N(R_x(i), \sigma_{Rx}^2) \qquad (66)$$

The above condition holds under the assumption that a common σ_{Rx}^2 applies for all BSs involved in the MT positioning which is the assumption used in the current paper. Based on the assumption provided in the previous section it can be concluded that in the steady state the resulting series of filtered RSS measurements will follow a Normal distribution with the same mean and a lower variance:

Optimization of Kalman Filtering Performance in Received Signal Strength... 29

$$\hat{R}_{xA}(i) \sim N\left(R_x(i), f_{Rx} \cdot \sigma_{Rx}^2\right) \qquad (67)$$

Figure 9 provides the Kalman Filter gain factor f_{Rx} versus σ_{Rx} as it has been produced based on simulation as well as based on the theoretical steady state Kalman filter performance.

Based on the above analysis it is feasible to consider an MT position calculation process which is equivalent to RSS Kalman filtering option (see Figure 10). As it can be seen from this figure, in the simplified process the impact of the Kalman filter has been replaced with the proper statistical characteristics of equivalent RSS measurements. The validity of the equivalent model has been verified through simulation. In the next section we will use this model in order to produce the revised MT-BS distance estimators which fit the RSS Kalman filtering option.

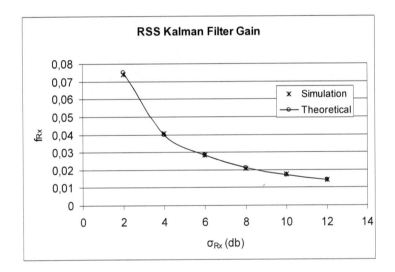

Figure 9. The RSS Kalman filtering gain versus σ_{Rx}.

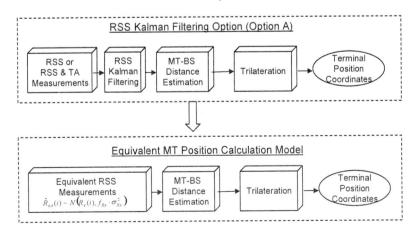

Figure 10. The RSS Kalman filtering Equivalent MT Position Calculation Model.

Option A: The revised MT-BS Distance Estimators

In the context of the above analysis and the analysis provided in [1] the MT-BS distance estimators in the case of RSS Kalman filtering will be the following:

(a) Estimator 0: Basic estimator for MT-BS distance: the one provided by Eq. (2).
(b) Estimator 1: The Unbiased MT-BS distance estimator:

$$\hat{d}_{1A}(i) = e^{-\beta(i) \frac{f_{Rx} \cdot \sigma_{Rx}^2}{2}} \cdot \hat{d}_o(i) \tag{68}$$

(c) Estimator 2: The Minimum MT-BS distance MSE estimator:

$$\hat{d}_{2A}(i) = e^{-\beta(i) \frac{3 f_{Rx} \sigma_{Rx}^2}{2}} \cdot \hat{d}_o(i) \tag{69}$$

(d) Estimator 3: The Unbiased MT coordinates estimator:

$$\hat{d}_{3A}(i) = e^{-\beta(i) \cdot f_{Rx} \sigma_{Rx}^2} \cdot \hat{d}_o(i) \tag{70}$$

(e). Estimator 4: The sub-optimal MT coordinates MSE estimator:

$$\hat{d}_{4A}(i) = e^{-3\beta(i) \cdot f_{Rx} \cdot \sigma_{Rx}^2} \cdot \hat{d}_o(i) \tag{71}$$

It should be noted that for $f_{Rx}=1$ corresponding to the case that Kalman filtering is not applied the estimators defined above become equal to the estimators defined for the plain RSS MT positioning case as defined [1].

6.4. Analysis of Option B: MT-BS Distance Kalman Filter

Option B: The pdf for the Kalman Filter output

The MT-BS distance is considered to be following a Lognormal distribution as described in [1]:

$$\begin{aligned} \hat{d}_o(i) &\sim LN\left(\mu_d(i), \sigma_d^2(i)\right) \\ \mu_d(i) &= \ln[d(i)] = \ln\left[10^{-\frac{R_x(i)+A(i)}{B(i)}}\right] \\ \sigma_d^2(i) &= \beta(i) \cdot \sigma_{Rx}^2 \end{aligned} \tag{72}$$

Based on the analysis presented in the previous section on the impact of Kalman filtering the pdf of the filter output can be approximated by the following:

$$\hat{d}_{oB}(i) \sim LN\left(\mu_{dB}(i), \sigma_{dB}^2(i)\right) \tag{73}$$

From the same analysis at the Kalman Filter steady state we should have:

$$E\left[\hat{d}_{oB}(i)\right] = E\left[\hat{d}_o(i)\right] \tag{74}$$

$$V\left[\hat{d}_{oB}(i)\right] = f_d(i) \cdot V\left[\hat{d}_o(i)\right] \tag{75}$$

Where $f_d(i)$ represents the Kalman filter gain.
From Eq. (74) we have:

$$e^{\mu_d(i) + \sigma_d^2(i)/2} = e^{\mu_{dB}(i) + \sigma_{dB}^2(i)/2} \tag{76}$$

And from Eq. (75):

$$e^{2\mu_d(i) + \sigma_d^2(i)} \cdot \left(e^{\sigma_d^2} - 1\right) = f_d(i) \cdot e^{\mu_{dB}(i) + \sigma_{dB}^2(i)/2} \cdot \left(e^{\sigma_{dB}^2} - 1\right) \tag{77}$$

Based on Eq. (76), Eq. (77) becomes:

$$f_d(i) \cdot \left(e^{\sigma_d^2(i)} - 1\right) = \left(e^{\sigma_{dB}^2(i)} - 1\right) \Rightarrow \sigma_{dB}^2(i) = \ln\left[1 + f_d(i) \cdot \left(e^{\sigma_d^2(i)} - 1\right)\right] \Rightarrow$$
$$\sigma_{dB}^2(i) = \ln\left[1 + f_d(i) \cdot \left(e^{\beta(i)\sigma_{Rx}^2} - 1\right)\right] \tag{78}$$

And based on Eq. (78) we get from Eq. (76):

$$\mu_{dB}(i) = \ln\left[d(i) \cdot \frac{e^{\beta(i) \cdot \sigma_{Rx}^2/2}}{\sqrt{1 + f_d(i) \cdot \left(e^{\beta(i) \cdot \sigma_{Rx}^2} - 1\right)}}\right] \tag{79}$$

The MT position calculation process which is equivalent to MT-BS distance Kalman filtering option is depicted in Figure 11. In the simplified process the Kalman filter has been replaced with the proper statistical characteristics of MT-BS distances. The validity of the equivalent model has been verified through simulation. In the next section we will use this model in order to produce the revised MT-BS distance estimators which fit the MT-BS Distance Kalman filtering option.

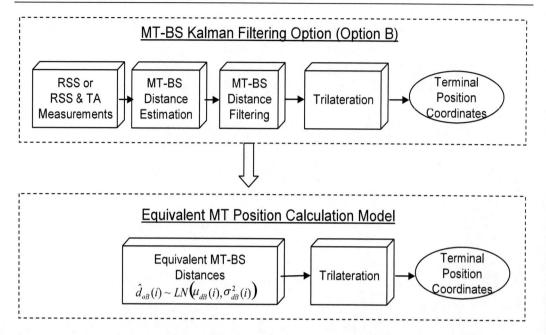

Figure 11. The MT-BS Distance Kalman filtering Equivalent MT Position Calculation Model.

Option B: The Revised MT-BS Distance Estimators

Taking into account the results of the previous section it is now possible to define the revised MT-BS distance estimators which are appropriate for the MT-BS Kalman filtering method.

(a) Estimator 0: Basic estimator for MT-BS distance: the one provided by Eq. (2).
(b) Estimator 1: The Unbiased MT-BS distance estimator is defined as follows:

$$d(i) = E\left[\hat{d}_{1B}(i)\right] = E\left[c_{1B}(i) \cdot \hat{d}_{oB}(i)\right] = c_{1B}(i) \cdot E\left[\hat{d}_{oB}(i)\right] = c_{1B}(i) \cdot E\left[\hat{d}_o(i)\right] = c_{1B}(i) \cdot e^{\beta(i)\frac{\sigma_{Rx}^2}{2}} \cdot d(i) \Rightarrow \quad (80)$$

$$c_{1B}(i) = e^{-\beta(i)\frac{\sigma_{Rx}^2}{2}} = c_1(i)$$

Therefore, the unbiased MT-BS distance estimator is identical to the one applied for the case that no Kalman filtering is applied (see Eq. (5)) due to the fact that the Kalman filter does not affect the mean value according to the assumption of the previous section (Eq. (63)).

(c) Estimator 2: The Minimum MT-BS distance MSE estimator: In this case we should identify an estimator that minimizes the following:

$$MSE\left(\hat{d}_{2B}(i)\right) = E\left(\left[\hat{d}_{2B}(i) - d(i)\right]^2\right) = c_{2B}^2(i) \cdot E\left(\hat{d}_{oB}^2(i)\right) - 2 \cdot d(i) \cdot c_{2B}(i) \cdot E\left(\hat{d}_{oB}(i)\right) + d^2(i) \quad (81)$$

The minimum MSE is achieved for the following estimator:

$$c_{2B}(i) = \frac{d(i) \cdot E(\hat{d}_{oB}(i))}{E(\hat{d}_{oB}^2(i))} = \frac{d^2(i) \cdot e^{\beta(i)\cdot\sigma_{Rx}^2/2}}{e^{2\mu'_d + 4\sigma'^2_d/2}} = \frac{d^2(i) \cdot e^{\beta(i)\cdot\sigma_{Rx}^2/2}}{d^2(i) \cdot e^{\beta(i)\cdot\sigma_{Rx}^2} \cdot \left[1 + f_d(i) \cdot (e^{\beta(i)\cdot\sigma_{Rx}^2} - 1)\right]} \Rightarrow$$

$$c_{2B}(i) = \frac{e^{-\beta(i)\cdot\sigma_{Rx}^2/2}}{\left[1 + f_d(i) \cdot (e^{\beta(i)\cdot\sigma_{Rx}^2} - 1)\right]} \quad (82)$$

(a) Estimator 3: The Unbiased MT coordinates estimator: From the analysis presented in [1] on trilateration process in order to achieve unbiased MT coordinates estimation the following condition should apply:

$$d^2(i) = E(\hat{d}_{3B}^2(i)) = c_{3B}^2(i) \cdot e^{2\mu'_d(i) + 4\sigma'^2_d(i)/2} = c_{3B}^2(i) \cdot d^2(i) \cdot e^{\beta(i)\cdot\sigma_{Rx}^2} \cdot \left[1 + f_d(i) \cdot \left(e^{\beta(i)\cdot\sigma_{Rx}^2} - 1\right)\right] \Rightarrow$$

$$c_{3B}(i) = \frac{e^{-\beta(i)\sigma_{Rx}^2/2}}{\sqrt{\left[1 + f_d(i) \cdot \left(e^{\beta(i)\sigma_{Rx}^2} - 1\right)\right]}} \quad (83)$$

(b) Estimator 4: The sub-optimal MT coordinates MSE estimator: According to the analysis presented in [1] on the trilateration process this estimator should minimize the following quantity:

$$\min\left\{E\left[\left[\hat{d}_{4B}^2(i) - d^2(i)\right]^2\right]\right\} = \min\left\{V\left(\hat{d}_{4B}^2(i) - d^2(i)\right) + E\left(\hat{d}_{4B}^2(i) - d^2(i)\right)\right\} \quad (84)$$

Based on the assumption that:

$$\hat{d}_{4B}(i) = c_{4B}(i) \cdot \hat{d}_{oB}(i) \quad (85)$$

it is easy to prove that the MSE minimization is achieved for:

$$c_{4B}(i) = d(i) \cdot \sqrt{\frac{E(\hat{d}_o^2(i))}{E(\hat{d}_o^2(i))^2 + V(\hat{d}_o^2(i))}} \Rightarrow$$

$$c_{4B}(i) = \frac{e^{-\beta(i)\cdot\sigma_{Rx}^2/2}}{\sqrt{\left[1 + f_d(i) \cdot \left(e^{-\beta(i)\cdot\sigma_{Rx}^2} - 1\right)\right]^5}} \quad (86)$$

It should be noted that for $f_d(i)=1$ (corresponding to the case that Kalman Filter is not applied at all and therefore no gain occurs) the above estimators become identical to the ones applicable for the non-filtered process (plain RSS based positioning) described in [1].

6.5. Analysis of Option C: MT Coordinates Kalman Filter

Option C: The pdf for the Kalman Filter output

Based on the analysis presented in [1] the MT coordinates (x,y) after the trilateration step can be expressed as follows:

$$\hat{z}_j = z + \frac{1}{D}\left[\sum_{i=1}^{3} a_z(i) \cdot \left[\hat{d}_j^2(i) - d^2(i)\right]\right], \text{ z=x,y} \qquad (87)$$

Where for (x_i, y_i) i=1,2,3 the coordinates of the BSs considered for the trilateration process:

$$D = 4 \cdot (x_1 - x_2) \cdot (y_1 - y_3) - 4 \cdot (x_1 - x_3) \cdot (y_1 - y_2) \qquad (88)$$

$$a_x(1) = -(y_2 - y_3), \; a_x(2) = (y_1 - y_3), \; a_x(3) = -(y_1 - y_2) \qquad (89)$$

$$a_y(1) = (x_2 - x_3), \; a_y(2) = -(x_1 - x_3), \; a_y(3) = (x_1 - x_2) \qquad (90)$$

The pdf of the MT coordinates is not known as it corresponds to a sum of Lognormal distributed parameters. Based on the assumptions adopted for the steady state performance of the Kalman Filter:

$$E(\hat{z}_{jC}) = E(\hat{z}_j), \text{ z=x,y} \qquad (91)$$

$$V(\hat{z}_{jC}) = f_z \cdot V(\hat{z}_j), \text{ z=x,y} \qquad (92)$$

In this Kalman filtering option we adopt the equivalent MT position calculation model depicted in Figure 12. As it can be seen from this figure it is assumed that the equivalent MT position calculation model is based on a set of equivalent MT-BS distances which follow a Lognormal distribution with proper mean and variance parameters. To derive the characteristics of the equivalent MT-BS distances the conditions provided by Eqs. (63) and (64) are exploited. In particular Eq. (91) leads to the following condition:

$$E(\hat{z}_{jC}) = E(\hat{z}_j) \Rightarrow \sum_{i=1}^{3} a_z(i) \cdot \left[E(\hat{d}_{oC}^2(i,z)) - E(\hat{d}_o^2(i))\right] = 0, \text{ z=x,y} \qquad (93)$$

And Eq. (92) leads to the following:

$$V(\hat{z}_{jC}) = f_z \cdot V(\hat{z}_j) \Rightarrow \sum_{i=1}^{3} a_z^2(i) \cdot \left[V(\hat{d}_{oC}^2(i,z)) - f_z \cdot V(\hat{d}_o^2(i))\right] = 0, \text{ z=x,y} \qquad (94)$$

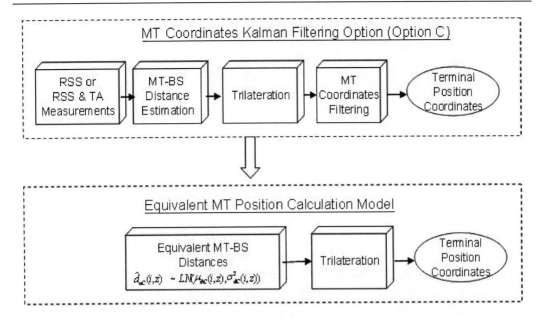

Figure 12. The MT Coordinates Kalman filtering Equivalent MT Position Calculation Model.

One solution that justifies both conditions provided by Eqs. (93) and (94) is the following:

$$E\left(\hat{d}_{oC}^2(i,z)\right) = E\left(\hat{d}_o^2(i)\right), \text{ z=x,y} \tag{95}$$

$$V\left(\hat{d}_{oC}^2(i,z)\right) = f_z \cdot V\left(\hat{d}_o^2(i)\right), \text{ z=x,y} \tag{96}$$

Assuming that $\hat{d}_{oC}(i,z)$ follows a Log-Normal distribution $LN(\mu_{dC}(i,z), \sigma_{dC}^2(i,z))$ we have based on Eq. (96):

$$V\left(\hat{d}_{oC}^2(i,z)\right) = f_z \cdot V\left(\hat{d}_o^2(i)\right) \Rightarrow E\left(\hat{d}_{oC}^2(i,z)\right) \cdot \left[e^{4\sigma_{dC}^2(i,z)} - 1\right] = f_z \cdot E\left(\hat{d}_o^2(i)\right) \cdot \left[e^{4\beta(i)\sigma_{Rx}^2} - 1\right], \text{ z=x,y} \tag{97}$$

Therefore:

$$\sigma_{dC}^2(i,z) = \ln\left[\sqrt[4]{1 - f_z \cdot \left(e^{4\beta(i)\sigma_{Rx}^2} - 1\right)}\right], \text{ z=x,y} \tag{98}$$

In this case Eq. (95) leads to the following:

$$e^{2\mu_{dC}(i,z)+4\sigma_{dC}^2(i,z)/2} = d^2(i) \cdot e^{2\beta(i)\sigma_{Rx}^2} \Rightarrow$$

$$\mu_{dC}(i,z) = \ln\left[\frac{d(i) \cdot e^{\beta(i)\sigma_{Rx}^2}}{\sqrt[4]{1-f_z \cdot \left(e^{4\beta(i)\sigma_{Rx}^2}-1\right)}}\right], \; z=x,y \qquad (99)$$

It is interesting to note that the equivalent MT-BS distances are a function of the BS i (i=1,2,3) as well as the MT coordinate z (z=x,y). As it will be shown in the next section this leads to different MT-BS distance estimators which should be used for the estimation of each of the MT coordinates.

Option C: The Revised MT-BS Distance Estimators

Based on the above analysis it is now feasible to proceed with the estimation of the revised MT-BS estimators:

(a) Estimator 0: Basic Estimator: the one provided by Eq. (2).
(b) Estimator 1: Unbiased MT-BS distance estimator: In this case the following condition should be satisfied:

$$E(\hat{d}_{1C}(i,z)) = d(i) \Rightarrow E(c_{1C}(i,z) \cdot \hat{d}_{oC}(i,z)) = d(i) \Rightarrow c_{1C}(i,z) = \frac{d(i)}{E(\hat{d}_{oC}(i,z))}, \; z=x,y \qquad (100)$$

Based on Eq. (98) and (99) we have:

$$c_{1C}(i,z) = e^{-\beta(i)\sigma_{Rx}^2} \cdot \sqrt[8]{1+f_z \cdot \left(e^{4\beta(i)\sigma_{Rx}^2}-1\right)}, \; z=x,y \qquad (101)$$

(c) Estimator 2: Minimum MT-BS distance MSE estimator: In this case the following condition should be minimized:

$$MSE(\hat{d}_{2C}(i,z)) = E\left[\left(\hat{d}_{2C}(i,z)-d(i)\right)^2\right] = c_{2C}^2(i,z) \cdot E(\hat{d}_{oC}^2(i,z)) + d^2(i) -$$
$$2d(i)c_{2C}(i,z) \cdot E(\hat{d}_{oC}(i,z))$$
$$, \; z=x,y \qquad (102)$$

It is easy to show that the above condition is minimized for the following estimator:

$$c_{2C}(i,z) = \frac{d(i) \cdot E(\hat{d}_{oC}(i,z))}{E(\hat{d}_{oC}^2(i,z))} = \frac{e^{-\beta(i)\sigma_{Rx}^2}}{\sqrt[8]{1+f_z \cdot \left(e^{4\beta(i)\sigma_{Rx}^2}-1\right)}}, \; z=x,y \qquad (103)$$

(d) Estimator 3: Unbiased MT-coordinates Estimator: The condition that needs to be satisfied in this case is the following:

$$E(\hat{z}_{3C}) = z \Rightarrow \sum_{i=1}^{3} a_z(i)\left[E(\hat{d}_{3C}^2(i,z)) - d^2(i)\right] = 0, \, z=x,y \quad (104)$$

An obvious solution for the above condition is the following:

$$E(\hat{d}_{3C}^2(i,z)) = d^2(i) \Rightarrow E(c_{3C}^2(i,z) \cdot \hat{d}_{oC}^2(i,z)) = d^2(i) \Rightarrow$$
$$c_{3C}(i,z) = \frac{d(i)}{\sqrt{E(\hat{d}_{oC}^2(i,z))}}, \, z=x,y \quad (105)$$

And based on Eqs. (98) and (99):

$$c_{3C}(i,z) = e^{-\beta(i)\sigma_{Rx}^2} = c_3(i), \, z=x,y \quad (106)$$

From the above equation it becomes clear that the Unbiased MT coordinates estimator is identical to the one applicable for plain RSS based positioning.

(e) Estimator 4: Sub-optimal MT coordinates MSE estimator: According to the analysis presented in [1] the following quantities should be minimized:

$$MSE(\hat{d}_{4C}(i,z)) = E\left[\left(\hat{d}_{4C}^2(i,z) - d^2(i)\right)^2\right] = c_{4C}^4(i,z) \cdot E(\hat{d}_{oC}^4(i,z)) + d^4(i) -$$
$$2d^2(i)c_{4C}^2(i,z) \cdot E(\hat{d}_{oC}^2(i,z))$$
$$z=x,y \quad (107)$$

It is easy to show that the minimization of the above quantities can be achieved for the following estimator:

$$c_{4C}(i,z) = d(i)\sqrt{\frac{E(\hat{d}_{oC}^2(i,z))}{E(\hat{d}_{oC}^4(i,z))}} = \frac{e^{-\beta(i)\sigma_{Rx}^2}}{\sqrt{1 + f_z \cdot (e^{4\beta(i)\sigma_{Rx}^2} - 1)}}, \, z=x,y \quad (108)$$

It is interesting to note that for $f_z=1$ corresponding to the case that Kalman filter is not applied the estimators defined above become equal to the estimators defined for the plain RSS case as defined in [1].

6.6. Performance Analysis

The simulation model described earlier in section 3 has been exploited for the analysis of the performance of the revised set of MT-BS distance estimators. In Figures 13, 14 and 15 we provide the MT positioning error for the three different Kalman Filtering options versus the standard deviation of the measured RSS (σ_{Rx}) and for both the original [1] and the revised MT-BS distance estimators (defined in this paper). The figures indicate that the MT-BS distance estimators provide the means to improve MT positioning accuracy in all Kalman filtering options. In particular the revised MT-BS distance estimators defined in this paper lead to the optimum performance (revised estimators 4A, 4B and 4C).

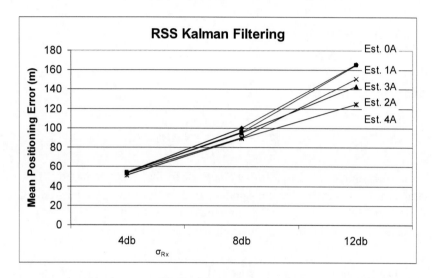

Figure 13. The performance of the revised MT-BS distance estimators for RSS Kalman Filtering Option.

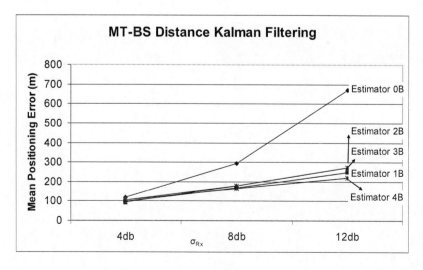

Figure 14. The performance of the revised MT-BS distance estimators for MT-BS distance Kalman filtering Option.

It is interesting to note that in the case of MT-BS distance Kalman filtering option (Option B) the performance of Estimator 1 which is identical to Estimator 1B (Unbiased MT-BS Distance Estimator) is very close to the optimum performing estimator (revised Estimator 4B). A similar observation applies for MT Coordinates Kalman filtering option (Option C) regarding the performance of Estimator 3 which is identical to Estimator 3C (MT Coordinates Unbiased Estimator). The reason for this result is the way that Kalman filtering affects the statistical characteristics of its input: Kalman filtering output maintains the input mean value and reduces the input variance. In the case that the input of the Kalman filtering is unbiased then the resulting positioning error becomes equivalent to the case of a lower σ_{Rx}. In such a case the performance of all estimators is converging as it can be seen from Figures 13, 14, 15 for e.g. σ_{Rx} =4db. On the other hand in the case that the input of Kalman filter is biased then the resulting output although it has lower variance (due to Kalman filter gain) it maintains the bias of the input leading possibly to a higher error. The best performing estimator (revised estimator 4) for all Kalman filter options actually provides a sub-optimal solution to MT coordinates MSE by balancing the resulting bias and the variance.

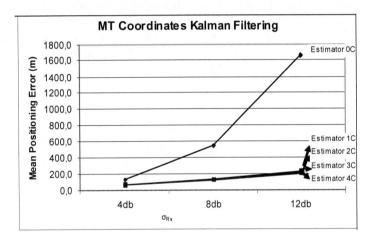

Figure 15. The performance of the revised MT-BS distance estimators for MT Coordinates Kalman filtering Option.

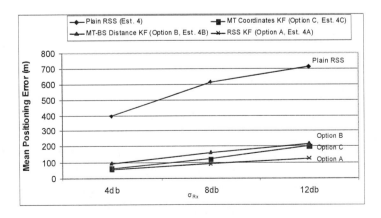

Figure 16. Comparative performance of the best performing estimators for the Kalman Filtering options applied to STAMP list and Plain RSS (based on a single RSS measurement).

Figure 16 compares the performance of the Kalman filter options based on their best performing estimator (4A, 4B and 4C respectively) as well as the performance of the plain RSS positioning (for Estimator 4). As it can be seen RSS Kalman filtering remains the best performing Kalman filtering option (this was the conclusion of [2] based on Estimator 0 analysis). However, as it can be seen from this figure the resulting MT positioning accuracy for all Kalman filter options is in a range (e.g., 100m-200m positioning error for σ_{Rx} =12db) which is valid for commercial exploitation.

6.7. Applicability Issues

An issue that should be considered is the ability to provide a good estimation of the parameters required for the application of the revised estimators and especially the Kalman filter gain factors. The analysis presented in this paper provides the means to estimate these parameters based on the steady state Kalman filter characteristics. In a commercial application such parameters can be calculated off-line in order to reduce the real-time processing effort. Another solution which is applicable to MT-BS distance Kalman filter (Option B) and MT coordinates Kalman filter (Option C) is to use the Estimators 1B and 3C respectively which do not require the estimation of Kalman filter gain parameters and according to the results presented in the previous section they exhibit a performance very close to the best performing estimators.

As mentioned in Section 3, the application of more than one Kalman filter options can not deliver any further improvement. This result is due to the fact that in the case that the output of a Kalman filter is used as an input to another Kalman filter then the second Kalman filter will not deliver the expected gain. This is because of the fact that its input does not exhibit one of the characteristics that allows Kalman filter to deliver gain: the series of input measurements should have measurement noise which is independent in each sample. However, the output of the first Kalman filter definitely does not have this property.

CONCLUSION

The current paper addresses an interesting application of Kalman filtering related to accuracy optimization in RSS based mobile terminal positioning. The existing research results on the definition of MT-BS statistical estimators as well as the definition of alternative options for the introduction of Kalman filtering in the MT position calculation process has led to a new research challenge addressed in this paper i.e. the definition of a set of properly revised MT-BS estimators which match the Kalman filtering options towards the optimization of MT positioning accuracy. To address this research problem a novel approach in treating Kalman filtering has been adopted. Considering steady state Kalman filter performance and based on a set of assumptions it is feasible to consider that the impact of Kalman filter in the MT positioning calculation process can be simplified by considering the impact of Kalman filtering on the statistical behaviour of the involved random variables. In this context, the impact of Kalman filtering on the mean value, the variance and the pdf of the random variables involved in the MT positioning process like the RSS, the MT-BS distance or the

MT coordinates is considered. This approach has allowed for the identification of a revised set of MT-BS distance estimators which not only led to further improvement of the MT positioning accuracy but it also provided a predictable performance for all the Kalman filtering options indicating the fact that the novel method presented in this analysis is valid and valuable for real applications.

It is interesting to highlight here that the whole analysis is based on simulation model which corresponds rather an ideal situation. In real mobile networks the nature of RSS error is such that the application of the theoretical results presented here are relevant, however, they do not deliver the expected benefits. It appears that there are additional aspects of the real propagation environment that need to be taken into account in order to deliver substantial MT positioning accuracy improvement in real network applications.

Beyond the specific application in MT positioning the analysis presented in this paper can be easily generalized for error optimization in complex calculation processes at the presence of Kalman filtering (e.g. GPS based positioning).

REFERENCES

[1] Markoulidakis, JG. *"Received Signal Strength Based Mobile Terminal Positioning Error Analysis and Optimization"*, *Elsevier Computer Communications*, Volume 33, Issue 10, 15 June 2010, Pages 1227-1234.

[2] Markoulidakis, JG; Dessiniotis, C; Nikolaidis, D. "Part Two: Kalman Filtering Options for Error Minimization in Statistical Terminal Assisted Mobile Positioning", *Elsevier Computer Communications*, Vol. 31, Issue 6, Pages, 1138-1147.

[3] Hata, M. Empirical Formula for Propagation Loss in Land Mobile Radio Services, IEEE *Transactions on Vehicular Technology*, 1980, Vol. VT-29, no. 3, August.

[4] COST 231. Urban transmission loss models for mobile radio in the 900 and 1800 MHz bands. Technical Report TD (91) 73, *European Cooperation in the Field of Scientific and Technical Research (COST)*, September, 1991.

[5] Markoulidakis, JG; Dessiniotis, C. "Statistical Terminal Assisted Mobile Positioning Technique", *IET Communications, June*, 2007, Vol. 1, Issue 3, 325-332.

[6] Hellebrandt, M; Mathar, R. "Location Tracking of Mobiles in Cellular Radio Networks", *IEEE Transactions on Vehicular Technology*, 1999, Vol. 48, No. 5, Sep., 1558-1562.

[7] IST MOTIVE project, FP6-IST 27659, Deliverable 2.1: Overview of MOTIVE concept applications, May 2006.

[8] GSM Association, Permanent Reference Document SE.23, *Location Based Services, January*, 2003, v.3.1.0.

[9] Jakub Borkowski, Jukka Lempiainen, *"Pilot correlation positioning method for urban UMTS networks"*, Proc. of European Wireless Conf., Nicosia, Cyprus, April 2005.

[10] Lee, JS; Miller, LE. *CDMA Systems Engineering Handbook*, Boston, Artech House, 1998.

[11] 3GPP TS 25.305 v.7.2.0 (2006-03), 3rd Generation Partnership Project, Technical Specification Group – Radio Access Network, Stage 2 Functional Specification of User *Equipment Positioning* in UTRAN (Release 7).

[12] Bao Long Le, K; Ahmed, and H. Tsuji, "Mobile Location Estimator with NLOS Mitigation Using Kalman Filtering," *IEEE Wireless Communications and Networking*, Vol. 3, 1969-1973, March 2003.

[13] Catrein, D; Hellebrandt, M; Mathar, R; Serrano, M. "Location tracking of mobiles: a smart filtering method and its use in practice", Vehicular Technology Conference, VTC 2004-Spring, 2004 IEEE 59th, Vol. 5, 2004, 2677–2681 Vol.5.

[14] 3GPP TS 25.304 v.6.8.0, *User Equipment procedures in idle mode and procedures for cell reselection in connected mode.*

[15] Laitetinen, H; Lahteenmaki, J; Nordstrom, T. "Database correlation method for GSM location", IEEE VTC Spring *Conf., May*, 2001, Rhodes, Greece.

[16] 3GPP TS 05.10 v.8.12.0, Radio Subsystem Synchronization.

[17] ETSI TS 100 911 V8.17.0 (2003-06), *Digital cellular telecommunications system* (Phase 2+); Radio Subsystem Link Control.

[18] Kalman, R. "A new approach to linear filtering and prediction problems", Transactions of the ASME *Journal of Basic Engineering, March*, 1960, 35-45.

[19] Open Mobile Alliance, OMA-TS-ULP-V1_0-20060127-C, User Plane Location Protocol, *Candidate Version*, 1.0-27 Jan 2006.

[20] Open Mobile Alliance, *Secure User Plane Location Architecture*. Version 2.0, April 2009.

[21] Markoulidakis, J; Nikolaidis, D; Desiniotis, C. "STAMP Accuracy Enhancement through Received Signal Strength Adaptive Kalman Filtering", accepted for publication to PIMRC 07, 18th Annual IEEE International Symposium on Personal, Indoor and *Mobile Radio Communications, Sept*. 2007, Athens, Greece.

[22] Catrein, D; Hellebrandt, M; Mathar, R; Serrano, M. "Location tracking of mobiles: a smart filtering method and its use in practice", Vehicular Technology Conference, VTC 2004-Spring, 2004, IEEE 59th, Vol. 5, 2004, 2677–2681 Vol.5.

[23] Najar, M; Vidal, J. "Kalman tracking for mobile location in NLOS situations", Personal, Indoor and Mobile Radio Communications, PIMRC 2003, *14th IEEE Proc.,* Vol. 3, 2003, 2203–2207 vol.3.

[24] Bertoni, HL. "Radio Propagation for Modern Wireless Systems", pp.16-23, Prentice Hall Professional *Technical Reference*, 1999.

[25] Welch, G; Bishop, G. "An Introduction to Kalman Filter", TR 95-041, Dept. of *Computer Science Univ. of North Carolina*, Apr 04.

In: Kalman Filtering
Editor: Joaquín M. Gomez

ISBN: 978-1-61761-462-0
© 2011 Nova Science Publishers, Inc.

Chapter 2

APPLICATION OF KALMAN FILTERING IN POWER SYSTEMS: HARMONIC DISTORTION AND VOLTAGE EVENTS

Julio Barros[*]*, Enrique Pérez, Ramón I. Diego and Matilde de Apráiz*
University of Cantabria, Dept. of Electronics and Computers,
Santander, Spain

ABSTRACT

Kalman filtering is a digital signal processing tool that has been extensively used in many electric power system applications. Voltage and current phasors, power system frequency, voltage flicker, high-impedance faults, harmonic distortion, voltage dips, voltage unbalance, high-frequency transients and other power system magnitudes can be successfully computed using Kalman filters. This chapter provides a general overview of the application of Kalman filters in electric power systems, focusing on their use for the analysis of two of the most important electric power quality disturbances in modern day power systems: harmonic distortion and voltage events. Section 2 gives a short introduction to Kalman filtering and a review of the main applications of Kalman filtering in electric power systems. Section 3 describes the practical application of Kalman filtering in the estimation of magnitude and phase angle of harmonics and interharmonics in voltage and current waveforms and section 4 explains how Kalman filters can be used for detection and analysis of voltage dips, short interruptions and overvoltages in voltage supply.

---

[*] Corresponding author: E-mail: barrosj@unican.es

1. INTRODUCTION

Electricity is generated and distributed as a set of three-phase pure sinusoidal voltages characterized by the following parameters:

Frequency
Magnitude
Waveform
Symmetry of three-phase voltage

These characteristics are subject to continuous variations during the normal operation of a supply system due to changes of load, disturbances generated by equipment and power system faults. These variations are to certain extent random and unpredictable, although some daily, weekly and seasonal trends can be observed.

As an example of voltage variations, figure 1 shows the weekly evolution of the three phase-to-phase voltages in the low-voltage distribution system of a building located on our campus.

European standard EN 50160 defines the main characteristics of the voltage at the customer's supply terminals in public low-voltage and medium-voltage electricity distribution systems under normal operating conditions [1]. This standard defines the nominal values and the range for accepted variations of power system frequency, supply voltage, voltage harmonic distortion and supply voltage unbalance.

Any deviation of voltage characteristics outside these ranges is considered a power quality disturbance. According to [2] power quality disturbances can be classified as variations and events. Variations are small deviations of voltage characteristics that can be measured at any moment in time. Frequency and voltage variations, harmonic distortion and voltage unbalance are examples of power quality variations. On the other hand, events are large and sudden variations that occur occasionally, such as voltage dips or voltage interruptions. Events are disturbances that start and end with a threshold crossing. These power quality disturbances reduce the reliability and the efficiency of the electric power system and must be measured exactly.

In this general context, the definitions and the use of the suitable signal processing tools for the complete characterization of voltage supply is of great importance in power systems. The root mean square voltage for estimation of voltage magnitude and variations, the Fourier analysis for estimation of the harmonic components in voltage supply and the symmetrical components for analysis of voltage unbalance are the main signal processing tools defined in international power quality standards for estimation of voltage supply characteristics [3]. Although all of these magnitudes are defined for stationary signals, they are also used in transient conditions as is the case of power quality disturbances. To improve the performance of the standard methods for the analysis of non-stationary signals, other signal processing tools, such as the Short Time Fourier Transform, Kalman filtering, wavelet analysis and others have been proposed in the literature in recent years [2, 4].

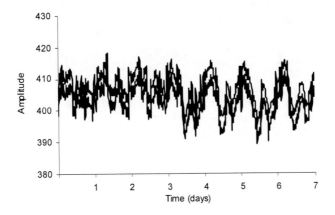

Figure 1. Weekly evolution of the three phase-to-phase voltages in a low-voltage distribution system.

The purpose of this chapter is to present a general overview of the application of Kalman filtering in electric power systems, focusing on its application for the analysis of two of the most important electric power quality disturbances in present day power systems: harmonic distortion and voltage events.

2. KALMAN FILTERING IN ELECTRIC POWER SYSTEMS

The Kalman filter is a set of mathematical equations that provides a recursive method to estimate the state of a process in a way that minimises the mean square error. The filter is initialized with an initial estimate of the system and its error covariance, using the measurements to update and refine this estimate. The update estimate is successively improved using new measurement data until, eventually, a steady-state condition is reached where no further improvement is obtained.

The Kalman filter is especially suited to on-line and real-time digital signal processing because the input data with noise (measurements) is processed recursively.

2.1. Discrete Kalman Filter

Kalman filtering addresses the problem of estimating the state x of a discrete-time process governed by a linear stochastic equation in the form:

$$x_{k+1} = \Phi_k x_k + w_k$$

where
x_k is the nx1 state vector at instant t_k
Φ_k is the nxn state transition matrix
w_k is an nx1 vector uncorrelated sequence with known covariance

The measurement of the process is assumed to be made at discrete instants of time according to the following equation:

$$z_k = H_k x_k + v_k$$

Where
z_k is the mx1 vector measurement at instant t_k
H_k is the mxn matrix given the ideal relation (without noise) between the measurement and the state vector
v_k is the mx1 measurement error vector assumed to an be uncorrelated sequence with known covariance.

w_k and v_k are the process and measurement error vectors respectively whose covariance matrices are defined by:

$$E[w_k w_i^t] = \begin{cases} Q_k & i = k \\ 0 & i \neq k \end{cases}$$

$$E[v_k v_i^t] = \begin{cases} R_k & i = k \\ 0 & i \neq k \end{cases}$$

where t means the transpose.

Starting from an initial estimation of the state vector x'$_k$ and its associated error covariance P'$_k$, the Kalman filter uses the measurement to update the initial estimation. A linear combination of the initial estimation and the noise measurement is chosen in accordance with the following equation:

$$x_k = x_k' + K_k(z_k - H_k x_k')$$

where
x'_k is the initial estimate
x_k is the estimate updated at t_k
K_k is the filter coefficient at instant t_k.

The difference ($z_k - H_k x'_k$) is called the residual and reflects the difference between the predicted measurement $H_k x'_k$ and the actual measurement z_k. A residual of zero means that the two are in complete agreement.

Making use of the state transition matrix Φ_k the filter is projected ahead, using the measurement at the instant t_{k+1} to obtain the new estimation for this instant x'_{k+1} and its error covariance P'_{k+1}:

$$x_{k+1}' = \Phi_k x_k + w_k$$

$$P'_{k+1} = \Phi_k P_k \Phi^t_k + Q_k$$

where Φ^t_k is the transpose of matrix Φ_k.

The filter coefficients K_k are time variable and are computed to minimise the mean square error between the actual values of the state vector and their estimates. These coefficients are computed recursively using the following equations:

$$K_k = \frac{P'_k H'_k}{H_k P'_k H'_k + R_k}$$

$$P_k = (I - K_k H_k) P'_k$$

where I is the identity matrix and H^t_k is the transpose of matrix H_k. Figure 2 shows the recursive algorithm of the discrete Kalman filter.

2.2. Extended Kalman Filter

The discrete Kalman filter is applied to estimate the state of a discrete time linear process. However when the process to be estimated or the relationship between the measurement and the process are non-linear, as would be the case in many electric power system applications, the Extended Kalman filter should be applied.

The Extended Kalman filter is a modified version of the linear Kalman filter that is applied in systems with non-linear process and measurement equations. In each step of the recursive algorithm, the non-linear equations are linearized at the latest estimate, using a first order Taylor series, to form a linear process, and then the linear Kalman filter model is applied.

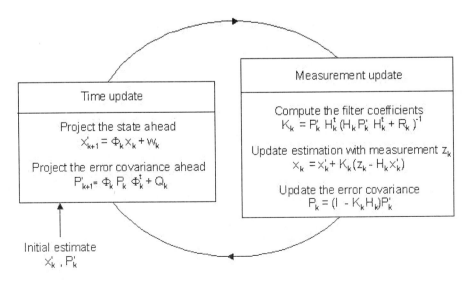

Figure 2. Recursive algorithm of a linear Kalman filter.

Considering a non-linear system with the following discrete process and measurement equations:

$$x_{k+1} = f[x_k, w_k]$$

$$z_k = h[x_k, v_k]$$

where x_k and z_k are the state vector and the measurement vector respectively at instant t_k and $f[x_k,w_k]$ and $h[x_k,v_k]$ are non-linear vector functions. P_k, Q_k and R_k are the covariance matrices of vectors x_k, w_k and v_k at instant t_k respectively.

In an Extended Kalman filter, the discrete process and the measurement equation are linearized using a first order Taylor series in the following way:

$$\Phi_{ij,k} = \frac{\partial f_i[x_k, w_k]}{\partial x_j}$$

and

$$H_{ij,k} = \frac{\partial h_i[x_k, v_k]}{\partial x_j}$$

where the state transition matrix, Φ_k, and the measurement matrix, H_k, are the Jacobian matrices of partial derivatives of function f with respect to x and function h with respect to x at instant t_k respectively, and f_i and h_i are the i-th elements of f and h.

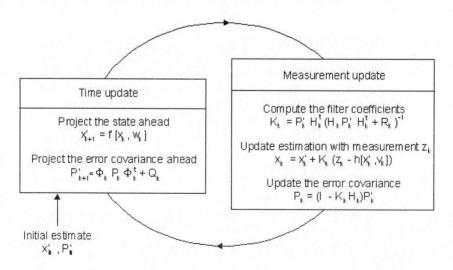

Figure 3. Recursive algorithm of an Extended Kalman filter.

Figure 3 shows the recursive algorithm of the Extended Kalman filter. As was the case in the linear Kalman filter, the recursive algorithm starts with an initial estimation of the state vector x'$_k$ and its associated error covariance P'$_k$, the Kalman filter uses the measurement to update the initial estimation. Then making use of the state transition matrix Φ_k the filter is projected ahead to obtain the new estimation for instant t$_{k+1}$, x'$_{k+1}$, and its error covariance P'$_{k+1}$.

2.3. Application of Kalman Filters in Power Systems

Kalman filtering has been extensively used in power system applications. Voltage and current phasors, power system frequency, voltage flicker, high-impedance faults, harmonic distortion, voltage events and other power system magnitudes can be successfully estimated using Kalman filters. This section presents a general overview of some of the main applications of Kalman filters in electric power systems.

Computer relaying applications

Kalman filters were first applied in power systems by Girgis and Brown for the estimation of voltage and current phasors in computer relaying applications [5, 6]. References [7, 8] present practical applications with two and three-state Kalman filters used for acquisition of voltage and current waveforms, modelling as noise harmonics and other undesirable frequency components present during the fault.

Sachdev et al. [9] introduce an eleven-state model including non-fundamental frequency components in the model of the input current for a better estimation of phasors. Murty and Smolinsky [10] use a five-state Kalman filter as a phasor estimator for fundamental and second order harmonic components for a three-phase transformer protection.

Reference [11] studies the characterization of Kalman filters in the frequency domain in order to improve their performance in relaying applications, and finally, reference [12] presents the practical issues involved in the implementation of a real-time fault detection and classification method in a low-cost 8-bit based microprocessor unit using two sets of Kalman filters, one initialized with values corresponding to fault condition and the other with the values corresponding to unfaulted condition. A probabilistic criterion is applied for fault confirmation.

Power system frequency estimation

Fundamental power system frequency can also be accurately estimated using Kalman filtering. Girgis and Hwang present in [13] the basic principles for application of Kalman filtering for estimation of power system frequency. They propose a three-state Extended Kalman filter model with the fundamental component and power system frequency as the state vector for estimation of the magnitude and phase angle of voltage supply and the frequency deviation.

Girgis and Peterson present in [14] a Kalman filtering-based technique for estimation of power system frequency deviation and its rate of change during emergency operation conditions. The method proposed is a two-stage algorithm that uses an Extended Kalman filter in series with a linear Kalman filter. The Extended Kalman filter is used to estimate the instantaneous frequency deviation and the linear Kalman filter estimates the mean frequency deviation and its rate of change.

Dash et al. propose in [15] an extended complex Kalman filter for the estimation of power system frequency in the presence of random noise and distortion. The method uses the sample values of three-phase voltages and converts this input using αβ-transform to a complex vector. An extended Kalman filter is then applied to this signal to estimate the power system frequency.

Routray et al. present in [16] a method based on the use of an extended Kalman filter for the measurement of the power system frequency. A hysteresis method is used for resetting the covariance matrix in the case of a sudden change of the signal which enables a faster tracking of the power system frequency.

Recently Reddy et al. [17] proposed the use of an unscented Kalman filter to overcome the difficulties of linearization and derivative calculations of signal in the extended Kalman filter for tracking the amplitude, phase, frequency and harmonics in power signals.

Voltage flicker

Kalman filtering was first introduced by Kamwa and Srinivasan for estimation of voltage flicker [18]. An extended Kalman filter algorithm is used for the computation of the instantaneous magnitude and phase of voltage supply, demodulating the amplitude of voltage for the estimation of the flicker level.

Girgis et al. present in [19] an algorithm based on Kalman filtering to estimate the instantaneous magnitude and frequency of the voltage flicker. A two-state Kalman filter is used to estimate the magnitude and phase angle of the fundamental power system voltage waveform. The envelope of this signal is separated into its constant and fluctuating elements and then an extended Kalman filter is used to estimate the magnitude and frequency of the instantaneous flicker level.

Elnady and Salama introduce a unified method for mitigation of voltage sags and voltage flicker using Kalman filtering [20]. In the case of the computation of instantaneous flicker level, the input signal is separated into a constant and a modulating signal, computing the amplitude and frequency of this modulating signal using an Extended Kalman filter. This extraction technique is also used in [21] for mitigation of voltage fluctuations

Detection and analysis of high-impedance faults

Girgis et al. first applied Kalman filtering to take into account the time-varying nature of the fundamental and harmonic components of current during a high-impedance fault [22]. The method proposed consists of two Kalman filtering-based schemes: a steady-state model looking for a transient in fundamental and harmonics during normal operating conditions and a fault-based scheme used when a transient is detected to classify or not this transient as a high-impedance fault.

Samantaray et al. present in [23] a method for the detection of high impedance faults in power distribution feeders using an Extended Kalman Filter and a probabilistic neural network. The Extended Kalman filter is used to estimate the different harmonic components in high-impedance fault and no-fault current signals that are used as features to train and test the probabilistic neural network for detection of the fault condition.

Samantaray and Dash propose in [24] a method using an Extended Kalman filter and a support vector machine for detection of high-impedance faults in power distribution feeders. The Extended Kalman filter is used for estimation of magnitude and phase angle of fundamental and odd harmonic components up to the thirteenth order using these magnitudes as feature inputs to the support vector machine.

Harmonic analysis

Kalman filtering can be used for measurement and tracking of harmonic distortion in voltage and current waveforms. Reference [26] presents the basic principles of Kalman filtering in tracking the time variation of power system harmonics, comparing the results with those obtained applying Fourier analysis.

References [25, 27] propose the use of Kalman filtering for harmonic state estimation. Ma and Girgis show in [28] how the Kalman filter could be used to solve the two basic problems in harmonic source identification: the optimal location of a limited number of harmonic meters and the dynamic estimation of harmonic injections and their locations. Kamwa et al. in [29] combine fundamental frequency tracking with a Kalman filter self-synchronized harmonic tracker. First, fundamental frequency is estimated with a Kalman filter then using this value to update the state matrix of a Kalman filter-based harmonic analyzer, which then becomes self-synchronizing.

Moreno and Barros present in [30] an 8-bit based microprocessor unit for real-time tracking of harmonics in voltage and current waveforms, using a 12-state linear Kalman filter to obtain the instantaneous magnitude of a maximum of six harmonic components.

S. Liu proposes in [31] an adaptive method for dynamic estimation of harmonic signals. In the case of a transient in harmonic parameters, the error covariance matrix P is reset to a predefined high level to increase the sensitivity of the Kalman filter. References [35, 36] propose the use of two values of the noise covariance matrix Q for dynamic harmonic tracking estimation, one for steady-state estimation and the other for transient estimation.

Dash et al. propose in [32] a hybrid combination of Kalman filter and Fourier linear combiner for estimation of time varying harmonics in the presence of frequency changes. Once the frequency is estimated using an extended Kalman filter the Fourier linear combiner computes the time varying amplitudes and phase jumps within a shorter time window of 1 to 2 cycles. Kennedy et al. explore in [33] the practical application of different possible implementations of Kalman filters to the analysis of harmonics in power systems, also examining the effect of fundamental frequency variation.

Barros and Pérez propose in [34] the use of a 24-state linear Kalman filter to compute in real-time the reference compensating current in single-phase shunt active power filters. The method proposed improves the dynamic performance of the active filter and can be used in two different working modes: the global harmonic compensation mode, where all harmonic components in the load current up to the 23rd order could be compensated, or the specific harmonic compensation mode where a specific harmonic pattern could be compensated.

Recently, Chen et al. present in [38] an extended real model of Kalman filter with a resetting mechanism for accurately tracking time-varying harmonic components. The model proposed includes the fundamental and harmonic components and frequency deviation in the state vector.

Detection and characterization of voltage events

Kalman filtering uses the change in magnitude of the fundamental component in voltage supply to detect and to analyze voltage events in power supply systems.

The detection properties of linear Kalman filtering and its accuracy in the estimation of the magnitude and duration of voltage events depend on the model of the system used and on the specific characteristics of the voltage event. References [39-43] discuss the Kalman filtering modelling issues and compare the performance of linear Kalman filters of different order in the detection and analysis of voltage dips. An important advantage of Kalman filtering over other methods is that it gives information about the magnitude and phase-angle jump associated with the voltage event and also about the point on the wave where it begins.

Reference [39] studies three different options for estimating the amplitude of voltage supply during a voltage dip using Kalman filtering: a two-state linear Kalman filter, modelling only the fundamental component of voltage supply, a low-pass filter followed by a two-state linear Kalman filter and a 20-state linear Kalman filter adding harmonic components to fundamental component. The conclusion is that the low-pass two-state Kalman filter shows the best performance of the three.

References [40, 41] present an expert system for automatic classification of three-phase power quality recordings, classifying voltage dips according to their causes. The mean square of the residuals of a Kalman filter is used for segmentation of the different stages of voltage supply during the event.

References [42, 43] propose the use of a three 12-state Kalman filters for real-time detection and analysis of voltage events in a three-phase voltage supply. The fundamental component and odd harmonic components from third to eleventh order are used to provide an accurate representation of the voltage supply during the event. A new value of fundamental magnitude and phase angle of voltage supply is obtained with each new sample of the voltage waveform.

One of the most critical issues related to the use of Kalman filters is the adequate selection of the noise covariance matrixes Q and R. These error covariance matrixes act as tuning parameters to balance the dynamic response of the filter against the sensitivity to noise. The theoretical values of Q and R can be computed mathematically, but in many cases and especially for non-linear systems, these theoretical values do not produce the most accurate results. It has been demonstrated that the filter response depends more on the Q/R ratio than on the values of Q and R [33].

An important aspect to be considered in the use of Kalman filtering is how the filter responds to abrupt changes, as is the case in voltage events, which can make the filter lose the ability to track these changes. Different solutions have been proposed to compensate for this problem. Liu proposes in [31] resetting the error covariance matrix P after detection of an abrupt change of the state variables, whereas references [35, 36] propose the use of two values of the noise covariance matrix Q, one for steady-state estimation and the other for transient estimation.

Another solution for overcoming the limitations of linear Kalman filters in the detection and analysis of voltage events is the use of an Extended Kalman filter (EKF) to better estimate the non-linear process associated with a voltage event.

E. Pérez presents in [44] the performance of different EKF algorithms in the detection and estimation of voltage events. Reference [45] proposes a combination of wavelet analysis and an Extended Kalman filtering for the classification and measurement of voltage events.

Wavelet analysis is used for detection and preliminary classification of voltage events, whereas the Extended Kalman filter is used for the estimation of amplitude, frequency and phase angle of voltage supply during the event.

Pérez and Barros propose in [46] an Extended Kalman filter approach for detection and analysis of voltage dips, including the fundamental power system frequency in the state vector of the system. In the method proposed in [47], Pérez and Barros propose the simultaneous use of discrete wavelet transform and an Extended Kalman filter for detection and classification of voltage events. The discrete wavelet transform is used for detection and estimation of the time-related parameters of the voltage event whereas the Extended Kalman filter is used for confirmation of the event and for estimation of the magnitude and phase-angle of voltage supply during the event.

3. KALMAN FILTERING AND HARMONIC DISTORTION

3.1. What Is Harmonic Distortion?

Waveform distortion in voltage or current is produced by frequency components other than the fundamental power system frequency. Table 1 provides a mathematical definition of waveform distortion in terms of its spectral components, where f_1 is the fundamental power system frequency and n is the harmonic order.

Harmonics in voltage and current waveforms are frequency components that are integer multiples of fundamental frequency, interharmonics are frequency components that are not integer multiples of the fundamental frequency and sub-harmonics are frequency components less than the power system frequency.

Waveform distortion is produced by the connection of non-linear loads and loads not-pulsating synchronously with the fundamental power system frequency. These loads draw on non-linear currents which in combination with the line impedances cause a harmonic voltage drop and disturb the sinusoidal voltage supply affecting equipment and the power system itself. Harmonic distortion cannot be avoided in modern power systems and its magnitude is increasing due to the extensive and growing use of non-linear and power electronic loads.

The fundamental component is dominant in most power system applications, especially for the case of voltage waveform, but this is not the case of current waveforms of non-linear power electronics loads, where low order harmonic components could be as high as the fundamental component. As an example of common voltage and current waveforms, figure 4.a shows one cycle of the waveform of voltage supply in p.u. in the low-voltage distribution system in our building while figure 4.b shows one cycle of the non-sinusoidal input current waveform of a personal computer in our laboratory.

Table 1. Mathematical definition of harmonics, interharmonics and subharmonics

Harmonic	$f = n*f_1$ where n is an integer > 0
Interharmonic	$f \neq n*f_1$ where n is an integer > 0
Sub-harmonic	$f > 0$ Hz and $f < f_1$

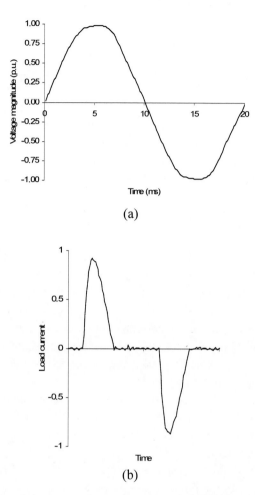

Figure 4. (a) Waveform of voltage supply in a low voltage distribution system, b) Non-sinusoidal input current of a personal computer.

As can be seen the voltage waveform in figure 4.a is not a pure sinusoidal waveform mainly due to third and fifth order harmonic components produced by the presence of non-linear loads in the distribution network. On the other hand, personal computers and other information technology equipment, some of the most common non-linear loads in many residential and commercial distribution networks, all show similar input current waveforms to the one in figure 4.b.

The effects of harmonics on equipment depend on the magnitude of harmonic distortion and on the sensitivity of equipment to this disturbance. Harmonics could severely affect protection systems, electrical machines and transformers, cables and measurement equipment, also producing interference in communication systems. The lifetime of devices might be reduced, the power factor decreased and sensitive loads could malfunction or even be damaged [48].

3.2. Assessment of Harmonic Distortion

IEC standard 61000-4-7 defines the instrument for the measurement of harmonics and interharmonics for power supply systems [49]. This instrument applies Discrete Fourier Transform (DFT) to the samples of voltage or current waveforms taken in a time window width of 10 periods (in 50 Hz systems) of the power system frequency using rectangular weighting. The output bins of the DFT analysis, with a 5 Hz resolution, are grouped to compute the harmonics and interharmonics in voltage and current waveforms.

DFT analysis produces exact results when the signal under analysis is stationary and periodic and the sampling window used is exactly synchronized with the fundamental frequency of the signal, otherwise three major pitfalls affect the successful application of Fourier analysis: aliasing, picket fence effect and spectral leakage [50].

- Aliasing is phenomenon by which high-frequency components of a signal can be translated into low frequencies if the sampling rate is too low (lower than the Nyquist frequency).
- The picket fence effect is produced if the input waveform has frequency components that are not integer multiples of the sampling window. These frequency components are not computed using DFT producing a spreading of energy to the adjacent discrete frequency components.
- Leakage is a spreading of energy from one frequency to adjacent frequencies that arises when the sampling window is not exactly synchronized with the fundamental frequency of the signal under analysis.

According to European standard EN 50160 [1] harmonic voltages can be evaluated in two different ways: individually by their relative amplitude (U_n) related to the fundamental voltage U_1, where h is the harmonic order

$$u_h = \frac{U_n}{U_1}$$

or globally, for example by the total harmonic distortion factor THD, computed using the following expression:

$$THD = \sqrt{\sum_{h=2}^{40} u_n^2}$$

Standard EN 50160 also defines the limits for individual voltage distortion and for total voltage distortion in percent under normal operating conditions.

At present and because of the continuous switching on/off of loads in the distribution system, THD in voltage supply waveforms fluctuates continuously, showing a daily pattern with maximum magnitude during the afternoon and minimum at night.

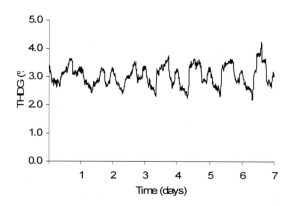

Figure 5. Weekly evolution of THD in voltage supply.

As an example, Figure 5 shows the weekly evolution of the magnitude of THD in voltage supply using 10-minute aggregation time, measured in the load-voltage distribution system of a building in our campus [51]. The building is supplied by a 12 kV/380 V three-phase distribution transformer. A large percentage of the load consists of lighting, computers and other information technology equipment. The THD shows the typical daily variation with the 5th order harmonic dominant in voltage supply distortion, showing maximum values close to or higher than 4% of fundamental component during the afternoons of different weekdays.

3.3. Application of Kalman Filtering for Estimation of Harmonic Distortion

This section first describes the general issues involved in the application of Kalman filtering in the estimation of harmonic distortion in power systems and then follows with two specific applications: the application of Kalman filtering for real-time tracking of power system harmonics and the use of Kalman filtering for estimation of the reference compensating current in shunt active power filters.

The application of Kalman filtering for harmonic analysis overcomes some of the previously mentioned drawbacks of Fourier analysis. It is not necessary to have an integer number of samples in an integer number of cycles of the input signal and Kalman filtering is not limited to stationary signals.

The Kalman filter allows the harmonic signal to be represented in different ways depending on different assumptions. The most common model used in the literature for estimation of harmonic distortion considers the fundamental frequency of the signal to be constant and takes the in-phase and quadrature components of the fundamental and desired harmonic components with respect to its respective rotating reference as the state vector. In this case each frequency component requires two state variables, the total number of state variables to represent n harmonic components being 2n.

Considering the voltage or current input signal with n harmonic components:

$$z(k) = \sum_{i=1}^{n} A_i \sin(iwk\Delta t + \theta_i)$$

where i is the harmonic order, A_i and θ_i are the amplitude and phase angle of each harmonic component, Δt is the sampling period and $w = 2\pi f$, f being the power system frequency. The following state vector is selected to model the system:

$$x_{1,k} = A_{1,k} \cos \theta_{1,k}$$
$$x_{2,k} = A_{1,k} \sin \theta_{1,k}$$
$$\vdots$$
$$x_{2n-1,k} = A_{n,k} \cos \theta_{n,k}$$
$$x_{2n,k} = A_{n,k} \sin \theta_{n,k}$$

where $x_{1,k}, x_{2,k}, \ldots x_{2n,k}$ are the in-phase and quadrature components of fundamental and harmonic components to order n at instant t_k.

In this case, the measurement matrix H_k is the following:

$$H_k = [\cos(wk\Delta t) \ -\sin(wk\Delta t) \ldots \cos(nwk\Delta t) \ -\sin(nwk\Delta t)]$$

where w is the fundamental power system frequency and Δt is the sampling interval.

The state transition matrix is the identity matrix.

$$\Phi = \begin{pmatrix} 1 & 0 & . & . & 0 \\ 0 & 1 & . & . & 0 \\ . & . & . & . & . \\ 0 & 0 & . & 1 & 0 \\ 0 & 0 & . & 0 & 1 \end{pmatrix}$$

The amplitude and phase-angle of each harmonic component can be calculated from the state variables as:

$$A_{n,k} = \sqrt{(x_{2n-1,k})^2 + (x_{2n,k})^2}$$

$$\theta_{n,k} = \arctan\left(\frac{x_{2n,k}}{x_{2n-1,k}}\right)$$

The correct application of Kalman filtering in power systems or in any other application requires a previous knowledge of the system in order to model the system as accurately as possible. The main drawback in the application of Kalman filtering for the estimation of harmonic distortion is that the filter produces as outputs the magnitude and phase angle of the harmonic (or interharmonic) components previously included in the model of the system. All other harmonic components present in the signal under analysis and not included in the model of the system are not computed and are included in the noise of the system.

3.3.1. Application of kalman filtering for tracking of power system harmonics

This section describes the application of Kalman filtering for real-time tracking of power system harmonics [30]. A 12-state linear Kalman filter has been used in an eight-bit microprocessor-based data acquisition unit to track in real time the magnitudes of a group of six harmonic components in voltage waveforms. This data acquisition unit is part of a distributed system for monitoring and supervisory control of an electric power system.

The sampling frequency used in the data acquisition unit is 1.6 kHz, making it possible to compute up to the 16th order harmonic component in the input signal. The in-phase and quadrature components of fundamental and 3rd, 5th, 9th, 11th and 13th order harmonic components of the voltage waveform are taken as the state variables.

To reduce the computation time, Kalman gains K_k and other filter coefficients independent of measurements H_k are calculated off-line. As can be seen in the equations of the discrete Kalman filter described in section 2.1, these values are independent of the measurements and are stored in memory to be used in real-time calculations. In addition, and as can be seen in [30] both K_k and H_k are repeated cyclically after the initialization time of the filter, reducing considerably the size of memory necessary for their processing.

Figure 6 shows the implementation of the method proposed in the data acquisition unit. The discrete Kalman filter is applied on the input signal using the off-line filter coefficients stored in memory. The amplitude and phase-angle of each harmonic component, $A_{n,k}$, $\theta_{n,k}$ at instant t_k, are calculated from the in-phase and quadrature components of fundamental and harmonic components selected in the model of the system.

The magnitude of each harmonic component and the total harmonic distortion are compared with the maximum individual and total harmonic distortion levels defined in electric power quality standards, generating an alarm signal if any defined threshold is surpassed.

As an example of the results obtained in the real-time tracking of harmonic distortion, figure 7 shows four cycles of the voltage waveform in a 130 kV bus measured using the data acquisition unit and the time evolution of the magnitude of fundamental component and magnitudes of 3rd, 5th, 9th, 11th and 13th order harmonic components computed using the 12-state Kalman filter previously described.

As can be seen in the figure, after a short initialization time the filter computes a new magnitude of the state variables in a time shorter than the sampling interval, showing the time evolution of harmonic distortion in the voltage waveform. The results obtained show a 1.06% and 1.47% third and fifth order harmonic component respectively with the rest of harmonic voltages computed being negligible.

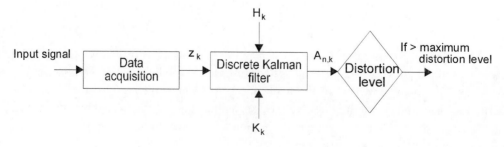

Figure 6. Implementation of the 12-state discrete Kalman filter in the data acquisition unit.

Application of Kalman Filtering in Power Systems

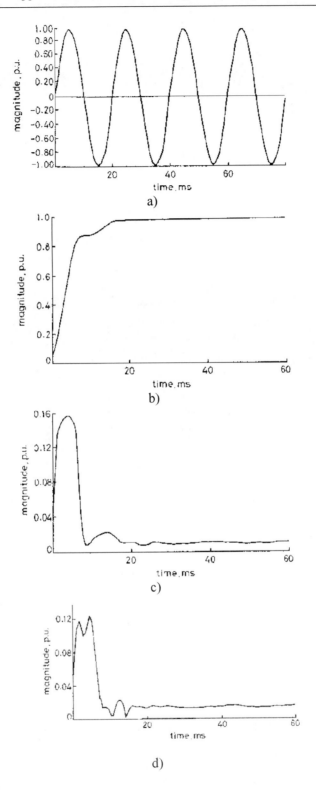

Figure 7. Continued.

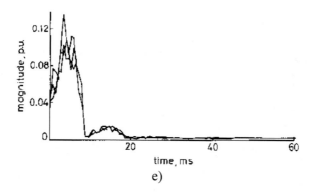

e)

Figure 7. a) Voltage waveform in a 130 kV bus, b) fundamental component, c) 3rd order harmonic component in voltage waveform, d) 5th order harmonic component, e) 9th, 11th and 13th order harmonic components.

3.3.2. Application of Kalman filtering in harmonic shunt active power filters

Conventional L-C passive filters have been used in power systems as a way of eliminating harmonic currents and improving the power factor, but they have the drawbacks of their large size, fixed compensation and resonance. Active power filters have been introduced to overcome the drawbacks of passive filters.

Figure 8 shows the general structure of a shunt active power filter. The filter injects the compensating current i_C to cancel the harmonics contained in the load current i_L. The result is that the current supplied by the power source i_S remains sinusoidal.

The heart of the active power filter is the development of the control strategy. This strategy affects the rating, steady-state and dynamic performance of the filter. Several control methods have been proposed, both for the generation of the reference current and for control of the filter.

This section presents an adaptive method for estimation of the reference compensating current in shunt active power filters based on the use of a linear Kalman filter, to obtain the reference compensating current in real-time [34]. This current, with the same harmonic components as the load current but in opposition, is used to obtain the reference current for the converter of the filter.

A 24-state Kalman filter has been used for signal processing. Using the linear constant fundamental frequency model described in the previous section, the fundamental component and the odd harmonics from 3rd order to 23rd order of the load current are taken as the state vector. It is unnecessary to increase the number of state variables because harmonic orders over 23rd in load currents are normally negligible. To reduce the computing time, gains and other filter coefficients independent of measurements are calculated off-line and are stored in memory to be used in real-time calculations.

The method proposed can be used in two different working modes: the global harmonic elimination mode (GHE) or the specific harmonic elimination mode (SHE). In the GHE mode all harmonic components in the load current up to the order 23rd could be compensated. In the SHE mode, the specific harmonic pattern to be compensated can be selected.

A DSP-based laboratory set-up has been used for simulation. Four channels of 16-bit, 200-kHz ADC and DAC converters are used for data acquisition and filter control. Both converters are interfaced to the synchronous serial port of the DSP. The sampling rate used for data acquisition is 3.2 Khz (64 samples/cycle in a 50 Hz power system). A 312.5 μs time interval is available for data acquisition, signal processing and filter control. The acquisition and conversion time of the converters are both 5 μs and the transport delay of the analog data from input to output is 10 μs. The rest of the time is available for signal processing and filter control.

The processing time for determining the reference compensating current in the GHE mode is only 91 μs, allowing the calculation of this current in the sampling interval of the system. The processing time in the SHE mode is always lower than in the GHE mode. As an example, this time is only 50 μs when only half of the harmonic components in the load current from 3rd order to 23rd order are selected for compensation.

Figures 9 to 11 show the results obtained when a load current, with the harmonic amplitudes shown in Table 2, is used. The total harmonic distortion (THD) of the load current is 11.59%. After 60 milliseconds a 50% and 42% increase is produced in the amplitude of the third and fifth order harmonic components of the load current. The new THD in this interval is 16.04%.

Table 2. Harmonic amplitudes of the load current

Harmonic order	1st	3rd	5th	7th	9th	11th	13th
Amplitude (p.u)	1	0.05	0.095	0.035	0.014	0.01	0.02

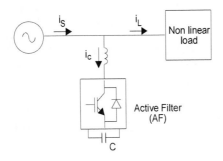

Figura 8. General structure of a shunt active power filter.

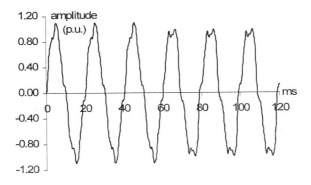

Figure 9. Waveform of the load current before compensation.

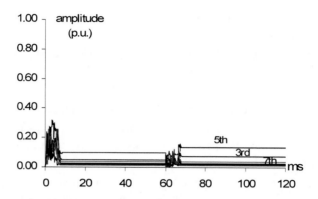

Figure 10. Amplitudes of harmonic components computed using the 24-state Kalman filter.

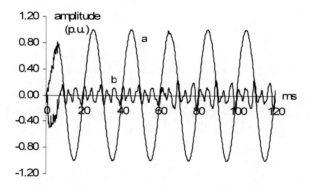

Figure 11. a) Load current after compensation and b) reference compensation current in the global harmonic compensation mode.

Figures 9 and 10 show, respectively, the waveform of the load current before compensation and the harmonic amplitudes in the load current as they are obtained by the Kalman filter.

As can be seen from Fig. 10, after a short initialization time, the active filter continuously tracks the harmonic components in the load current. Fig. 11 shows the waveform of the reference compensating current and the waveform of the load current after compensation in the GHE mode. The THD of the load current after compensation is less than 0.1%. As can also be seen in Fig. 11, the active filter responds in only a few milliseconds to a sudden variation in the load current.

4. KALMAN FILTERING AND VOLTAGE EVENTS

4.1. What Are Voltage Events?

A voltage event is a sudden and temporary variation in the magnitude of voltage supply. Depending on their magnitude voltage events are classified as:

Voltage dips
Supply interruptions
Overvoltages

According to European standard EN 50160 [1] a voltage dip is a sudden reduction in voltage supply to a value between 90% and 1% of nominal voltage followed by a return to the initial level with a duration between 10 milliseconds and 1 second. A supply interruption is defined as a condition in which the voltage supply is lower than 1% of the nominal voltage. The interruption is classified as a short interruption if its duration is less than 3 min; otherwise the interruption is classified as a long interruption. Finally, an overvoltage is an increase in voltage supply greater than 110% of nominal voltage of relatively long duration, if the duration of the overvoltage is only a few milliseconds it is defined as a transient overvoltage.

Voltage events are unpredictable and their frequency of occurrence varies greatly depending on the type of supply system and the point of observation, their distribution over the year being very irregular. They are one of the most important power quality disturbances in the present day, owing to their frequency of occurrence and their effect on equipment.

A voltage event is characterized by a pair of values: magnitude and duration. The magnitude of the voltage dip or an interruption is defined as the lowest root mean square (r.m.s.) voltage magnitude measured during the event (the highest r.m.s. magnitude in the case of an overvoltage) and the duration is defined as the time difference between the beginning and the end of the voltage event. As an example, Figure 12 represents the waveform of a single-phase voltage dip (49.1% magnitude, 63.4 ms duration) measured in the low-voltage distribution network in our building.

4.2. Effects of Voltage Events

Voltage events are usually associated with system faults, switching of heavy loads, starting of large motors, transformer energization and capacitor switching. They cause numerous process disruptions, producing malfunctions, shutdown or failure of electronic and lighting equipment. Furthermore, electrical and electronic equipment is becoming more sensitive to voltage events due to its increasing complexity.

A way to characterize the sensitivity of equipment to voltage fluctuations is the use of voltage-tolerance curves. These curves, also known as power acceptability curves, are plots of bus voltage deviation versus time duration. They divide the voltage deviation - time duration plane in two regions: "acceptable power" and "unacceptable power".

Figure 13 shows the ITIC curve [52]. It describes the AC input voltage envelope which typically can be tolerated (no interruption in function) by most information technology equipment. Although this curve is defined only for single-phase, 120-volt, 60-Hz, information technology equipment, it has become a standard for measuring the performance of many other devices.

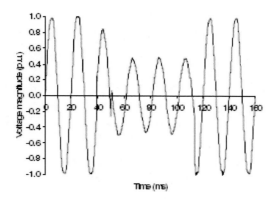

Figure 12. Waveform of voltage supply during a voltage dip.

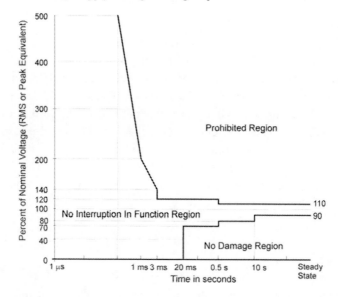

Figure 13. The ITIC curve.

4.3. Assessment of Voltage Events

The most common method used for the majority of power quality instruments for detection and analysis of voltage events is the calculation of the r.m.s. voltage. In a digital system the r.m.s. magnitude of voltage supply is obtained using the following equation:

$$V_{rms} = \sqrt{\frac{1}{N}\sum_{i=1}^{N} v_i^2},$$

where v_i refers to the voltage samples and N is the number of samples taken in a window.

The r.m.s. magnitude is simple and easy to compute but it has the drawback of its dependency on the window length and on the time interval for updating the values. The window size can be selected from a half-cycle of the power system frequency up to any multiple of half-cycles. On the other hand, the r.m.s. magnitude could be updated with each new sample taken, using an overlapping window, or it could be updated each sampling window when a non overlapping window is employed.

IEC Standard 61000-4-30 defined the $U_{rms(1/2)}$ magnitude as the basic measurement of a voltage event [3]. This magnitude is defined as the r.m.s. voltage measured over 1 cycle, commencing at a fundamental zero crossing, and refreshed each half-cycle. According to this standard, in single-phase systems a voltage dip begins when the $U_{rms(1/2)}$ magnitude falls below the dip threshold (90% of nominal voltage) and ends when this magnitude is equal to or above the dip threshold.

The $U_{rms(1/2)}$ method shows a limited performance in the detection and estimation of voltage events, mainly in the case of short-duration and less severe events. As is reported in [53], the error in the computation of the magnitude of these events using the $U_{rms(1/2)}$ method can be of such a magnitude that the voltage event might go undetected, because the detection threshold might not be surpassed.

Using the $U_{rms(1/2)}$ value the magnitude of a voltage event is exactly computed only when the duration of the event is longer than the window width used (1 cycle). Furthermore, as the $U_{rms(1/2)}$ magnitude is computed each half-cycle of the fundamental component, the duration of a voltage event is given in integer multiples of half-cycles and thus, depending on the point on the wave where the voltage event begins and on its magnitude, the error in the duration could be very important, especially for short duration and low magnitude voltage events. Finally, there is always a delay in the detection of the beginning of a voltage event that depends on the magnitude and the point on the wave where the voltage event starts.

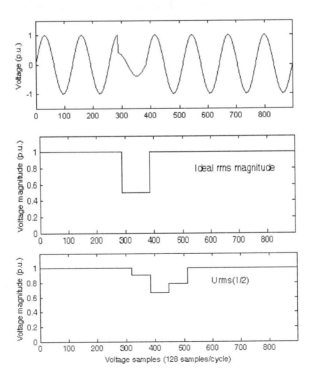

Figure 14. Simulated voltage dip and corresponding r.m.s. ideal and $U_{rms(1/2)}$ magnitudes.

As an example of the limitations in the detection and analysis of voltage events using the standard method, Figure 14 shows a 50% magnitude, 15 millisecond duration and 90° of point on the wave of the beginning simulated voltage dip and the corresponding ideal r.m.s. and $U_{rms(1/2)}$ magnitudes.

The shape and the magnitude of the r.m.s. value obtained using the $U_{rms(1/2)}$ method is very different to the real magnitude. Furthermore, the voltage dip is not immediately detected and the duration does not correspond to the real value.

4.4. Application of Kalman Filtering for Detection and Analysis of Voltage Events

This section describes two specific applications of detection and analysis of voltage events in a low-voltage distribution system using linear and Extended Kalman filtering respectively.

4.4.1. Linear kalman filtering approach

This section presents an automatic system for real-time detection and analysis of voltage events using a set of three linear Kalman filters to detect when a voltage event begins and to estimate the three-phase voltage supply during the event [43].

Three 12-state linear Kalman filters, taking the fundamental component and the odd harmonic components from third to eleventh order of voltage supply as the state vector, have been used to estimate the fundamental magnitude of the three-phase voltage supply, as is shown in Figure 15. Higher order harmonic components in voltage supply are normally too small and it is not necessary to include them in the model of the system. Each frequency component requires two state variables, the in phase and quadrature components with respect to its respective rotating reference, the total number of state variables to represent fundamental and five harmonic components being twelve.

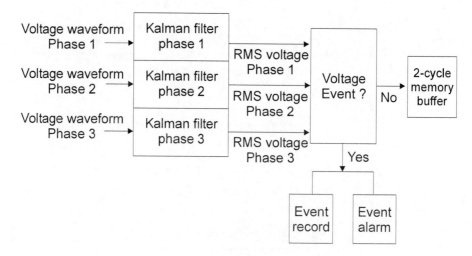

Figure 15. Voltage event detection using three Kalman filters.

For each voltage phase the state vector selected is the following:

$$x_{1,k} = A_{1,k} \cos \theta_{1,k}$$
$$x_{2,k} = A_{1,k} \sin \theta_{1,k}$$
$$\vdots$$
$$x_{11,k} = A_{11,k} \cos \theta_{11,k}$$
$$x_{12,k} = A_{11,k} \sin \theta_{11,k}$$

where $A_1, A_3, \ldots, A_{11}, \theta_1, \theta_3, \ldots, \theta_{11}$ are the amplitude and phase angle of fundamental and odd harmonic components up to eleventh order respectively.

The measurement matrix H_k is the following:

$$H_k = [\cos(wk\Delta t) \ -\sin(wk\Delta t) \ \ldots \cos(nwk\Delta t) \ -\sin(nwk\Delta t)]$$

where $w = 2\pi\ 50$ and Δt is the sampling interval.

The state transition matrix is the identity matrix:

$$\Phi = \begin{pmatrix} 1 & 0 & . & . & 0 \\ 0 & 1 & . & . & 0 \\ . & . & . & . & . \\ 0 & 0 & . & 1 & 0 \\ 0 & 0 & . & 0 & 1 \end{pmatrix}$$

The amplitude and phase-angle of the fundamental component of the three-phase voltage supply at instant t_k, $A_{1,k}$ and $\theta_{1,k}$ respectively, can be calculated from the state variables as:

$$A_{1,k} = \sqrt{(x_{1,k})^2 + (x_{2,k})^2}$$

$$\theta_{1,k} = \arctan\left(\frac{x_{2,k}}{x_{1,k}}\right)$$

The equations of the discrete Kalman filter are described in Section 2.1

The daily mean magnitudes of harmonic components and their mean square values are taken as the initial estimation of the state vector and the initial error covariance respectively. These values were obtained by monitoring harmonics in voltage supply in the low-voltage distribution network for several days. Finally, a Q/R ratio of 0.25 gives a good response of the Kalman filters used in this specific distribution network.

The event detection method proposed compares the magnitude of the voltage supply, as it is obtained in real-time by the three Kalman filters, with the threshold of accepted supply voltage variation defined in [1]. If the magnitude of voltage supply is outside the range of ±10% of the nominal voltage for three consecutive values in any of the three phases, a voltage event is detected and an event record and an alarm signal are generated (Figure 15). The voltage event finishes when the fundamental voltage magnitude returns to within the limits of acceptable voltage supply variation. In case of no-event detection in voltage supply, the system stores the waveforms and the magnitude and phase angle of the three-phase voltage supply in a two-cycle memory buffer.

The event record contains two cycles of the waveforms and the magnitude and phase angle of the three-phase voltage supply recorded before the start of the event, the waveforms and the magnitude and phase angle of voltage supply during the event and the same magnitudes for two cycles after the end of the event. The magnitude and duration of the voltage event are calculated and are also added to the event record.

The same DSP-based laboratory set-up as the one described in Section 3.3.2 has been used to implement the system. Three Hall-effect voltage transducers and three input channels of 16-bit, 200-kHz ADC and DAC converters are used for data acquisition and signal processing (Figure 16). The DAC converter is used to generate an alarm signal in the case of detection of a voltage event in the voltage supply.

A software module in C language has been implemented for data acquisition and real-time estimation of the three-phase voltage supply. The sampling rate used for data acquisition is 3.2 kHz (312.5 µs sampling period). To reduce the processing time, Kalman gains (K_k) and other filter coefficients independent of measurements (H_k and P_k) are calculated off-line and are stored in memory as constants. The processing time of the algorithm developed is only 67 µs per voltage phase, making it possible to calculate the fundamental magnitude and phase angle of the three-phase voltage supply in the sampling interval available.

As an example of the performance of the method proposed in comparison with the standard method for detection and analysis of voltage events, Figure 17.a shows the waveform of a simulated rectangular voltage dip of 70% magnitude and 17.4 ms duration. The voltage event begins after two cycles of the voltage supply and at zero crossing of the voltage waveform. Figure 17.b shows the r.m.s. magnitude of voltage supply computed using the $U_{rms(1/2)}$ method and the 12-state Kalman filter method proposed. As can be seen the Kalman filter clearly improved the results obtained with the standard method. Reference [43] presents the complete characterization of this method in the detection and estimation of the magnitude and duration of voltage events.

References [42, 43] report the results of six months monitoring of voltage events in a three-phase low-voltage distribution system using the system proposed. As an example of the voltage events detected during the monitoring period, figures 18 and 19 show the waveform and the fundamental component of voltage supply during a three-phase complex voltage event. A two-step voltage dip is observed in one phase of voltage supply and a simultaneous two-step overvoltage in the other two voltage phases. The maximum magnitude of the voltage supply during the overvoltage is 354.39 volts with a of 30.9 millisecond duration and the minimum magnitude of the voltage supply during the voltage dip is 50.9 volts with the same duration.

Application of Kalman Filtering in Power Systems

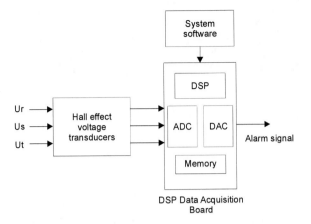

Figure 16. Hardware for voltage event detection and analysis.

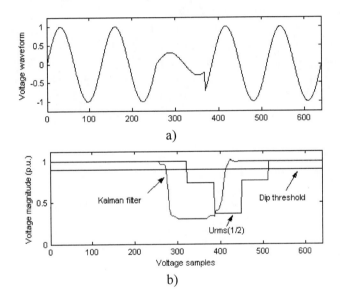

Figure 17. Voltage waveform of a simulated voltage dip, b) $U_{rms(1/2)}$ magnitude and fundamental component of voltage supply computed using a 12-state linear Kalman filter.

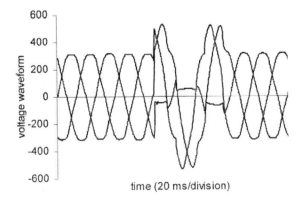

Figure 18. Waveform of the three-phase voltage supply during a voltage event.

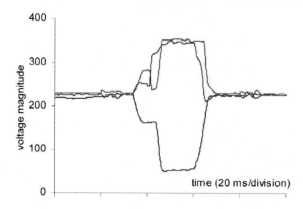

Figure 19. Fundamental component of the three-phase voltage supply during the event.

4.4.2. Extended kalman filtering approach

In the new approach proposed in paper [46], an Extended Kalman Filter (EKF) algorithm is used for the detection and the complete characterization of voltage dips. The use of EKF increases the complexity of the filter and obviously the computational load in the estimation of the magnitude and phase angle of voltage supply during the voltage dip, but this greater complexity does not mean there is any drawback in the method proposed, because the algorithm has been optimized to be implemented in a DSP system in real-time, computing a new value of the magnitude of voltage supply with each new voltage sample within the sampling interval used in the system.

In fact, the use of EKF significantly improves the performance of the method in comparison with the IEC standard method and other digital signal processing methods proposed in the literature, improving the detection properties and the estimation of the magnitude and the time-related parameters of voltage dips.

Considering again the voltage input signal with n harmonic components

$$z(k) = \sum_{i=1}^{n} A_i \sin(iwk\Delta t + \theta_i)$$

where i is the harmonic order, A_i and θ_i are the amplitude and phase angle of each harmonic component, Δt is the sampling period and $w = 2\pi f$, f being the power system frequency. In order to detect and estimate the parameters of a voltage dip in voltage supply, the following state vector is selected to model the system,

$$\begin{aligned}
x_{1,k} &= A_{1,k} \cos \theta_{1,k} \\
x_{2,k} &= A_{1,k} \sin \theta_{1,k} \\
&\vdots \\
x_{11,k} &= A_{11,k} \cos \theta_{11,k} \\
x_{12,k} &= A_{11,k} \sin \theta_{11,k} \\
x_{13,k} &= wk\Delta t
\end{aligned}$$

where $x_{1,k}$, $x_{3,k}$, $x_{5,k}$, $x_{7,k}$, $x_{9,k}$ and $x_{11,k}$ are the in-phase components of fundamental and odd harmonic components from 3rd to 11th order of voltage supply $x_{2,k}$, $x_{4,k}$, $x_{6,k}$, $x_{8,k}$, $x_{10,k}$ and $x_{12,k}$ are the in-quadrature components and finally $x_{13,k}$ is the power system frequency at instant t_k;. The magnitude and phase angle of fundamental and odd harmonic components are extracted from the instantaneous rectangular components of each harmonic frequency considered in the model of the system.

Assuming a negligible change in the amplitude and phase angle between consecutive samples, it can be stated that

$$x_{1,k+1} \cong x_{1,k}$$
$$x_{2,k+1} \cong x_{2,k}$$
$$\vdots$$
$$x_{13,k+1} = w(k+1)\Delta t = wk\Delta t + w\Delta t = x_{13,k} + \frac{x_{13,k}}{k} = \frac{k+1}{k} x_{13,k}$$

and Φ_k and H_k have the following form

$$\Phi_k = \begin{bmatrix} 1 & 0 & 0 & 0 & 0 & \cdots & 0 & 0 \\ 0 & 1 & 0 & 0 & 0 & \cdots & 0 & 0 \\ 0 & 0 & 1 & 0 & 0 & \cdots & 0 & 0 \\ 0 & 0 & 0 & 1 & 0 & \cdots & 0 & 0 \\ 0 & 0 & 0 & 0 & 1 & \cdots & 0 & 0 \\ \vdots & \vdots & \vdots & \vdots & \vdots & \ddots & \vdots & \vdots \\ 0 & 0 & 0 & 0 & 0 & \cdots & 1 & 0 \\ 0 & 0 & 0 & 0 & 0 & \cdots & 0 & \frac{k+1}{k} \end{bmatrix}$$

$$H_k = [\sin x_{13,k} \quad \cos x_{13,k} \quad \sin 3x_{13,k} \quad \cos 3x_{13,k} \quad \cdots \quad \sin 11 x_{13,k} \quad \cos 11 x_{13,k}$$
$$x_{1,k} \cos x_{13,k} - x_{2,k} \sin x_{13,k} + \ldots + 11 x_{11,k} \cos 11 x_{13,k} - 11 x_{12,k} \sin 11 x_{13,k}]$$

The equations of the Extended Kalman filter are described in Section 2.2.

As an example of the performance of the 13-state EKF algorithm developed and to compare the results with those obtained using the standard $U_{rms(1/2)}$ method, Figure 20 shows the voltage waveform of a 73% magnitude, 20-ms duration and 90° point-on-wave voltage dip and the corresponding ideal r.m.s., $U_{rms(1/2)}$ magnitudes and the fundamental component of voltage supply computed using the 13-state EKF algorithm proposed. Using the method proposed, the fundamental component of voltage supply is compared with the dip threshold to determine when the voltage dip begins and ends and to compute its duration. As can be seen the results obtained are considerably better using the 13-state EKF method proposed than those obtained using the $U_{rms(1/2)}$ magnitude.

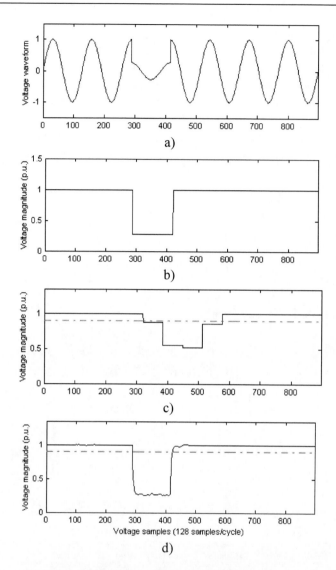

Figure 20. Voltage waveform of a simulated voltage dip, b) ideal r.m.s. voltage, c) $U_{rms(1/2)}$ magnitude, d) fundamental component of voltage supply computed using EKF.

References [44, 46] report the results obtained in the complete characterization of the method proposed in the detection and estimation of the magnitude and duration of voltage events comparing the results with those obtained using the standard method. As is reported in these references the results obtained show better performance than the standard method, overcoming the limitations of the standard method in the detection and analysis of short-duration voltage events and improving considerably the estimation of the instant of beginning and end and the time-related parameters of voltage events.

Finally, as an example of the performance of the method developed in the detection and analysis of real voltage events, Figures 21 shows the magnitude of the fundamental component of voltage supply during the voltage dip in Figure 12, detected in the low-voltage distribution system in a building located on our campus, computed using the 13-state EKF and the $U_{rms(1/2)}$ methods.

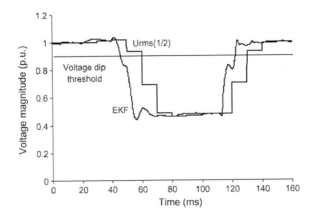

Figure 21. Fundamental component of voltage supply computed using EKF method and $U_{rms(1/2)}$ magnitude for the voltage supply in Figure 12.

The daily mean magnitude of harmonic components of voltage supply and their mean square values are taken as the initial estimation of the state vector and the initial error covariance respectively in the EKF method. These values were obtained from several days' monitoring of harmonics in voltage supply. The initial value of the power system frequency selected is 50 Hz and the Q/R ratio selected for the system modelled was 10.

As can be seen in Figure 21, the shape of the fundamental component and the r.m.s magnitude of voltage supply during the voltage dip is very different using both methods. The voltage dip is detected (its magnitude goes below the dip threshold, 0.9 p.u. of the nominal voltage) at instant 46.094 ms using the 13-state EKF method, whereas using the $U_{rms(1/2)}$ magnitude the voltage dip is detected at instant 60.156 ms, with a delay of 14.062 ms with respect to the EKF method. On the other hand, the magnitude (the lowest magnitude measured during the dip) and the duration (time difference between the beginning and the end) are similar using both methods, the duration being slightly lower using the EKF method.

REFERENCES

[1] European Standard EN 50160, *Voltage characteristics of electricity supplied by public distribution systems*, Cenelec, 1999.

[2] Bollen, MHJ; Gu, IYH. Signal processing of power quality disturbances, IEEE Press *Wiley-Interscience*, 2006, ISBN, 978-0-471-73168-9.

[3] International Electrotechnical Commission, IEC 61000-4-30. *Electromagnetic compatibility (EMC) - Part 4-30: Testing and measurement techniques - Power quality measurement methods*, Switzerland, Second edition 2008.

[4] Bollen, MHJ; Gu, IYH; Santoso, S; McGranaghan, M; Crossley, PA; Ribeiro, MV; Ribeiro, PF. Bridging the gap between signal and power, *IEEE Signal Processing Magazine*, 2009, vol. 26, no. 4, 12-31.

[5] Girgis, AA; Brown, RG. Application of Kalman filtering in computer relaying, IEEE *Transactions on Power Apparatus and Systems*, 1981, vol. PAS-100, no. 7, 3387-3395.

[6] Girgis, AA. Application of Kalman filtering in computer relaying of power systems, PhD *Dissertation*, Iowa State University, 1981.

[7] Girgis, AA. A new Kalman filtering-based digital distance relay, IEEE Transactions on Power *Apparatus and Systems*, 1982, vol. PAS-101, no. 9, 3471-3480.

[8] Girgis, AA; Brown, RG. Modelling of fault-induced noise signals for computer relaying applications, *IEEE Transactions on Power Apparatus and Systems*, 1983, vol. PAS-102, no. 9, 2834-2841.

[9] Sachdev, MS; Wood, HC; Johnson, NG. Kalman filtering applied to power system measurement for relaying, IEEE *Transactions on Power Apparatus and Systems*, 1985, vol. PAS-104, no. 12, 3565-3571.

[10] Murty, YVVS; Smolinski, WI. Design and implementation of a digital differential relay for a 3-phase power transformer based on Kalman filtering theory, *IEEE TPWRD*, 1988, vol. 3, no. 3, 525-533.

[11] Benmouyal, G. Frequency-domain characterization of Kalman filters as applied to power system protection, *Ieee T Power Deliver*, 1992, vol. 7, no. 3, 1129-1136.

[12] Barros, J; Drake, JM. Real-time fault detection and classification in power systems using microprocessors, IEE Proceedings - Generation, *Transmission and Distribution*, 1994, vol. 141, no. 4, 315-322.

[13] Girgis, AA; Hwang, D. Optimal estimation of voltage phasors and frequency deviation using linear an non-linear Kalman filtering: Theory and limitations, *IEEE Transactions on Power Apparatus and Systems*, 1984, vol. PAS.103, no. 10, 2943-2949.

[14] Girgis, AA; Peterson, WL. Adaptive estimation of power system frequency deviation and its rate of change for calculating power system overloads, *Ieee T Power Deliver*, 1990, vol. 5, no. 2, 585-591.

[15] Dash, PK; Pradhan, AK; Panda, G. Frequency estimation of distorted power system signal using Extended complex Kalman filter, *Ieee T Power Deliver*, 1999, vol. 14, no. 3, 761-766.

[16] Routray, A; Pradhan, AK; Rao, KP. A novel Kalman filter for frequency estimation of distorted signals in power systems, *Ieee T Instrum Meas*, 2002, vol. 51, no. 3, 469-479.

[17] Reddy, JBV; Dash, PK; Samantaray, R; Moharana, AK. Fast tracking of power quality disturbance signals using an optimized unscented filter, *IEEE T INSTRUM MEAS*, 2009, vol. 58, no. 12, , 3943-3952.

[18] Kamwa, I; Srinivasan, K. A Kalman filter-based technique for combined digital estimation of voltage flicker and phasor in power distribution systems, *ETEP*, 1993, vol. 3, no. 2, 131-142.

[19] Girgis, AA; Stephens, JW; Makram, EB. Measurement and prediction of voltage flicker magnitude and frequency, *IEEE T POWER DELIVER*, 1995, vol. 10, no. 3, 1600-1604.

[20] Elnady, A; Salama, MMA. Unified approach for mitigation voltage sag and voltage flicker using the *DSTATCOM, IEEE T POWER DELIVER*, 2005, vol. 20, no. 2, 992-1000.

[21] Elnady, A; Salama, MMA. Mitigation of the voltage fluctuations using an efficient disturbance extraction technique, *EPSR*, 2007, vol. 77, no. 3-4, 266-275.

[22] Girgis, AA; Chang, W; Makram, EB. Analysis of high-impedance fault generated signals using a Kalman filtering approach, *IEEE T POWER DELIVER*, 1990, vol. 5, no. 4, 1714-1720.

[23] Samantaray, SR; Dash, PK; Upadhyay, SK. Adaptive Kalman filter and neural network, *IJEPES*, 2009, vol. 31, no. 4, 167-172.

[24] Samantaray, SR; Dash, PK. High impedance fault detection in distribution feeders using Extended Kalman filter and support vector machines, *ETEP*, 2009, doi: 10.1002/etep.321.

[25] Dash, PK; Sharaf, AM. A Kalman filtering approach for estimation of power system harmonics, 3rd *International Conference on Harmonics in Power Systems*, 1988, 34-40.

[26] Girgis, AA; Chang, WB; Makram, EB. A digital recursive measurement scheme for on-line tracking of power system harmonics, *IEEE T POWER DELIVER*, 1991, vol. 6, no. 3, 1153-1160.

[27] Beides, HM; Heydt, GT. Dynamic state estimation of power system harmonics using Kalman filter methodology, *IEEE T POWER DELIVER*, 1991, vol. 6, no. 4, 1663-1670.

[28] Ma, H; Girgis, AA. Identification and tracking of harmonic sources in a power system using a Kalman filter, *IEEE T POWER DELIVER*, 1996, vol. 11, no. 3, 1659-1665.

[29] Kamwa, I; Gronding, R; McNabb, D. On-line tracking of changing harmonics in stresses power systems: application to Hydro-Québec Network, *IEEE T POWER DELIVER*, 1996, vol. 11, no. 4, 2020-2027.

[30] Moreno, V; Barros, J. Application of Kalman filtering for continuous real time tracking of power system harmonics, IEE Proceedings - Generation, *Transmission and Distribution*, 1997, vol. 144, no. 1, 13-20.

[31] Liu, S. An adaptive Kalman filter for dynamic estimation of harmonic signals, 8th International *Conference on Harmonics and Quality of Power*, 1998, vol. 2, 636-640.

[32] Dash, PK; Pradhan, AK; Panda, G; Jena, RK; Panda, SK. On-line tracking of time varying harmonics using an integrated complex Kalman filter and Fourier linear combiner, *IEEE Power Engineering Society Winter Meeting*, 2000, vol. 3, 1575-1580.

[33] Kennedy, K; Lightbody, G; Yacamini, R. Power system harmonic analysis using the Kalman filter, *IEEE Power Engineering Society General Meeting*, 2003, vol. 2, 752-757.

[34] Barros, J; Pérez, E. An adaptive method for determining the reference compensating current in single-phase shunt active power filters, *IEEE T POWER DELIVER*, 2003, vol. 18, no. 4, 1578-1580.

[35] Yu, KKC; Watson, NR; Arrillaga, J. An adaptive Kalman filter for dynamic harmonic state estimation and harmonic injection tracking, *IEEE T POWER DELIVER*, 2005, vol. 20, no. 2, 1577-1584.

[36] Rosendo, JA; Gómez, A. Self tunning of Kalman filters for harmonic computation, *IEEE T POWER DELIVER*, 2006, vol. 21, no. 1, 501-503.

[37] Mostafa, MA. Kalman filtering algorithm for electric power quality analysis: harmonics and voltage sag problems, *Conference on Large Power Engineering Systems*, 2007, 159-165.

[38] Chen, CI; Chang, GW; Hong, RC; Li, HM. Extended real model of Kalman filter for time-varying harmonics estimation, *IEEE T POWER DELIVER*, 2010, vol. 25, no. 1, 17-26.

[39] Styvaktakis, E; Gu, IYH; Bollen, MHJ. Voltage dip detection and power system transients, *IEEE Power Engineering Society Summer Meeting*, 2001, 683- 688.

[40] Styvaktakis, E; Bollen, MHJ; Gu, IYH. Expert system for voltage dip classification and analysis, IEEE *Power Engineering Society Summer Meeting*, 2001, 671- 676.

[41] Styvaktakis, E; Bollen, MHJ; Gu, IYH. Expert system for classification and analysis of power system events, *IEEE T POWER DELIVER*, 2002, vol. 17, no. 2, 423-428.

[42] Barros, J; Pérez, E. Measurement and analysis of voltage events in a low voltage distribution network, 12th *IEEE MELECON,* 2004, Dubrovnik, Croatia, 2004, 1083-1086.

[43] Barros, J; Pérez, E. Automatic detection and analysis of voltage events in power systems, *IEEE T INSTRUM MEAS*, 2006, vol. 55, no. 5, 1487-1493.

[44] Pérez, E. New method for real-time detection and analysis of events in voltage supply using a combined wavelet - Extended Kalman filtering model, PhD Thesis, *University of Cantabria, Santander*, Spain, 2006.

[45] Dash, PK; Chilukuri, MV. Hybrid S-transform and Kalman filtering approach for detection and measurement of short duration disturbances in power networks, *IEEE T INSTRUM MEAS*, 2004, vol. 53, no. 2, 588-595.

[46] Pérez, E; Barros, J. An Extended Kalman filtering approach for detection and analysis of voltage dips in power systems, EPSR, 2008, vol. 78, no. 4, 618-625.

[47] Pérez, E; Barros, J. A proposal for on-line detection and classification of voltage events in power systems, *IEEE T POWER DELIVER*, 2008, vol. 23, no. 4, 2132-2138.

[48] IEEE Task Force on Effects on Harmonics on Equipment, Effects of harmonics on equipment, *IEEE T POWER DELIVER*, 1993, vol. 8, no. 2, 672-680.

[49] International Electrotechnical Commission, IEC 61000-4-7. Electromagnetic compatibility (EMC) - Part 4-7: Testing and measurement techniques – General guide on harmonics and interharmonics measurement and instrumentation, for power supply systems and equipment connected thereto, Second edition, 2002.

[50] Girgis AA; Ham, FM. A quantitative study of pitfalls in the FFT, IEEE Transactions *Aerospace and Electronic Systems,* 1980, vol. AES-16, no. 4, 434-439.

[51] Barros, J; de Apráiz, M; Diego, RI. Measurement of subharmonics in power system voltages, IEEE PowerTech2007, *Lausanne, Switzerland*, 2007, 1736-1740.

[52] ITI (CBEMA) curve application note, Information Technology Industry Council (ITIC), Washington, USA, (available) http://www.itic.org/technical/iticurv.pdf.

[53] Barros, J; Pérez, E. Limitations in the use of r.m.s. value in power quality analysis, *IEEE IMTC2006*, Sorrento, Italy, 2006, 2261-2264.

In: Kalman Filtering
Editor: Joaquín M. Gomez

ISBN: 978-1-61761-462-0
© 2011 Nova Science Publishers, Inc.

Chapter 3

STATISTICAL STATE-SPACE MODELING VIA KALMAN FILTRATION

*Marek Brabec**

Academy of Sciences of the Czech Republic, Institute of Computer Science, Department of Nonlinear Modeling, Pod Vodarenskou vezi 2, 182 07 Praha 8, Czech Republic; *and* National Institute of Public Health, Department of Biostatistics and Informatics, Srobarova 48, 100 42 Praha 10, Czech Republic

ABSTRACT

In this paper, we will start with a brief review of the theory underlying the Kalman filter (KF) applications in statistical modeling based on the state-space approach. In particular, we will stress the prediction error decomposition (achievable through the KF application to the analyzed time series) as a highly effective way of computing the likelihood function, useful when maximum likelihood estimate (MLE) of certain structural parameters is attempted. One-step prediction errors (evaluated at the MLE of structural parameters) are useful for other purposes as well, including the model diagnostics. Similarly, KF (evaluated at the MLE of structural parameters) is used to estimate state variables.

Next, we will illustrate how the state-space modeling and KF can be useful for solving practical problems from several interesting real-life applications.

Firstly, the state-space approach and subsequent (extended) KF estimation will be shown as a valuable tool for estimation of time-varying parameters (radon entry rate and air exchange rate) describing radon concentrations in houses. The model here will be built on two underlying differential equations summarizing the radon and tracer dynamics, similarly as in Brabec, Jilek (2007a). As such, the methodology can be viewed as a flexible way to approach functional data analysis, Ramsay, Silverman (1997).

Secondly, we will show how the Kalman filtration can be useful for estimation of underlying (de-noised) growth curve of children. As demonstrated in Brabec (2004), KF is useful here not only for direct filtering of an individual growth curve, but also for

---

[*] Corresponding author: mbrabec@cs.cas.cz

effective information retrieval from relatively complex (semi-longitudinal) data containing irregular measurements for many individual children.

In addition to the previous examples oriented to univariate time series modeling (but with multivariate state-space), we will consider also multivariate approach useful for individualized natural gas consumption modeling, expanding the approach of Brabec et al. (2007b). There, it is useful to account for gradual evolution of individual long-term consumption averages, or to allow for (individual scaling factor)*(shape of consumption dynamics in daily resolution) interaction which is not present in the current generation of gas consumption models, Brabec et al. (2009b).

INTRODUCTION

Kalman filter (KF) is a well known tool for effective state prediction/estimation, especially in linear state-space systems. That is in systems described by the linear (potentially time-varying) observation equation(s):

$$z_t = H_t x_t + \varepsilon_t \tag{1a}$$

linear (potentially time-varying) state equation(s):

$$x_t = F_t x_{t-1} + v_t \tag{1b}$$

where $\{H_t\}_{t=1}^T, \{F_t\}_{t=1}^T$ are known system matrices. They can reduce to scalars in case of univariate observation z_t and/or univariate state x_t. We start from initial state:

$$x_0 \sim N(a_0, P_0) \tag{1c}$$

(where $\sim N(a, B)$ stands for a normal distribution with mean vector a and variance matrix B), independently from $\{\varepsilon_t\}_{t=1}^T, \{v_t\}_{t=1}^T$. $\begin{pmatrix} \varepsilon_t \\ v_t \end{pmatrix} \sim N(0, diag(R_t, Q_t))$ are independent across t and variance matrices $P_0, \{R_t\}_{t=1}^T, \{Q_t\}_{t=1}^T$ are known (as well as the initial mean a_0).

If we denote conditional averages and variances relevant for prediction and filtering as: $\hat{x}_{t|t} \equiv E(x_t | z_1, z_2, \ldots, z_t), P_{t|t} \equiv Var(x_t - \hat{x}_{t|t} | z_1, z_2, \ldots, z_t)$, $\hat{x}_{t|t-1} \equiv E(x_t | z_1, z_2, \ldots, z_{t-1})$, $P_{t|t-1} \equiv Var(x_t - \hat{x}_{t|t-1} | z_1, z_2, \ldots, z_{t-1})$, the Kalman filter algorithm can be written recursively as a series of steps for $t = 1, \ldots, T$, starting from initial conditions a_0, P_0. In fact, each step consists from two sub-steps, namely from the *prediction*:

$$\hat{x}_{t|t-1} = F_t \hat{x}_{t-1|t-1} \tag{2a}$$

$$P_{t|t-1} = F_t P_{t-1|t-1}(F_t)' + Q_t \qquad (2b)$$

(where ' stands for the transposition)
and the *update*:

$$\hat{x}_{t|t} = \hat{x}_{t|t-1} + K_t \tilde{y}_t \qquad (3a)$$

$$P_{t|t} = (I - K_t H_t) P_{t|t-1} \qquad (3b)$$

where I is the identity matrix of appropriate dimension. $\tilde{y}_t$ is the (observation) prediction error (or just the prediction error, for short),

$$\tilde{y}_t = z_t - H_t \hat{x}_{t|t-1} \qquad (3c)$$

further, K_t is the so called Kalman gain (matrix),

$$K_t = P_{t|t-1}(H_t)'(S_t)^{-1} \qquad (3d)$$

with observation prediction error variance (matrix) S_t,

$$S_t = H_t P_{t|t-1}(H_t)' + R_t \qquad (3e)$$

KF popularity stems not only from its recursive nature, computational simplicity and efficiency, but also because it is well known as an answer to certain class of optimal filtering problems (namely under linearity of the state-space system and normality of both states and observations), Anderson, Moore (2005). Hence, it enjoys many appealing frequentist properties. It is also well known that KF can be motivated from the viewpoint of Bayesian sequential state estimation, Meinhold, Singpurwalla (1983).

When the system is linear but at least one of the random distributions of observation errors (in 1a), of structural errors (in 1b), or of initial conditions (1c) are not normal, the KF algorithm produces optimal estimates/predictions in the linear class, using the first two moments only, or linear Bayes estimates as in West, Harrison (1999). When the system is not linear, (1a, 1b) generalizes to the following (using possibly nonlinear (vector) functions h_t, f_t):

$$z_t = h_t(x_t) + \varepsilon_t \qquad (4a)$$

$$x_t = f_t(x_{t-1}) + v_t \qquad (4b)$$

Then the KF approximation based on local linearization and application of the delta theorem, DasGupta (2008), called the extended Kalman filter (EKF), Harvey (1990) is often employed. It amounts to using $F_t \equiv \frac{\partial f_t(x_{t-1})}{\partial (x_{t-1})'}|_{x_{t-1}=\hat{x}_{t-1|t-1}}$ and/or $H_t \equiv \frac{\partial h_t(x_t)}{\partial (x_t)'}|_{x_t=\hat{x}_{t|t-1}}$ when deriving (2b, 3b-3e).

All of this is good and powerful by itself, but it does not cover an important situation, often encountered in practice, when at least some of the structural parameters (denoting vectorization operator by $vec(.)$ and half-vectorization of a symmetric matrix by $vech(.)$):
$((vech(H_t)_{t=1}^T), (vech(F_t)_{t=1}^T), (vech(R_t)_{t=1}^T), (vech(Q_t)_{t=1}^T), (vech(P_0))', (a_0)')$ are not known and have to be estimated somehow, before the KF algorithm can ever be applied. Moreover, the unknown parameters have often a physical interpretation and can be of interest by themselves. Say that the unknown parameters are assembled into the vector θ.

While the forward run of the state-space model (1a-1c) is easy – amounting to the state estimation (prediction and update) using known θ and the Kalman filter machinery (2a-3e), the backward, or inverse direction from data to structural parameters θ is much more demanding both technically and conceptually. In fact, we are dealing here essentially with a version of a (relatively tame) inverse problem, Tarantola (2005).

One general and flexible approach to the estimation of unknown parameters is the maximum likelihood estimation (MLE). It has many appealing theoretical properties, Schervish (1995), related to asymptotic optimality in frequentist's sense and to the fact that its asymptotic distribution is normal and that it can be evaluated in a relatively straightforward manner. In general, the MLE, or the maximum likelihood estimate $\hat{\theta}$ of a parameter θ is obtained as $\hat{\theta} = \arg\max_\theta L(\theta)$, where $L(\theta) = \log(f(z_1, z_2, \ldots, z_T; \theta))$ is the log likelihood, or logarithm of the density of the observed sample ($z_1, z_2, \ldots, z_T$), taken as a function of parameters, evaluated at θ. It is well known from general statistical theory that under relatively mild (regularity) conditions, one has:

$$\sqrt{T}(\hat{\theta} - \theta) \approx N\left(0, \left(-\frac{d^2 L(\hat{\theta})}{d\theta^2}\right)^{-1}\right) \tag{5}$$

having practical consequence that, in addition to the MLE estimates themselves, one can, at least in principle, rather easily assess their uncertainty as approximate (asymptotic, in the sense of improvement being obtained with increasing T) standard errors (and covariances), once the second derivatives of the log likelihood function evaluated at the MLE are known. This means that approximate confidence intervals and tests of (smooth) hypotheses (i.e. of hypotheses about smooth enough functions of θ) can be obtained in a straightforward manner. For real world problems, MLE estimates can rarely be obtained in a closed form. Typically, they have to be obtained via numerical maximization of the log likelihood. Efficient evaluation of the log likelihood function is then an imperative. If needed for maximization, the log likelihood derivatives can be either derived analytically or evaluated by numerical means. Note from (5) that the Hessian at the optimum has much more extensive use than just in the optimization itself. Its (negative) inversion gives asymptotic covariance matrix of the MLE estimates.

A complication occurs with the MLE computation in the context of models having latent (i.e. unobservable) random variables with complicated structure, such as the state-space model (1a-1c) here. This happens because MLE, at least in its standard form, is based on maximization of the logarithm of the marginal data density $f(z_1, z_2, \ldots, z_T; \theta)$. That is, the density of the observed data, with unobservable random variables being integrated out. Such a direct marginalization can be difficult and/or costly to obtain (especially beyond the normal linear model class). An alternative approach is to use the EM (expectation-maximization) algorithm, Dempster, Laird, Rubin (1977) which iterates between the local marginalization at the current θ estimate (E-step) and maximization to obtain improved θ estimate (M-step).

In our, state-space model (1a-1c) context, the E-step amounts to the Kalman filter run at the current estimate of θ (say at $\theta_{current}$), yielding one-step-ahead prediction errors $\{\tilde{y}_{current,t}\}_{t=1}^{T}$ which are uncorrelated but heteroscedastic in general, Durbin, Koopman (2001) and can be used to evaluate the log likelihood at a value $\theta_{current}$ as:

$$L(\theta_{current}) = -\frac{kT2\pi}{2} - \frac{1}{2}\sum_{t=1}^{T}\det(S_t) - \frac{1}{2}\sum_{t=1}^{T}(\tilde{y}_{current,t})(S_t^{-1})\tilde{y}_{current,t} \qquad (6)$$

where k is the dimension of the observation vector and $\det(.)$ stands for the determinant (assuming that the $\{S_t\}_{t=1}^{T}$'s evaluated at $\theta_{current}$ are nonsingular). The way of writing the log likelihood function used in (6) is called the *prediction error decomposition* of the (log) likelihood and it is where the Kalman filter comes in as an effective tool for (otherwise complicated) log likelihood evaluation. The M-step then amounts to maximization of $L(\theta)$ with respect to θ. Nice discussion of the EM algorithm in time series estimation context can be found in Shumway, Stoffer (1982), while further details about the MLE estimation theory and methods in the state-space context can be found in Harvey (1990).

Practical implementation of the MLE estimation should use an efficient numerical maximization procedure. In our experience, the BFGS algorithm, Fletcher (1970) gives nice results even for relatively cumbersome problems, being stable and fast. Moreover, it can produce log likelihood's Hessian at the optimum in order to estimate Fisher information matrix (and hence asymptotic covariance matrix of the estimates via inversion).

Note that all of the above discussion is based on the split between state estimation (estimation of the latent variables) and structural parameter (θ) estimation. While, in the EM style algorithm, the two are being switched iteratively, the uncertainty induced by the fact that state parameters are not known but only estimated (and just plugged-in, in place of the unknown θ) is not acknowledged. This is in line with the prevailing practice in many fields of Statistics, including the classical time series analysis, Brockwell, Davis (1991) and models with random effects in general, Davidian, Giltinan (1995), Pinheiro, Bates (2000). In fact, such a plug-in approach is in the spirit of the so called Empirical Bayes approach, Casella (1985). In the linear mixed model context, the approach amounts to the EBLUP (estimated

BLUP, or estimated structural parameters best linear unbiased prediction). Although it is known that such and approach typically underestimates the uncertainty in the state estimates somewhat, it is also well known that fixing this caveat amounts to a very hard problem (at least when being approached from frequentist's and not from Bayesian side), Kackar, Harville (1984).

APPLICATIONS OF THE MLE ESTIMATION FOR STATE-SPACE MODELS, BASED ON KALMAN FILTERING

Room Air Exchange Rate Modeling and Radon Entry Rate Estimation

Room air exchange model description

One relatively simple situation where one might want to apply online estimation of states in the Kalman filter style is when air exchange rate (ACH) estimate is needed for a single room with closed windows and doors, based on a tracer experiment. Such an estimate is useful for air tightness assessment related to energy balance assessment, for assessment of ventilation itself, as well as for estimation of sources of radon and other noxes. A simple, physically founded description of the room ventilation based on the tracer gas measurement can be formulated as follows:

$$\frac{dc_\tau}{d\tau} = Q - c_\tau k_\tau \tag{7}$$

where τ is the (continuous) time within the interval covered by the tracer gas experiment ($\tau \in \Im = [t_1, t_2]$). c_τ is the tracer concentration (in ppm) at time τ. Q is the time-invariant tracer entry rate (in ppm/h). Since the tracer is supplied artificially through a judiciously controlled process, Q is known. k_τ is the unknown air exchange rate, or ACH (in 1/h) to be estimated. Note that ACH has the meaning of the room volume proportion that is exchanged within an hour. In general, ACH is influenced by various unstable/random external influences and hence it has to be assessed properly as a time-varying quantity. In our illustrative data from a campaign conducted by the National Radiation Protection Institute in Praha and kindly provided by K. Jilek, N_2O was used as the tracer gas.

Unfortunately, relationship (7) is not usable directly since the tracer concentrations are taken only at discrete time points (placed equidistantly, 1 minute apart). They are measured with an (observation) error. This calls for separation of the structural and observation error variability, and hence for a (discrete time, continuous space) state-space model. One possible formulation is:

$$\begin{aligned} c_t &= \gamma_t + \varepsilon_t \\ \gamma_t &= \gamma_{t-1} + \Delta.Q - \Delta.\gamma_{t-1}.k_{t-1} + \eta_{1t} \\ k_t &= k_{t-1} + \eta_{2t} \end{aligned} \tag{8}$$

where $t = 1,...,T$ indexes the time of tracer observation. While c_t is the observed concentration, γ_t is its true ("denoised") version. Observation errors $\varepsilon_t \sim N(0,\sigma^2)$ form a white noise process and are independent from the structural errors $\begin{pmatrix} \eta_{1t} \\ \eta_{2t} \end{pmatrix} \sim N\left(\begin{pmatrix} 0 \\ 0 \end{pmatrix}, diag(\sigma_1^2, \sigma_2^2) \right)$ and from the initial condition $\begin{pmatrix} \gamma_0 \\ k_0 \end{pmatrix} \sim N\left(\begin{pmatrix} 0 \\ 0 \end{pmatrix}, diag(v_1^2, v_2^2) \right)$. Structural errors are also assumed to be independent temporarily (i.e. to form a bivariate white noise process). Δ is a conversion factor that needs to be introduced in order to account for the distance between the measurement instances. In our case when measurements were taken one minute apart, we take $\Delta = 1/60$.

The model (8) is almost a linear state-space model with gaussian (normally distributed) errors (1a-1c), except for the nonlinearity introduced in the first state equation via multiplication of the state vector elements. This is not the most severe form of the nonlinearity and we will deal with it via the Extended Kalman Filter (EKF). As expected, the ACH as the (time-varying) quantity of interest enters the model as a latent variable (i.e. as a component of the state vector). In order to cope with the measurement error, one needs also the other state vector element which might or might not be of direct interest. It can be useful for model checking and other purposes, however.

The structural parameters $\theta' = (\sigma^2, \sigma_1^2, \sigma_2^2, v_1^2, v_2^2)$ needed for any application of the EKF are not known and hence have to be estimated, first. We use the MLE estimation here. To this end, we reparametrized to log variances (both for numerical stability and in order to deal with the positivity constraint) before using the quasi Newton-Raphson procedure BFGS. Log likelihood function was evaluated via prediction error decomposition, as described in the Introduction. On the other hand, the γ_0 and k_0 parameters are not estimated (even if they are unknown as well, strictly speaking). They are fixed at their expert estimate $k_0 = 0.5$ and tracer value observed just before the start of the analyzed period ($\gamma_0 = 33.98$). This means that their values might not be entirely perfect and hence they might influence the fit of the model at the very beginning.

Figure 1 compares the fit of the model (8) to the data. Clearly, the tracking ability of the model is quite good, despite its simplicity. Note that the one-step-ahead prediction errors are distributed quite symmetrically around zero, as they should. Figure 2 shows this in a more detail.

By far, the point estimates are not the only quantities we should be looking at. Model (8) is a fully specified statistical model and as such, it allows for standard errors (for the one-step-ahead predicted observations) and for the confidence intervals, as well. The approximate 95% confidence intervals are shown in Figure 3. Obviously, the approximation status arises due to the local linearization used in the EKF. Note that the confidence intervals are constructed in a pointwise fashion (time point by time point) and hence that they cannot guarantee simultaneous nominal coverage. Nevertheless, the percentage of the points located beyond the confidence limits is very close to the nominal. In particular, 2.48% of datapoints are above the upper limit and 2.41% are below the lower limit. Note that estimates, standard errors and confidence limits can easily be produced for the filtered quantity as well.

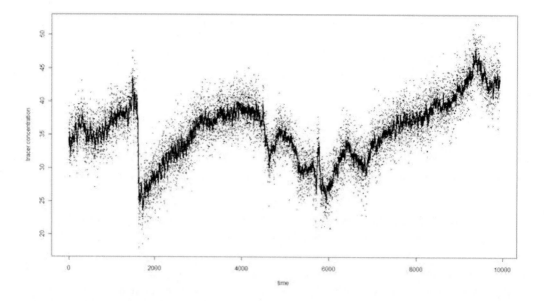

Figure 1. Predictor of the tracer concentration $\hat{y}_{t|t-1} \equiv E(y_t \mid c_1, c_2, \ldots, c_{t-1})$ (line) compared to the measurements c_t (points).

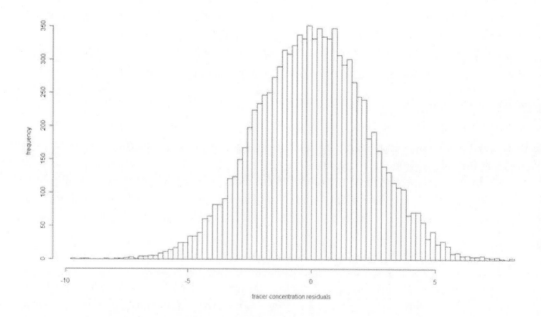

Figure 2. Histogram of the one-step-ahead prediction residuals.

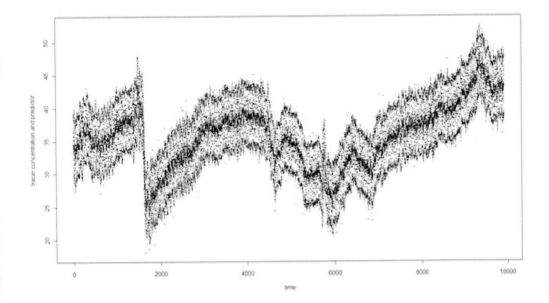

Figure 3. Predicted tracer concentration $\hat{\gamma}_{t|t-1} \equiv E(\gamma_t \mid c_1, c_2, \ldots, c_{t-1})$ (solid line), approximate 95% confidence limits (dotted lines), compared to the measurements c_t (points).

Figure 4 shows the filtered estimate of the ACH, together with the approximate 95% confidence limits. Once again, they are constructed in pointwise manner. Notice that the influence of the possible initial condition misspecification, if any, diminishes quite quickly. Initial uncertainty is more than covered by the width of the confidence intervals for the early measurement times. For both state vector elements, the filter-predictor difference, corresponding to the current observation update, is not large. Quite nicely, it is not systematic – as it should when in accord with the model assumptions.

Figure 5 shows how the filtered estimates of the two elements of state vector are correlated. Note that this correlation is induced just by the estimation process, since both elements are estimated from the same data - the model assumes their independence. Situation with the predicted estimates is quite similar. While they are somewhat correlated at earlier times, correlation of their estimates becomes very weak later. The correlation is negative, as expected from the form of the structural part of the model (8). Note also the wiggly appearance of the correlation curve. This is caused by occasional missing data points caused by the measuring device not being able to recalibrate quickly enough to be ready for the next measurement, hence skipping the particular measurement but taking the subsequent one all right. Dealing with such missing completely at random (MCAR), Little, Rubin (2002) cased is easy within the KF or EKF framework. When an observation is missing, the update is skipped, both in the filtered estimate and in the filter variance computations. Therefore, filter estimate becomes the same as the predictor estimate and the predictor variance is not reduced as in the usual update, for such a missing data case.

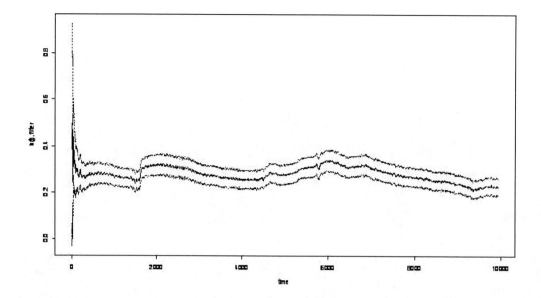

Figure 4. Filtered ACH, $\hat{k}_{t|t} \equiv E(k_t \mid c_1, c_2, \ldots, c_t)$ (solid line) and approximate 95% confidence limits (dotted lines).

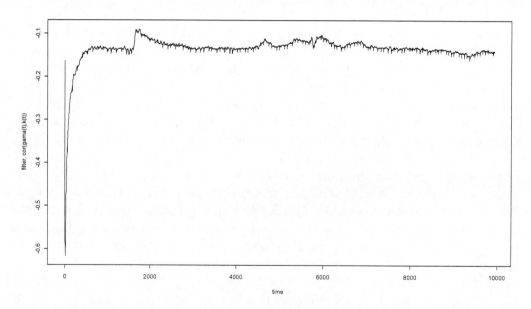

Figure 5. Correlation between filtered estimates of the two state vector components, $cor(\hat{\gamma}_{t|t}, \hat{k}_{t|t})$.

Figure 6 looks into the EKF's kitchen a bit more closely. For the Kalman filter gain vector, which has two elements in the case of model (8), it shows the first element. That illustrates influence of the one-step-ahead prediction residual when a new observation is assimilated during the filter, or update step. It becomes quite stabilized rather quickly – except for the local behavior caused by the occasional missings mentioned previously when discussing the correlations of estimates.

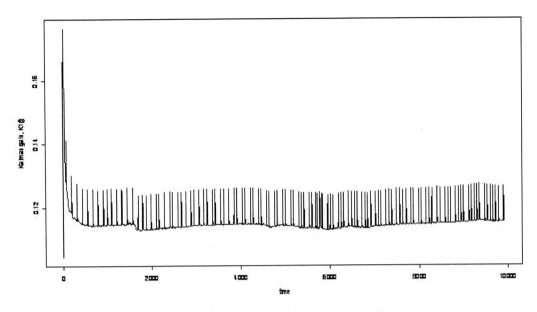

Figure 6. The first element of the Kalman gain vector, corresponding to the weight of the one-step-ahead prediction error for the true tracer concentration update.

Since the model (8) is nonlinear, one might be curious about the stability of the filter. In other words, one might worry about the behavior of the transition matrix F_t from (1b) obtained by the linearization of the state equations mentioned above in the connection with the EKF implementation. The 2x2 matrix has one eigenvalue equal to 1, all the time. Figure 7 shows the dynamics of the other eigenvalue. One can see that it is quite safely bounded away from the unstable region. This is a nice feature having implications for practical usability of the EKF here, for its standard errors as well as for its ability to cope with outliers. This is not of theoretical-only concern. In fact, the data originally contained a gross outlier (corresponding to an impossible tracer concentration of -18.24 obtained when the measurement was not calibrated properly) that we deliberately deleted (denoting it as a missing value) and did all analyses so far without it. It is reassuring to see what would happen if it would not be deleted (which can easily happen in a routine analysis) in Figure 8. We see that the filter behavior is very nice, letting the observation to influence the state estimates only very locally (in line with the Kalman gain entries shown in Figure 6). It just singles out the erroneous observation without spoiling other estimates substantially. Letting the outlier in has also other, more subtle, consequences, however. It leads to a larger prediction error variability and hence to wider confidence intervals. Then they are wider than they should be, having much larger coverage than nominal. This is not ideal, but at least the effect goes in the conservative direction.

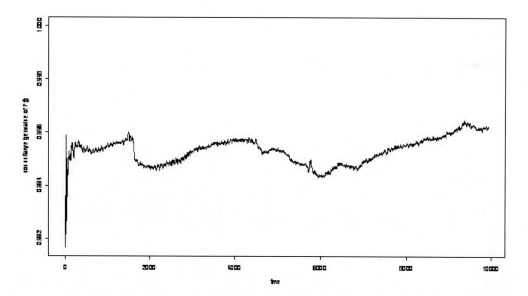

Figure 7. Non-unitary eigenvalue of the transition matrix obtained by the local linearization of the state equations involved in EKF implementation.

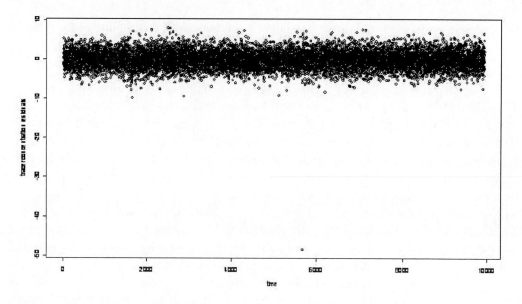

Figure 8. One-step-ahead prediction residuals from the model (8) when the single gross outlier is not deleted.

Radon entry rate estimation

One context in which the air exchange rate (ACH) estimation discussed in the previous paragraph is of substantial importance is when assessing radon entry rate (RER) for a room/house. Then, two series of measurements are taken simultaneously: tracer and radon concentrations in the same room are obtained. This experimental setup is discussed and motivated in Brabec, Jilek (2007a). Here, we will deal with a simplified model useful for measurement period that is not very close to the experiment's beginning:

$$\begin{aligned}
c_t &= \gamma_t + \varepsilon_t \\
r_t &= \rho_t + \omega_t \\
\gamma_t &= \gamma_{t-1} + \Delta.Q - \Delta.\gamma_{t-1}.k_{t-1} + \eta_{1t} \\
k_t &= k_{t-1} + \eta_{2t} \\
\rho_t &= \rho_{t-1} + \Delta.G_{t-1} - \Delta.\rho_{t-1}.k_{t-1} + \delta_{1t} \\
G_t &= G_{t-1} + \delta_{2t}
\end{aligned} \quad (9)$$

where $t = 1,\ldots,T$ indexes the tracer observation location in time. While c_t and r_t are the observed tracer and radon concentrations (in ppm and Bq/m^3), γ_t and ρ_t are their true versions, respectively. Observation errors $\begin{pmatrix} \varepsilon_t \\ \omega_t \end{pmatrix} \sim N\left(\begin{pmatrix} 0 \\ 0 \end{pmatrix}, diag(\sigma^2, A.\rho_t)\right)$ form a bivariate white noise process. They are independent from the structural errors $\begin{pmatrix} \eta_{1t} \\ \eta_{2t} \\ \delta_{1t} \\ \delta_{2t} \end{pmatrix} \sim N\left(\begin{pmatrix} 0 \\ 0 \\ 0 \\ 0 \end{pmatrix}, diag(\sigma_1^2, \sigma_2^2, \sigma_3^2, \sigma_4^2)\right)$ and from the (random) initial condition $\begin{pmatrix} \gamma_0 \\ k_0 \\ \rho_0 \\ G_0 \end{pmatrix} \sim N\left(\begin{pmatrix} a_1 \\ a_2 \\ a_3 \\ a_4 \end{pmatrix}, diag(v_1^2, v_2^2, v_3^2, v_4^2)\right)$. Structural errors are also assumed to be independent temporarily. Once again, Δ is a conversion factor that needs to be introduced in order to account for the distance between the measurement instances. In our case when measurements were taken one minute apart, we take $\Delta = 1/60$. Q is the (time-invariant and known) trace entry rate (in ppm/h), G_t is the radon entry rate, or RER (in Bq/m^{3}h), k_t is the air exchange rate, or ACH (in ppm/h). Notice that both ACH and RER are time varying, unlike in some simplified approximative models mentioned in Brabec, Jilek (2007a). While the tracer measurement error is homoscedastic – in agreement with (8) and real data properties, the radon measurement error is heteroscedastic. In fact, it is assumed to be proportional to the square root of the true measured value to reflect the near-Poisson character of the radioactive decay counts on which the radon measurement is based. There is a substantial data preprocessing within the measurement device so that the concentrations are not exactly Poisson, however. Notice that this amounts to a particular case of the so called conditional gaussian model, Harvey (1990), in which the observation error variance matrix R_t can be computed from the predicted state ($\hat{\rho}_{t|t-1}$) when it is needed for (3e).

There are two observation equations (or the observation is bivariate) and four state variables in the model (9). In a typical radon expert view, the only quantity of interest is the RER. The other ones are often perceived either as pure nuisance variables (whose presence is dictated only by the technical necessity for being able to pose a complete model from which RER can be correctly estimated), or as the quantities of secondary interest. As in the case of model (8), the more comprehensive model (9) is almost a linear state-space model with gaussian (normally distributed) errors, except for the nonlinearity introduced by multiplications of certain state vector elements. Again, we will deal with it using the Extended Kalman Filter (EKF) algorithm.

At least in principle, one can think of estimating k_t and G_t simultaneously from the radon measurement data only (and hence from one observation equation and three state equations obtained as an obvious extension of a model (8) analogue). This is not a preferable approach, however, since the radon data are typically obtained with smaller time resolution and smaller precision. Analysis of the coupled radon-tracer experiment via model like (9) is the recommended and preferable tactic. Yet another tempting simplification based on the triangular estimating equations, Small, Wang (2003) - that is on estimating k_t first from (8) and then plugging it into a model obtained by retaining only 2nd, 5th and 6th equation of (9) - is not to be recommended either. The plug-in approach does not utilize the measurement information efficiently and even more importantly, it does not acknowledge the estimation error in the $\hat{k}_t$ (either in $\hat{k}_{t|t-1}$, or in $\hat{k}_{t|t}$) to be plugged in the second step.

The structural parameters $\theta' = \left(\sigma^2, \sigma_1^2, \sigma_2^2, \sigma_3^2, \sigma_4^2, v_1^2, v_2^2, v_3^2, v_4^2, A\right)$ are not known. We use the MLE to estimate them. To that end, we reparametrize to $\log(\theta)$, i.e. to log variances and $\log(A)$. This is motivated by both numerical stability and by the positivity constraints. For optimization, we utilize the quasi Newton-Raphson procedure BFGS, evaluating the log likelihood function via prediction error decomposition. On the other hand, the starting values of the state variables (which are not known either) are not estimated. They are set to the expert estimates ($G_0 = 94$, $k_0 = 0.5$) and to the values obtained form measurements taken just prior to the first observation used in the analysis ($\gamma_0 = 33.98$ and $\rho_0 = 300$). This is possible because he measurements were taken a while before the analysis period started in order to avoid problems and instabilities that are often encountered at the start of the experiment, see Brabec, Jilek (2007a).

Typically, in the tracer-radon experiments, the time resolutions of the tracer and of the radon are different. Also here, the tracer was measured with equidistant 1 minute spacing (apart from occasional missings), while radon was measured with 1 hour spacing, for metrologic and sensitivity reasons, Jilek, Thomas, Brabec (2008). A simple way to deal with this obstacle is to work with the finer (1 minute, equidistant) time steps and deal with (quite a few) radon missings later in the EKF. The only new facet here is that, as the observation is bivariate, we can have 3 types of missings at a particular time t: i) both c_t and r_t are missing, ii) c_t missing, but r_t observed, iii) c_t observed, but r_t missing. These three cases have to be reflected separately when computing the filter/update. i) is straightforward – analogously to the univariate case mentioned in the previous paragraph, the update is just skipped here, both in the filtered estimate and in the filter variance computations. Therefore, filter estimate becomes the same as the predictor estimate and the predictor variance is not reduced as in the usual update, for such a missing data case. ii) needs to compute K_t in (3d), but differently (using different, just scalar variance of the radon-only observation, S_t) than for the fully observed data case. In fact, Kalman gain now becomes a vector (instead of matrix) and that will be reflected later in (3c, 3a, 3b). Case iii) is analogous.

MLE estimates of structural parameters were:

$$\hat{\sigma}^2 = 4.5097, \hat{\sigma}_1^2 = 0.0671, \hat{\sigma}_2^2 = 0.0000, \hat{\sigma}_3^2 = 0.0006, \hat{\sigma}_4^2 = 0.0034,$$
$$\hat{v}_1^2 = 0.6198, \hat{v}_2^2 = 0.0366, \hat{v}_3^2 = 2699.2805, \hat{v}_4^2 = 130.6736, \hat{A} = 1.2936$$

Asymptotic variance matrix of $\log(\hat{\theta})$ is rather decent. It gives asymptotic correlations that stay reasonably low, as seen from the Table 1. In other words, the structural parameters are well separated by the data (their estimates *are not* collinear).

Figure 9 shows how the model (9) fits the radon data when it uses the available information from both tracer and radon measurements simultaneously. Not only that the fit is decent (center line corresponding to the one-step-ahead prediction or to the conditional average), but also the coverage of the (approximate, pointwise constructed) 95% confidence limits for the predictions, computed from (3e) and (2a), is OK. In fact, the coverage stays on conservative side, being a bit larger than nominal, being 0.18% larger than the nominal 5%. It is quite symmetric, having about the same percentage of observations falling above and below the upper and lower confidence bounds, respectively. Analogous figure for the tracer looks quite similar to Figure 1. Figure 10 shows the difference between the one-step-ahead predictions of the same quantity (of the true tracer concentration) from the model (9) and the from model (8). Considering the levels of the tracer (as seen in Figure 1), they are rather small (the difference makes less than 0.55% of the tracer level). Also the uncertainty of the tracer predictions is quite similar in the two models. This suggests that even in the simultaneous model (9), the information from the two sources is not mixed substantially. The tracer measurements contribute mainly to the estimation of the tracer-related quantities and similarly the radon measurements contribute mainly to the radon-related quantities. This idea is supported by other characteristics as well. For instance, Figure 11 shows how the correlation between the true tracer and true radon one-step-ahead predicted concentrations evolve in time. These two quantities are just predictions of particular state elements and hence their correlation is easily accessible from the prediction variance matrix (2b) obtained in the course of the EKF recursions. Clearly, the correlations keep rather low all the time. While they start at zero by model (9) assumption, they increase a little bit temporarily at the beginning of the filtering (when the amount of information accumulated from the data is still low) but then decrease again. Separation between the two quantities is very good, indeed. Correlation between predictions of other state vector component is different but generally stays reasonably low after a possible intermittent increase at the beginning of the filtering (these plots are not shown).

Figure 12 shows predictors of the time varying RER (as of the quantity of the main interest) and of the ACH (as of a practically important byproduct of the analysis). We can see that mainly ACH changes relatively abruptly several times over the course of the time covered by the observations. As expected, RER is smoother. Widths of the confidence intervals stabilize relatively quickly. ACH interval width decreases and stabilizes for earlier time indexes because there are about 60 times more tracer observations per time unit than radon observations.

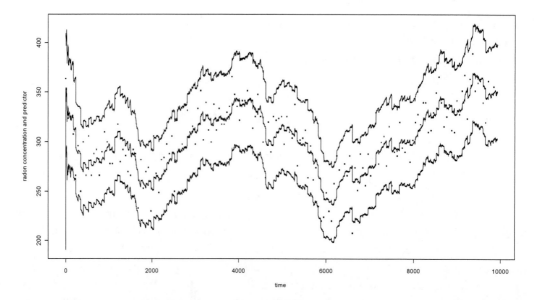

Figure 9. Predicted radon concentration $\hat{\rho}_{t|t-1} \equiv E(\rho_t \mid c_1, c_2, \ldots, c_{t-1}; r_1, r_2, \ldots, r_{t-1})$ (solid line), approximate 95% confidence limits (dotted lines), compared to the measurements ρ_t (points) obtained from the simultaneous model (9).

Figure 10. Differences of predicted tracer concentration obtained from model (9) and from model (8), $E(\gamma_t \mid c_1, c_2, \ldots, c_{t-1}; r_1, r_2, \ldots, r_{t-1}) - E(\gamma_t \mid c_1, c_2, \ldots, c_{t-1})$.

Table 1. Asymptotic correlation matrix of the MLE for structural parameters

	$\log(\hat{\sigma}^2)$	$\log(\hat{\sigma}_1^2)$	$\log(\hat{\sigma}_2^2)$	$\log(\hat{\sigma}_3^2)$	$\log(\hat{\sigma}_4^2)$	$\log(\hat{v}_1^2)$	$\log(\hat{v}_2^2)$	$\log(\hat{v}_3^2)$	$\log(\hat{v}_4^2)$	$\log(\hat{A})$
$\log(\hat{\sigma}^2)$	1	-0.28	0.01	0.00	0.04	-0.01	0.01	0.00	0.02	0.00
$\log(\hat{\sigma}_1^2)$	-0.28	1	-0.04	0.00	-0.16	0.03	-0.02	0.01	-0.07	0.02
$\log(\hat{\sigma}_2^2)$	0.01	-0.04	1	-0.02	-0.03	0.01	-0.01	0.02	-0.02	-0.16
$\log(\hat{\sigma}_3^2)$	0.00	0.00	-0.02	1	-0.02	0.00	0.00	0.00	-0.01	-0.02
$\log(\hat{\sigma}_4^2)$	0.04	-0.16	-0.03	-0.02	1	-0.04	0.05	0.03	0.32	-0.04
$\log(\hat{v}_1^2)$	-0.01	0.03	0.01	0.00	-0.04	1	-0.04	0.02	-0.02	-0.01
$\log(\hat{v}_2^2)$	0.01	-0.02	-0.01	0.00	0.05	-0.04	1	-0.02	0.03	0.01
$\log(\hat{v}_3^2)$	0.00	0.01	0.02	0.00	0.03	0.02	-0.02	1	0.02	-0.03
$\log(\hat{v}_4^2)$	0.02	-0.07	-0.02	-0.01	0.32	-0.02	0.03	0.02	1	0.00
$\log(\hat{A})$	0.00	0.02	-0.16	-0.02	-0.04	-0.01	0.01	-0.03	0.00	1

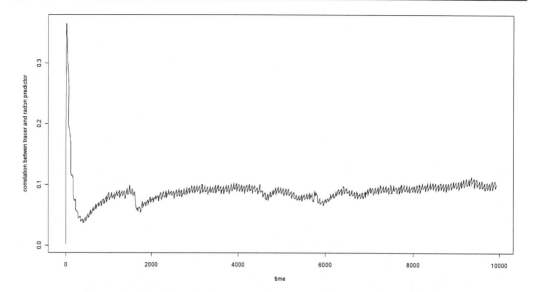

Figure 11. Correlation between one-step-ahead predictors of the true tracer and the true radon concentrations (i.e. $cor(\hat{\gamma}_{t|t-1}, \hat{\rho}_{t|t-1})$).

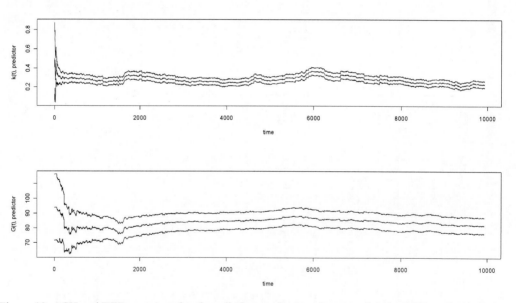

Figure 12. ACH and RER one-step-ahead predictions and approximate, pointwise 95% conf. Intervals.

So far, we took the radon data as point observations taken one hour apart. This is just a simplification, since they, in fact, represent hourly averages. Due to the sensitivity problems and relatively low radioactivity of the air, the radon measurements have to be done in integrated fashion. Therefore, strictly speaking, a radon observation r_t at time t does not represent a noisy observation of true concentration at time t only, as assumed before ($r_t = \rho_t + \varepsilon_t$). Instead, it corresponds to a noisy observation of the average concentration over the past hour:

$$r_t = \left(\frac{1}{60}\right)\sum_{j=0}^{59} \rho_{t-j} + \varepsilon_t \qquad (10)$$

where we now have $\varepsilon_t \sim N\left(0, \left(\frac{A}{60}\right)\sum_{j=0}^{59}\rho_{t-j}\right)$. Conceptually, it is not difficult to deal with the complication still within the EKF framework. All what it needs is to:

(i) expand the state vector from the length of 4 to the length of 63 (keeping the state vector the same, except for replacing the ρ_t element and hence ρ_t equation in model (9) by 60 equations, starting with the original one, being followed by the equations for ρ_{t-1}, ..., ρ_{t-59}, each having no structural error and developing according to a subdiagonal F_t submatrix),

(ii) add more initial conditions for the expanded part (we use the same starting value for all sixty ρ-related state element),

(iii) modify observation operator matrix H_t. It expands (to 2x63 from 2x4) but stays time invariant. While the first row, corresponding to tracer part is similar as before, the second row, corresponding to the radon observation contains $1/60$ in the positions corresponding to the ρ-related state elements.

Structural parameters remain the same and do not expand, in this case. Hence the optimization is complicated only by the more difficult evaluation of the log likelihood and not by the parameter space expansion. Hence the expansion causes "only" computational problem. The state dimension, and more importantly the dimension of the state variance matrices increase by 59. That increases computational load and storage requirements in (2a, 2b, 3a, 3b, 3c, 3d). It is doable, however, and it goes through in our R implementation, R core (2010), albeit more slowly than before. Note that it is not enough to use the model (9) in its original form and integrate/average $\hat{\rho}_{t+j|t}, j=1,2,...,60$ (among other things, this would not provide correct variances and hence weights to the filtering algorithm) - more laborious route we just described has to be taken instead. In a more general case, we can have a functional relationship between RER and ACH, e.g. as:

$$G_t = \int_0^L k_{t-s} u_s \, ds \qquad (11)$$

for a given maximum lag L and a given weight function (or kernel) u. There we would proceed in a similar way, paying the price of the state dimension increase. When u would not be known completely, we might think of restricting it to a parametric family (as $u(\vartheta)$) and estimate the expanded structural parameter vector $(\theta', \vartheta')'$ by the MLE again.

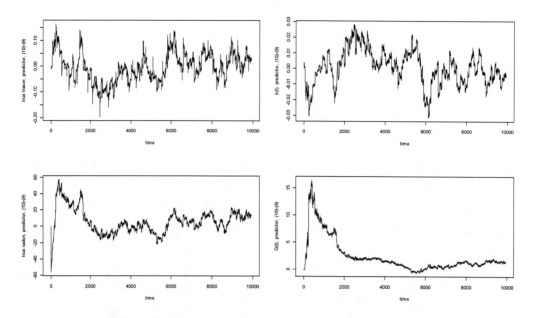

Figure 13. Difference between the one-step-ahead predictions obtained from model (10) and model (9) for RER, ACH, true tracer and true radon.

It is interesting to see how the one-step-ahead predictions of the quantities of interest (RER, ACH, true tracer, true radon) change between the model (10) with correct recognition of the functional nature of the radon observations and between the model (9) which can be perceived as a cheap approximation. Figure 13 shows that the difference is not entirely negligible, but typically, it is not huge, either. This is the case especially in the relative sense, when taking into account substantially larger resources needed for (10). The approximation can provide a first orientation in the problem relatively easily.

Individualized Growth Curve Modeling

Children growth has been of both practical and theoretical interest of anthropologists and pediatricians for more than a century. One of the more practical reasons for such popularity is related to the fact that various body size characteristics and velocities of their change are related to the current or past health status of an individual child in the period when it grows to reach mature size. Therefore, monitoring the characteristics can serve as a cheap tool for health screening widely accessible even at a population level. Growth charts based on marginal quantiles (i.e. for quantiles estimated from cross-sectional data corresponding to standards valid for one check at any given time and not for repeated checks) are used for this purpose routinely (and even inappropriately – for repeated checks of a single child). They are available for various countries, and used in clinical practice, see e.g. Kuczmarski, et al. (2000), Vignerova et al. (2006). Growth curves have been extensively analyzed also from scientific, structuralistic point of view, very often using nonparametric estimators, see e.g. Sheehy et al (1999). Predictive point of view is much less frequent, despite the fact that it corresponds more to the idea of conditional (or individualized) standards given the history of past measurements, suitable for repeated use and even using the accumulated info in an appropriate way. One attempt to construct such a standard for height, which is one of the most popular anthropometric indicators, can be found in Brabec (2004).

Here we will show a simplified construction of such a standard using a linear state space model and Kalman filter. As before, the KF will be used as a tool for likelihood evaluation needed for optimization involved in maximum likelihood estimation of structural parameters. Unlike before, we will deal here with a data setup that is more complicated than that assumed in a standard purely time series oriented analysis. Here, we have a longitudinal dataset coming from a semilongitudinal study, Blaha et al. (2006), where a large number of children (more than 900 boys and more than 900 girls) were measured repeatedly. Five replicated measurements approximately half year apart were planned originally. Due to the non-compliance, some of them were randomly skipped, however. The study was organized into three cohorts (1, 2, 3) with starting age of about 6.5, 9.5 and 11.5 years, respectively. That leads to the planned coverage of the 6.5 to 13 age interval. Several anthropometric characteristics were recorded, see Blaha et al. (2006) for a detailed description of the study. Here, were will use only the height data for boys (in cm). Figure 14 shows the situation. Our task will be to find some structural features in the data cloud. Note that there is an obvious increasing trend dictated by the growth and correlation among the measurements of the same individual.

For our purposes, we record the time by 0.1 year. We stress that the timing of the measurements is not at all regular. The spacing between subsequent measurements is roughly 6 months but the exact measurement date is shifted moderately, for convenience and technical reasons. Therefore, we will formulate a model that runs in discrete time (indexed by t) with step of 0.1 year and deal with missings given by both the design a irregularities easily within the KF by the procedure mentioned in the previous paragraph.

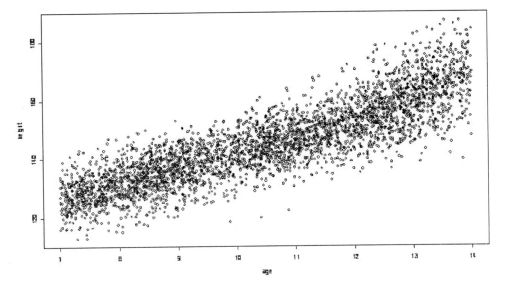

Figure 14. Boys height data from the semilongitudinal study (a child contributes up to 6 observations to the data cloud).

For height measurements of an individual child (say the i-th), we pose the following simple model (which falls into the linear state space category (1a, 1b, 1c)):

$$z_{it} = c_0 + c_1 \tau_t + x_{it} + \varepsilon_{it}$$
$$x_{it} = x_{i,t-1} + b_{i,t-1} \quad (12)$$
$$b_{it} = b_{i,t-1} + \eta_{it}$$

where τ_t is the age corresponding to the t-th time in years (time 1 corresponds to age of 7 years, i.e. $\tau_1 = 7, \tau_2 = 7.1, \ldots$). $\varepsilon_{it} \sim N(0, \sigma^2)$, (independently across time), is the measurement/observation error which is known to be homoscedastic in the case of height measurements, according to long-term anthropometric experience. The first equation is the observation equation and the other two are state equations. Such a model corresponds to a special case of the locally linear model used widely e.g. in econometrics and discussed in detail in Harvey (1990), West, Harrison (1999). Specifically, this model is called a smooth trend model, Koopman et al. (2000), reflecting the fact that there is no error in the first state equation. This fact imposes substantial smoothness into the process trajectories and hence also into the KF estimates. The feature is used in order to reflect well growth process smoothness that is well known empirically. It is quite simple model, well suited for this time range and the empirical data behavior. If a wider time intervals were covered by the measurements, possibly with increased time resolution, more ambitious models would be possible/needed, especially when the so called pubertal growth spurt were covered more extensively. An alternative, more complicated model suitable for research, rather than for routine purposes, can be found in Brabec (2004). The $\eta_{it} \sim N(0, \sigma_1^2)$ term describes the structural (dynamic-, or signal-related) variability. It is assumed that the structural errors are independent across time. They are also independent from observation errors and from the initial condition $\begin{pmatrix} x_{i0} \\ b_{i0} \end{pmatrix} \sim N\left(\begin{pmatrix} 0 \\ 0 \end{pmatrix}, diag(v_1^2, v_2^2) \right)$, which in turn, is assumed to be independent from the observation errors as well. Due to the way the model (12) is specified, the initial condition corresponds to the height at 7, for all individuals. Clearly, there is a linear trend in the model growth, whose departures are modeled as smooth trend. This, in total, gives 6 unknown structural parameters, $\theta' = (\sigma^2, \sigma_1^2, v_1^2, v_2^2, c_0, c_1)$ to be estimated from the data. Most certainly, that would not be feasible from a set of 5 measurements of a particular individual. Not all parameters would be even estimable, without at least some external information if we would attempt to estimate θ_i separately for each individual.

Instead, we assume structural similarity among the individual children. That is, we assume that θ is free of i. By far, that does not mean that the growth of different children is the same. It just has the same statistical properties. For "stitching" the information coming from various individuals together, we assume the most simple structure given by the independence across n individuals in the sample, i.e. across i's. The inter-individual similarity is concentrated to the common structural parameters. Empirically, this is a rather reasonable model with substantial computational advantages. In particular, the KF algorithm and prediction error decomposition can be very easily used to evaluate log likelihood function in a very straightforward way. The KF is evaluated individual by individual with the same parameters. Note that due to the linearity and normality, the filter can be derived in advance - running the recursions (2b, 3b)), with no data. Such a feature could offer further computational simplifications if the measurement schedule were more regular, i.e. having only a few fixed timing patterns. That is not the case here, but in any case, the total log likelihood of the whole sample of individuals is then obtained simply by summing individual log likelihood contributions, $L(\theta) = \sum_{i=1}^{n} L_i(\theta)$, without increasing complexity of the computations or state dimension for individual evaluation. Once again, we reparametrize to logs (except for c_0) for numerical stability in loglikelihood optimization.

The MLE estimates of model (12) structural parameters are: $\hat{\sigma}^2 = 4.6858e-1, \hat{\sigma}_1^2 = 6.2421e-4, \hat{v}_1^2 = 34.196, \hat{v}_2^2 = 1.3141e-5, \hat{c}_0 = 83.33, \hat{c}_1 = 5.818$. Asymptotic correlation among these MLEs are generally low, except for the intercept-slope correlation $cor(\hat{c}_0, \log(\hat{c}_1))$, which comes out as -0.89, negative as expected. Note that the $\hat{\sigma}$ estimate comes quite close to 0.5, commonly cited as the measurement error for the human height assessment. There is substantial variability in the initial height, allowing for significant assimilation of the measurement information. Figure 15 illustrates how the model works on a particular individual, once the structural parameters are estimated. Time corresponds to the age, but runs in 0.1 year steps, as mentioned in the model (12) specification. Figure 15 compares individual measurements with one-step-ahead predictions. They are accompanied by the 95% prediction intervals as with assessments of the prediction uncertainty implied by the model (12). All of these are computed for four different individuals. As implied by the (zero) initial condition in the specification of the model (12), we always start from the common linear growth trend (since that contains all the information available before measurement for a particular individual starts). That might be considerably off – vertical placement typically needs to be adjusted individually. Top right panel shows a large adjustment for a 5-observation individual, from the second cohort. The adjustment is almost additive (predicted trajectory is shifted downwards, but almost parallel to the population-average linear trend). Note how quickly the adjustment occurs. The prediction beyond the area covered by the data is linear, with the slope being equal to the last slope update. Note also how quickly the prediction intervals narrow in the region covered by the data and how they get wider gradually beyond that region. Top right panel shows similar type of adjustment but occurring earlier, for individual from the first cohort. Bottom left panel illustrates model behavior for an individual which is data-poor (he skipped quite a few planned measurements). The prediction is much closer to the population average, as it should when the individualized information is scarce. Bottom right panel shows a more complicated adjustment pattern, where slope is readjusted, probably in connection with the early pubertal spurt onset.

Figure 16 shows how the predictor and the filter of the local slope differ in case of one particular individual. The filter for a time of a particular observation already assimilated the information from this observation. That information is then propagated into the predictor of further observations until a further observation comes in and hence until the new update/filter is available. Note that the growth velocity is given per year in c_1, and per the time-step (0.1 year) for b, so that the two differ by factor of 10.

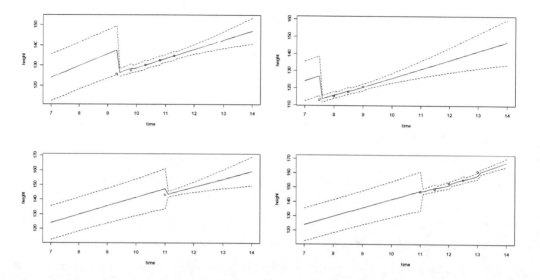

Figure 15. Measurements (dots), height one-step-ahead predicted from model (12) (solid line) and pointwise 95% confidence intervals for the prediction computed for four different individuals.

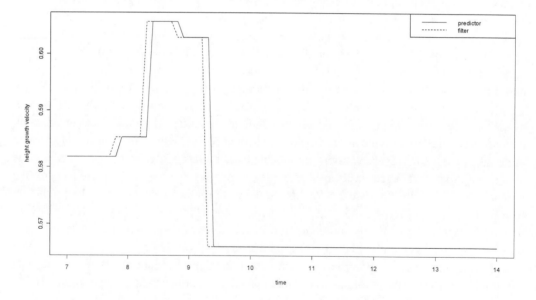

Figure 16. Predictor and filter for growth velocity (i.e. $\hat{b}_{i,t|t-1} = Eb_{it} \mid z_{i1},\ldots,z_{i,t-1}$ and $\hat{b}_{i,t|t} = Eb_{it} \mid z_{i1},\ldots,z_{i,t}$, respectively) for one individual.

Taken all together, once the structural parameters are estimated, the model (12), despite its structural simplicity can provide nice conditional (individualized) predictions that account for all available information in an elegant way, allowing for measurement schedule irregularity often seen in pediatric practice. Moreover, a model like (12), together with structural parameter MLEs, can be used for various design considerations – for instance when a new study or children growth surveillance program is to be launched (e.g. looking for measurement spacing just close enough to satisfy uncertainty requirements given a priori), see Brabec (2004) for more discussion of this issue.

Annual Gas Consumption Modeling (How to Assess the Individual*time Interaction)

In gas industry, there are various models that aim at modeling consumption of small- to medium-size customers (SMC), both household and commercial. The modeling here is driven mostly by practical needs of substantial economic importance. While for the large customers, consumption is tracked (almost) continuously, the smaller ones are too numerous to be followed individually in sufficient time resolution. Their meters are typically red in larger time periods, providing only consumption sums (or averages) per longer intervals (e.g. annual). Moreover, not all SMC meters are red at once. They are red more or less regularly throughout year (reading a portion of the total SMC pool each month). On the other hand, for various purposes, including balancing, planning, transport and storage capacity considerations, marketing, etc., the gas utility company needs to have an estimate of the amount of gas consumed for a particular day, month (or any other time interval) by all SMC customers (or by some subset of the total SMC customer pool). In order to be able to construct such estimates, one needs to disaggregate the interval sums/averages into daily consumptions. There are several models of varying complexity that do just this. Since the gas consumption is strongly temperature-dependent, a reasonable model should take that into account. There are various standardized load profile (SLP) and other models available, which do adjust for temperature as well as for other effects related to holidays, type of the weekday, as well as for the character of a particular customer, Brabec at al. (2009b), Vondracek et al. (2008). In those models that are statistically derived, one often relies on the multiplicative decomposition of the nonrandom part of the consumption model that separates individual but constant part and time-varying but common part.

$$p_i f(T_t, t; \lambda) \qquad (13)$$

for i-th individual, time t, nonlinear function (or even functional) f of temperature T_t, time, and some vector of parameters. $f(T_t, t; \lambda)$ captures the common dynamics driven by temperature, calendar irregularities etc. In practice, it is often normalized, somehow. In order for this to work, $f(T_t, t; \lambda)$ is applied to a realatively homogeneous segment of SMC customers (the parameters are estimated and the model used upon stratification by segment). On the other hand, p_i captures individual size of the consumption. It is often termed as planned (individual) annual consumption, expected consumption (EYC) and alike. EYC is typically estimated via various ad hoc procedures from historical readings of the particular customer prior to the period of interest. Perhaps the most typical is the simple CLOF strategy (carry the last observation forward).

Of course, in the analysis of variance (ANOVA) terminology, Graybill (1976), (13) amounts just to the assumption of additivity (or lack of interaction) on the log scale. The (log scale) additivity or (original scale) multiplicativity is convenient, parsimonious and allows for capturing substantial portion of the consumption variability. A more elaborate model can be of use especially in the less stable periods when the EYC changes more rapidly and individually (e.g. in times of economic crisis, or when thermal insulation is massively improved as a result of governmental subsidy programs, etc.). Obviously, the full (saturated) individual*time interaction model is out of question here. It would not even be estimable. On the other hand, very simple and restrictive forms of interaction, allowing e.g. for different linear trends (and other individual-specific but time global parametric models) for different customers would be estimable easily, but not of much use, because we are after changes that are of many different shapes taking place among different individuals. What we want is something between no interaction model (13) and full interaction, but less inclined to the "no interaction" than the parametric restrictions just mentioned (something with substantially more degrees of freedom for the interaction part of the model). State space model in the residuals seems to be one possible route to take. Such a model would be advantageous also from operational point of view. EKF updates can be done in an on line way, as soon as the meter readings come in, via the Kalman filtering.

Here we will illustrate the approach on the historical reading data kindly provided by the West Bohemian Gas Utility Company. They are related to a particular household customer segment consisting of those that use the gas for heating, as well as other purposes. We have reading histories of various length for 5819 individual customers, there. The model we use here is a refinement of the model (13). For an individual i, and its t-th reading of the annual consumption, we have (assuming independence among the individuals):

$$\log(z_{it}) = \log(f(T_t, t, \lambda)) + x_{it} + \varepsilon_{it}$$
$$x_{it} = x_{i,t-1} + \eta_{it}$$
(14)

note that here, the time is indexed individually, by the reading occasion, $t = 0,1,2,\ldots,m_i$ (m_i varying from 1 to 7), not globally as in the previous examples. $f(T_t, t; \lambda)$ is of similar structure as in Brabec et al. (2009b) or Brabec et al. (2009a). Importantly, we take its parameters λ as known and fixed. In fact, they were estimated on a rather large sample of continuously followed customers previously. This is the way how the f-like model, Brabec et al. (2009b) is applied routinely in practice in both Czech Republic and Slovakia, Brabec et al. (2009c). The (log) reading error is assumed to be normal, $\varepsilon_{it} \sim N(0, \sigma^2)$, homoscedastic, independent both across time and from the structural errors. That means that the original reading error is lognormal, effectively. Structural errors are also assumed to be independent across time, and $\eta_{it} \sim N(0, \sigma_1^2)$. Initial condition is $x_{i0} \sim N(a_i, v_1^2)$, where a_i is constructed from the consumptions red just before the first observation used for the analysis (from observation indexed by 0, i.e. from z_{i0}'s). Therefore, we have the following vector of unknown structural parameters (common to all $i = 1,2,\ldots,n$ individuals), $\theta' = (\sigma^2, \sigma_1^2, v_1^2)$. Reparametrizing to $\log(\theta)$ for the purpose of numerical log likelihood optimization, these will be estimated via Kalman filter application and prediction error decomposition. Because of the assumed inter-individual independence, we obtain the total likelihood conveniently as the sum of individual components, $L(\theta) = \sum_{i=1}^{n} L_i(\theta)$, similarly as in the previous example.

It is immediately clear that the model (14) corresponds to the lognormal random walk with drift given by f mediating some of the systematic external influences upon the consumption. On the original scale, we have a time-varying and individually specific factor modifying the original SLP model f. Notice also that it is the latent variable x_{it} that is related to the EYC, as to the variable of interest. In fact, unlike in (13) the p_i is allowed to be time-varying and we have $p_{it} = \exp(x_{it})$ in model (14).

MLE of θ is obtained quickly and easily. It comes as $\hat{\sigma}^2 = 0.0328$, $\hat{\sigma}_1^2 = 0.1251$, and $\hat{v}_1^2 = 0.1063$. Asymptotic correlation of $\log(\hat{\theta})$ is given in the Table 2, with asymptotic standard errors on the diagonal. This is reasonable and precise estimate. Notice that the signal to noise ratio is about 1.95 in this case, but the initial uncertainty is high. That throws some suspicion on the idea of using simple CLOF. Since the CLOF strategy is currently very popular in practice, we will explore this in a more detail. Figure 17 shows the behavior of the two predictors of the consumption in the next roughly annual interval, as compared to the data (consisting from a consumption reading from a standard gas meter) obtained from several individual customers (displayed on the individual panels). There, we can see that, possibly after adaptation of one or two periods, the model (14) based predictor tends to be better than CLOF in tracking the measured data. This impression is confirmed by overall performance statistics based on all available data and not only four individuals shown. Table 3 compares the overall performance characteristics. From there, we can clearly see that while both predictors overestimate (which might be related to a small general downward trend of gas consumption connected to insulation, ineffective appliance replacements etc.), the model (14) predictor is systematically better than the popular CLOF. It is better both in terms of difference and ratio (relative) residual means, as well as in terms of variability of residuals. This is due to the fact that, unlike CLOF, model (14) takes into account variability of the initial estimate. It is also open to the new information coming with an additional observation in just a right way, given by the derivations underlying the Kalman filter (2a-3e). Figure 18 shows 95% pointwise confidence intervals computed for four individual customers. It illustrates the generally good tracking ability of the model (14). Due to the multiplicative nature of the model (14), the prediction intervals are asymmetric. Coverage is large enough – in fact, it is even somewhat larger than nominal (the prediction intervals tend to be conservative, especially with respect to the upper limit). It might be worthwhile to expand the model (14) in future, in order to split the residual variability into several components and to cut the intervals shorter.

Table 2. asymptotic correlation matrix of the MLE for structural parameters

	$\log(\hat{\sigma}^2)$	$\log(\hat{\sigma}_1^2)$	$\log(\hat{v}_1^2)$
$\log(\hat{\sigma}^2)$	0.0343	-0.75	0.12
$\log(\hat{\sigma}_1^2)$	-0.75	0.0167	-0.27
$\log(\hat{v}_1^2)$	0.12	-0.27	0.0472

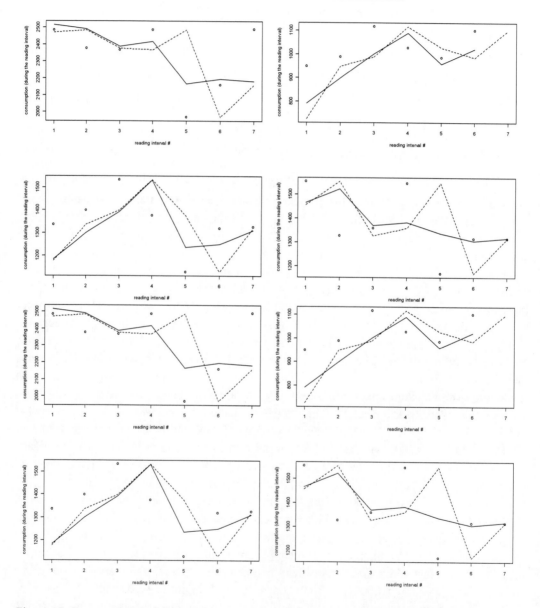

Figure 17. Consumption during roughly annual intervals for several customers. Comparison of data (dots), one-step-ahead (i.e. one reading interval ahead) predictor obtained from model (14) as $\hat{z}_{t|t-1} = E(z_t \mid z_0, z_1, \ldots, z_{t-1})$ (solid line) and the CLOF (carry the last observation forward, dotted line).

Table 3. performance characteristics for the model (14) based predictor and for the CLOF predictor. $I(.)$ is an indicator function, equal to 1 if the condition in the argument is true and to 0 otherwise

	Model (14) based predictor	CLOF predictor
$M = \dfrac{\sum_{i=1}^{n}\left(z_{it} - \hat{z}_{it,predictor}\right)}{\sum_{i=1}^{n} m_i}$	-30.44	-36.88
$\sqrt{\dfrac{\sum_{i=1}^{n}\left(z_{it} - \hat{z}_{it,predictor} - M\right)^2}{\sum_{i=1}^{n} m_i}}$	414.36	464.53
$Mr = \dfrac{\sum_{i=1}^{n}\left(z_{it}/\hat{z}_{it,predictor}\right)}{\sum_{i=1}^{n} m_i}$	1.27	1.39
$\sqrt{\dfrac{\sum_{i=1}^{n}\left(\left(z_{it}/\hat{z}_{it,predictor}\right) - Mr\right)^2}{\sum_{i=1}^{n} m_i}}$	7.86	13.10
$100\,\dfrac{\sum_{i=1}^{n} I\left(z_{it} < \hat{z}_{it,predictor}\right)}{\sum_{i=1}^{n} m_i}$	54.68	51.93
$100\,\dfrac{\sum_{i=1}^{n} I\left(z_{it} > \hat{z}_{it,predictor}\right)}{\sum_{i=1}^{n} m_i}$	45.31	47.74

Computations related to the implementation of the model (14) are not heavy – due to its simplicity. Also storage requirements are very low due to the fact that the dimension of the state vector is not larger than the dimension of the observations and that different individuals can be processed separately. Conceptually, there is no problem in posing a model similar to model (14) which would allow for dependence among the structural errors across individuals, i.e. allowing for $cor(\eta_{it}, \eta_{i't}) \neq 0, i \neq i'$ (retaining independence among structural errors in different times). "All" what it would take is the increase of the state vector dimension – and all data would have to be processed simultaneously. When the number of individuals would be moderate, it would not mean much. Nevertheless, in the situation, not uncommon in practice, when a customer segment contains thousands or even tens of thousands of individuals, it becomes a real trouble. A simplification is possible, however, for instance when the reading schedule is regular (reading interval length is exactly the same for all customers in the segment) and when the correlation structure of the structural errors and of the initial conditions is exchangeable. That is when we have model analogous with (14), but with:

$$\begin{pmatrix} \eta_{1t} \\ \vdots \\ \eta_{nt} \end{pmatrix} \sim N\left(\begin{pmatrix} 0 \\ \vdots \\ 0 \end{pmatrix}, q_1 I + q_2 J \right)$$
$$\begin{pmatrix} x_{10} \\ \vdots \\ x_{n0} \end{pmatrix} \sim N\left(\begin{pmatrix} a_1 \\ \vdots \\ a_2 \end{pmatrix}, r_1 I + r_2 J \right)$$
(15)

for some positive q_1, q_2, r_1, r_2. I is the identity matrix, while J is the matrix composed of all ones (of appropriate dimension). Independence of measurement errors would be retained as in model (14). Under this setup, the exchangeability (which is the correlation structure assumed for both η_{it}'s and x_{it}'s in (15)) is preserved during the Kalman filter prediction and update steps (2a-3e). In fact, only the variance/covariance related computations (2b, 3b, 3d, 3e) are relevant here for the filter appearance, since we are dealing with normal linear model where the variances do not depend on the state. Then, one can easily derive explicit formulas that can be evaluated easily (using just addition/subtraction, multiplication/division, with no need to use matrix algebra) for large number of customers.

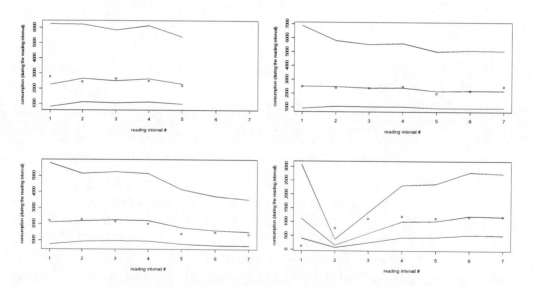

Figure 18. Observed consumption per roughly annual interval, one observation ahead prediction and its 95% (pointwise constructed) confidence interval.

CONCLUSION

Kalman filtering is very much like a regression. Even their theoretical bases are very similar. Having its flexibility, it can be used to tackle various nonstandard problems in a truly creative way (to derive relatively easy-to-compute estimates). In turn, KF or EKF algorithm is useful not only in implementation of the state-space model prediction/filtering with fixed structural parameters, but it is very much instrumental in structural parameter estimation, offering effective log likelihood function evaluation. All of this is based on the so called prediction error decomposition (based on orthogonality of prediction error to the space spanned by the previous to current observation). The log likelihood is then used to define structural parameter MLEs and their asymptotic variance, correlation, asymptotic distribution yielding confidence intervals, hypothesis tests etc, based on the general statistical theory. The estimated structural parameters are plugged into the KF/EKF in order to get state estimates (prediction and filter) in the EBLUP style (neglecting the variability of the structural parameter plug-ins).

KF/EKF algorithm offers vast possibilities for application in various interesting problems, both small and large in size (where size can be perceived in terms of state dimension, observation dimension, or both), both linear and nonlinear, both of time-series oriented and of more complicated (e.g. longitudinal) data framework. It can be perceived as a tool for estimation of time-varying latent variables in an on line manner. Importantly, not only point estimates of a dynamically evolving quantity are produced, they are accompanied by standard errors and possibly also by (pointwise constructed) confidence intervals (prediction- or filter-related). A big advantage, invaluable in practical situations is that the algorithm deals easily, almost automatically, with missing data (at least with the MCAR type of missingness).

We have shown KF/EKF applications from three vastly different areas and showed how they can be useful to deal with otherwise difficult practically motivated problems. Our illustrations stress development of the full statistical state-space model, from formal specification, to structural parameter estimation and finally to state vector prediction/filtering.

Some KF algorithm applications (especially to longitudinal data) are computer intensive, some are not. Occasionally, one can derive the filter update explicitly leading to vary simple and convenient computations that can be applied to large number of dynamically evolving units. In other cases, when the system is gaussian and linear, measurement schedule regular and common to all individuals, the simplifications are possible at least via computing the common filter form upfront (no data are actually needed in order to evaluate Kalman gain and variance matrices (2b, 3b, 3d, 3e)).

ACKNOWLEDGMENT

The work was partly supported by the Institutional Research Plan AV0Z10300504 'Computer Science for the Information Society: Models, Algorithms, Applications'.

We also would like to thank to those who kindly provided the data that we used for illustration of KF-related algorithms. Karel Jilek from the National Radiation Protection Institute (NIRP) in Praha, Czech Republic provided us the tracer and radon concentration measurements from a specialized campaign conducted by the NIRP in an experimental house located in the Czech Republic, as described in Brabec, Jilek (to appear in 2010). Jana Vignerova from the National Institute of Public Health (NIPH) in Praha, Czech Republic provided us height measurements from the unique semilongitudinal children growth study conducted by the NIPH, Blaha et al. (2006). People from the West Bohemian Gas Utility Company provided us some historical data about gas consumption related to a specific customer segment and specific region in the Czech Republic.

REFERENCES

Anderson, B. D. O. & Moore, J. B. (2005). *Optimal filtering.* Dover.

Blaha, P., Krejcovsky, L., Jiroutova, L., Kobzova, J., Sedlak, P., Brabec, M., Riedlova, J. & Vignerova, J. (2006). *Somatic development of contemporary Czech children.* Semilongitudinal study. Charles University and National Institute of Public Health. Praha.

Brabec, M. & Jilek, K. (2010). Simplified radon entry rate estimation methodology from tracer and radon concentrations based on functional data analysis approach. In: *Radioactivity Assessment.* Nova Publishers (to appear in)

Brabec, M., Konar, O., Maly, M. & Pelikan, E. (2009a). *Gas consumption and temperature. A detailed view and strategies of statistical modeling.* 29th International Symposium on Forecasting. Hong Kong, June 21-24.

Brabec, M., Konár, O., Malý, M., Pelikán, E. & Vondráček, J. (2009b). A statistical model for natural gas standardized load profiles. *JRSS C - Applied Statistics.*, 58, 123-139

Brabec, M., Konar, O., Maly, M., Pelikan, E. & Konar, O. (2009c). Statistical calibration of the natural gas consumption model. *WSEAS Transactions on Systems.*, 8, 902-912.

Brabec, M. & Jilek, K. (2007a). State-space dynamic model for estimation of radon entry rate, based on Kalman filtering, *Journal of Environmental Radioactivity*, 98, 285-297

Brabec, M., Konár, O., Malý, M., Pelikán, E. & Vondráček, J. (2007b). *State space model for aggregated longitudinal data.* Book of abstracts, ISF symposium on forecasting, page 46, New York.

Brabec, M. (2004). State-space model for individual growth curves. Folia Fac. Sci. Nat. Univ. Masaryk. *Brunensis, Mathematica*, 15, 15-28.

Brockwell, P. J. & Davis, R. A. (1991). *Time series: theory and methods.* Springer. New York.

Casella, G. (1985). An introduction to empirical Bayes data analysis. *American Statistician.*, 39, 83-87.

DasGupta, A. (2008). *Asymptotic theory of statistics and probability.* Springer. New York.

Davidian, M. & Giltinan, D. M. (1995). *Nonlinear models for repeated measurement data.* Chapman and Hall, New York.

Dempster, A. P., Laird, N. M. & Rubin, D. B. (1977). Maximum likelihood from incomplete data via the *EM algorithm. JRSS* B, 39, 1-38

Durbin, J. & Koopman, S. J. (2001). *Time series analysis by state-space methods.* Oxford University Press. Oxford.

Fletcher, R. (1970). A new approach to variable metric algorithms. *Computer Journal.*, 13, 317-322

Graybill, F. A. (1976). Theory and application of the linear model. Wadsworth & Brooks/Cole. Pacific Grove. California.

Harvey, A. C. (1990). *Forecasting, structural time series models and the Kalman filter.* Cambridge University Press. Cambridge,

Jilek, K., Thomas, J. & Brabec, M. (2008). QA programme for radon and its short-lived progeny measuring instruments in NRPI Prague. *Radiation Protection Dosimetry.*, 1-5

Kackar, R. N. & Harville, D. A. (1984). Approximations for standard errors of estimators of fixed and random effects in mixed linear models. *JASA*, 79, 853-862.

Koopman, S. J., Harvey, A. C., Doornik, J. A. & Shephard, N. (2000). *STAMP, structural time series analyzer, modeller and predictor.* Timberlake Consultants. London.

Kuczmarski, R. J., Ogden, C. L. & Grummer-Strawn, L. M. (2000). et al. *CDC growth charts: United states advance data from vital and health statistics.* No. 314. National Center for Health Statistics. Hyattsville, Maryland.

Little, R. J. & Rubin, D. B. (2002). *Statistical Analysis with Missing Data.* John Wiley, New York.

Meinhold, R. J. & Singpurwalla, N. D. (1983). Understanding the Kalman filter. *The American Statistician*, 37, 123-127.

Pinheiro, J. C. & Bates, D. M. (2000). *Mixed Effect Models in S and S-Plus.* Springer. New York.

R core: R. (2010). (http://cran.at.r-project.org/).

Ramsay, J. O. & Silverman, B. W. (1997). *Functional data analysis.* Springer. New York.

Schervish, M. J. (1995). *Theory of statistics.* Springer. New York.

Schumway, R. H. & Stoffer, D. S. (1982). An approach to time series smoothing and forecasting using the EM algorithm. *Journal of Time Series Analysis.*, 4, 253-263

Sheehy, A., Gasser, T., Molinari, L. & Largo, R. H. (1999). An analysis of variance of the pubertal and midgrowth spurts for length and width. *Annals of Human Biology*, 26, 309-331

Small, C. G. & Wang, J. (2003). *Numerical methods for nonlinear estimating equations.* Clarendon press. Oxford.

Tarantola, A. (2005). *Inverse problem theory and methods for model parameter estimation.* SIAM. Paris.

Vignerova, J., Riedlova, J., Blaha, P., Kobzova, J., Krejcovsky, L., Brabec, M. & Hrušková, M. (2001). *The 6th Nation-wide Anthropological Survey of Children and Adolescents,* (Czech republic). Přírodovědecká fakulta UK. Praha. 2006

Vondracek, J., Pelikan, E., Konar, O., Cermakova, J., Eben, K., Maly, M. & Brabec, M. A. (2008). statistical model for the estimation of natural gas consumption. *Applied Energy.*, 85, 362-370

West, M. & Harrison, J. (1999). *Bayesian forecasting and dynamic models.* Springer. New York. (2nd edition).

In: Kalman Filtering
Editor: Joaquín M. Gomez

ISBN: 978-1-61761-462-0
© 2011 Nova Science Publishers, Inc.

Chapter 4

FORECASTING THE WEEKLY US TIME-VARYING BETA: COMPARISON BETWEEN GARCH MODELS AND KALMAN FILTER METHOD

Taufiq Choudhry[1]* and Hao Wu[2]

School of Management, University of Southampton,
Southampton, United Kingdom

ABSTRACT

This paper investigates the forecasting ability of four different GARCH models and the Kalman filter method. The four GARCH models applied are the bivariate GARCH, BEKK GARCH, GARCH-GJR and the GARCH-X model. The paper also compares the forecasting ability of a non-GARCH model in the Kalman method. Forecast errors based on twenty US companies' weekly stock return forecasts (based on estimated time-varying beta) are employed to evaluate the out-of-sample forecasting ability of both the GARCH models and the Kalman method. The results when measuring forecast errors overwhelmingly support the Kalman filter approach. Among the GARCH models, the BEKK model appears to provide somewhat more accurate forecasts than the other bivariate GARCH models. However, the predominance of BEKK over the other GARCH models is not significant. Jel Classification: G1, G15

Keywords: Forecasting, Kalman Filter, GARCH, Time-varying beta, Volatility.

1. INTRODUCTION

The standard empirical testing of the Capital Asset Pricing Model (CAPM) assumes that the beta of a risky asset or portfolio is constant (Bos and Newbold, 1984). Fabozzi and

---

* Corresponding author: Phone: (44) 2380-599286, Fax: (44) 2380-593844, E-mail: T.Choudhry@soton.ac.uk

Francis (1978) suggest that a stock's beta coefficient may move randomly through time rather than remain constant.[1] Fabozzi and Francis (1978) and Bollerslev et al. (1988) provide tests of the CAPM that imply time-varying betas.

As indicated by Brooks et al. (1998), several different econometrical methods have been applied to estimate time-varying betas of different countries and firms.[2] Two of the most well known methods are the different versions of the GARCH models and the Kalman filter approach. The GARCH models apply the conditional variance information to construct the conditional beta series. The Kalman approach recursively estimates the beta series from an initial set of priors, generating a series of conditional alphas and betas in the market model. Brooks et al. (1998) provide several citations of papers that apply these different methods to estimate the time-varying beta.

Given that the beta is time-varying, empirical forecasting of the beta has become important for several reasons. Since the beta (systematic risk) is the only risk that investors should be concerned about, prediction of the beta value makes it easier for investors to make their investment decisions. The value of beta can also be used by market participants to measure the performance of fund managers through the Treynor ratio. For corporate financial managers, forecasts of the conditional beta not only benefits them in capital structure decisions but also in investment appraisal.

This paper empirically estimates and attempts to forecast (by means of four GARCH models and the Kalman filter technique) the weekly time-varying beta of twenty US firms. This paper thus empirically investigates the forecasting ability of four different GARCH models: the standard bivariate GARCH, the bivariate BEKK, the bivariate GARCH-GJR and the bivariate GARCH-X. The paper also studies the forecasting ability of the non-GARCH Kalman filter approach. A variety of GARCH models have been employed previously to model time-varying betas for different stock markets (see Bollerslev et al. (1988), Engle and Rodrigues (1989), Ng (1991), Bodurtha and Mark (1991), Koutmos et al. (1994), Giannopoulos (1995), Braun et al. (1995), Gonzalez-Rivera (1996), Brooks et al. (1998) and Yun (2002). Similarly, the Kalman filter technique has also been used by some studies to estimate the time-varying beta (see Black et al., 1992; Well, 1994).

Given the different methods available the empirical question to answer is which econometrical method provides the best forecast. Although a large literature exists on volatility forecasting models, no single model has yet emerged as superior. Akgiray (1989) finds the GARCH(1,1) model specification exhibits superior forecasting ability to traditional ARCH exponentially weighted moving average and historical mean models, using monthly US stock index returns. The apparent superiority of GARCH is also observed by West and Cho (1995) in forecasting exchange rate volatility for a one week horizon, although for a longer horizon none of the models exhibit forecast efficiency. In contrast, Dimson and Marsh (1990), in an examination of the UK equity market, conclude that the simple models provide more accurate forecasts than GARCH models.

More recently, empirical studies have emphasised the comparison between GARCH models and relatively sophisticated non-linear and non-parametric models. Pagan and

---

[1] According to Bos and Newbold (1984), the variation in the stock's beta could be due to the influence of microeconomics factors and/or macroeconomics factors. A detailed discussion of these factors is provided by Rosenberg and Guy (1976a, 1976b).

[2] Brooks et al. (1998) provide several citations of papers that apply these different methods to estimate the time-varying beta.

Schwert (1990) compare GARCH, EGARCH, Markov switching regime, and three non-parametric models for forecasting US stock return volatility. While all non-GARCH models produce very poor predictions, the EGARCH, followed by the GARCH models, perform moderately. As a representative applied to exchange rate data, Meade (2002) examines the forecasting accuracy of the linear AR-GARCH model versus four non-linear methods using five data frequencies, and finds that the linear model is not outperformed by the non-linear models. However, despite this debate, and the inconsistent evidence, as Brooks (2002, p. 493) says, it appears that conditional heteroscedasticity models are among the best that are currently available.

Franses and Van Dijk (1996) investigate the performance of the standard GARCH model and the non-linear Quadratic GARCH and GARCH-GJR models for forecasting the weekly volatility of various European stock market indices. Their results indicate that non-linear GARCH models cannot beat the original model. In particular, the GJR model is not recommended for forecasting. In contrast to their results, Brailsford and Faff (1996) find the evidence favours the GARCH-GJR model for predicting monthly Australian stock volatility, compared with the standard GARCH model. However, Day and Lewis (1992), using out of sample forecast comparisons, find limited evidence that, in certain instances, GARCH models provide better forecasts than EGARCH models.

Few papers have compared the forecasting ability of the Kalman filter method with the GARCH models. The Brooks et al. (1998) paper investigates three techniques for the estimation of time-varying betas: GARCH; a time-varying beta market model approach suggested by Schwert and Seguin (1990); and the Kalman filter. According to in-sample and out-of-sample return forecasts based on beta estimates, the Kalman filter is superior to the others. Faff et al. (2000) finds all three techniques are successful in characterising time-varying beta. Comparisons based on forecast errors support the theory that time-varying betas estimated by the Kalman filter are more efficient than other models. One of the main objectives of our paper is to compare the forecasting ability of the GARCH models against the Kalman method.

2. THE (CONDITIONAL) CAPM AND THE TIME-VARYING BETA

One of the assumptions of the capital asset pricing model (CAPM) is that all investors have the same subjective expectations on the means, variances and covariances of returns.[3] According to Bollerslev et al. (1988), economic agents may have common expectations on the moments of future returns, but these are conditional expectations and therefore random variables rather than constant.[4] The CAPM that takes conditional expectations into consideration is sometimes known as conditional CAPM. The conditional CAPM provides a convenient way to incorporate the time-varying conditional variances and covariances

---

[3] See Markowitz (1952), Sharpe (1964) and Lintner (1965) for details of the CAPM.
[4] According to Klemkosky and Martin (1975) betas will be time-varying if excess returns are characterised by conditional heteroscedasticity.

(Bodurtha and Mark, 1991).[5] An asset's beta in the conditional CAPM can be expressed as the ratio of the conditional covariance between the forecast error in the asset's return and the forecast's error of the market return, and the conditional variance of the forecast error of the market return.

The following analysis relies heavily on Bodurtha and Mark (1991). Let $R_{i,t}$ be the nominal return on asset i ($i= 1, 2, ..., n$) and $R_{m,t}$ the nominal return on the market portfolio m. The excess (real) return of asset i and market portfolio over the risk-free asset return is presented by $r_{i,t}$ and $r_{m,t}$, respectively. The conditional CAPM in excess returns may be given as

$$E(r_{i,t}|I_{t-1}) = \beta_{i|t-1} E(r_{m,t}|I_{t-1}) \qquad (1)$$

where,

$$\beta_{i|t-1} = \text{cov}(R_{i,t}, R_{m,t}|I_{t-1})/\text{var}(R_{m,t}|I_{t-1}) = \text{cov}(r_{i,t}, r_{m,t}|I_{t-1})/\text{var}(r_{m,t}|I_{t-1}) \qquad (2)$$

and $E(\cdot|I_{t-1})$ is the mathematical expectation conditional on the information set available to the economic agents during the last period (t-1), I_{t-1}. Expectations are rational based on Muth's (1961) definition of rational expectation where the mathematical expected values are interpreted as the agent's subjective expectations. According to Bodurtha and Mark (1991), asset I's risk premium varies over time due to three time-varying factors: the market's conditional variance; the conditional covariance between the asset's return; and the market's return and/or the market's risk premium. If the covariance between asset i and the market portfolio m is not constant, then the equilibrium returns $R_{i,t}$ will not be constant. If the variance and the covariance are stationary and predictable, then the equilibrium returns will be predictable.

3. GARCH MODELS

3.1. Bivariate GARCH

As shown by Baillie and Myers (1991) and Bollerslev et al. (1992), weak dependence of successive asset price changes may be modelled by means of the GARCH model. The multivariate GARCH model uses information from more than one market's history. According to Engle and Kroner (1995), multivariate GARCH models are useful in multivariate finance and economic models, which require the modelling of both variance and covariance. Multivariate GARCH models allow the variance and covariance to depend on the information set in a vector ARMA manner (Engle and Kroner, 1995). This, in turn, leads to the unbiased and more precise estimate of the parameters (Wahab, 1995).

---

[5] Hansen and Richard (1987) have shown that omission of conditioning information, as is done in tests of constant beta versions of the CAPM, can lead to erroneous conclusions regarding the conditional mean variance efficiency of a portfolio.

The following bivariate GARCH(p,q) model may be used to represent the log difference of the company stock index and the market stock index:

$$y_t = \mu + \varepsilon_t \qquad (3)$$

$$\varepsilon_t / \Omega_{t-1} \sim N(0, H_t) \qquad (4)$$

$$vech(H_t) = C + \sum_{j=1}^{p} A_j vech(\varepsilon_{t-j})^2 + \sum_{j=1}^{q} B_j vech(H_{t-j}) \qquad (5)$$

where $y_t = (r_t^c, r_t^f)$ is a (2x1) vector containing the log difference of the firm (r_t^c) stock index and market (r_t^f) index, H_t is a (2x2) conditional covariance matrix, C is a (3x1) parameter vector (constant), A_j and B_j are (3x3) parameter matrice, and vech is the column stacking operator that stacks the lower triangular portion of a symmetric matrix. We apply the GARCH model with diagonal restriction.

Given the bivariate GARCH model of the log difference of the firm and the market indices presented above, the time-varying beta can be expressed as:

$$\beta_t = \hat{H}_{12,t} / \hat{H}_{22,t} \qquad (6)$$

where $\hat{H}_{12,t}$ is the estimated conditional variance between the log difference of the firm index and market index, and $\hat{H}_{22,t}$ is the estimated conditional variance of the log difference of the market index from the bivariate GARCH model. Given that conditional covariance is time-dependent, the beta will be time-dependent.

3.2. Bivariate Bekk Garch

Lately, a more stable GARCH presentation has been put forward. This presentation is termed by Engle and Kroner (1995) the BEKK model; the conditional covariance matrix is parameterized as

$$vech(H_t) = C'C + \sum_{K=1}^{K} \sum_{i=1}^{q} A'_{Ki} \varepsilon_{t-i} \varepsilon'_{t-i} A_{ki} + \sum_{K=1}^{K} \sum_{i=1}^{p} B'_{Kj} H_{t-j} B_{kj} \qquad (7)$$

Equations 3 and 4 also apply to the BEKK model and are defined as before. In equation 7, A_{ki}, $i = 1,..., q$, $k = 1,... K$, and B_{kj} $j = 1, ... p$, $k = 1,..., K$ are all $N \times N$ matrices. This formulation has the advantage over the general specification of the multivariate GARCH in that the conditional variance (H_t) is guaranteed to be positive for all t (Bollerslev et al., 1994). The BEKK GARCH model is sufficiently general in that it includes all positive definite diagonal representations, and nearly all positive definite vector representation. The following presents the BEKK bivariate GARCH(1,1), with K=1.

$$H_t = C'C + A'\varepsilon_{t-1}\varepsilon'_{t-1}A + B'H_{t-1}B \qquad (7a)$$

where C is a 2x2 lower triangular matrix with intercept parameters, and A and B are 2x2 square matrices of parameters. The bivariate BEKK GARCH(1,1) parameterization requires estimation of only 11 parameters in the conditional variance-covariance structure, and guarantees H_t positive definite. Importantly, the BEKK model implies that only the magnitude of past returns' innovations is important in determining current conditional variances and co-variances. The time-varying beta based on the BEKK GARCH model is also expressed as equation 6. Once again, we apply the BEKK GARCH model with diagonal restriction.

3.3. Garch-GJR

Along with the leptokurtic distribution of stock returns data, negative correlation between current returns and future volatility have been shown by empirical research (Black, 1976; Christie, 1982). This negative effect of current returns on future variance is sometimes called the leverage effect (Bollerslev et al. 1992). The leverage effect is due to the reduction in the equity value which would raise the debt-to-equity ratio, hence raising the riskiness of the firm as a result of an increase in future volatility. Thus, according to the leverage effect stock returns, volatility tends to be higher after negative shocks than after positive shocks of a similar size. Glosten et al. (1993) provide an alternative explanation for the negative effect; if most of the fluctuations in stock prices are caused by fluctuations in expected future cash flows, and the riskiness of future cash flows does not change proportionally when investors revise their expectations, the unanticipated changes in stock prices and returns will be negatively related to unanticipated changes in future volatility.

In the linear (symmetric) GARCH model, the conditional variance is only linked to past conditional variances and squared innovations (ε_{t-1}), and hence the sign of return plays no role in affecting volatilities (Bollerslev et al. 1992). Glosten et al. (1993) provide a modification to the GARCH model that allows positive and negative innovations to returns to have different impacts on conditional variance.[6] This modification involves adding a dummy variable (I_{t-1}) on the innovations in the conditional variance equation. The dummy (I_{t-1}) takes the value one when innovations (ε_{t-1}) to returns are negative, and zero otherwise. If the coefficient of the dummy is positive and significant, this indicates that negative innovations have a larger effect on returns than positive ones. A significant effect of the dummy implies nonlinear dependencies in the returns volatility.

Glostern et al. (1993) suggest that the asymmetry effect can also be captured simply by incorporating a dummy variable in the original GARCH.

$$\sigma_t^2 = \alpha_0 + \alpha u_{t-1}^2 + \gamma u_{t-1}^2 I_{t-1} + \beta \sigma_{t-1}^2 \tag{8}$$

---

[6] There is more than one GARCH model available that is able to capture the asymmetric effect in volatility. Pagan and Schwert (1990), Engle and Ng (1993), Hentschel (1995) and Fornari and Mele (1996) provide excellent analyses and comparisons of symmetric and asymmetric GARCH models. According to Engle and Ng (1993), the Glosten et al. (1993) model is the best at parsimoniously capturing this asymmetric effect.

where $I_{t-1} = 1$ if $u_{t-1} > 0$; otherwise $I_{t-1} = 0$. Thus, the ARCH coefficient in a GARCH-GJR model switches between $\alpha + \gamma$ and α, depending on whether the lagged error term is positive or negative. Similarly, this version of the GARCH model can be applied to two variables to capture the conditional variance and covariance. The time-varying beta based on the GARCH-GJR model is also expressed as equation 6.

3.4. Bivariate Garch-X

Lee (1994) provides an extension of the standard GARCH model linked to an error-correction model of cointegrated series on the second moment of the bivariate distributions of the variables. This model is known as the GARCH-X model. According to Lee (1994), if short-run deviations affect the conditional mean, they may also affect conditional variance; and a significant positive effect may imply that the further the series deviate from each other in the short run, the harder they are to predict. If the error correction term (short-run deviations) from the cointegrated relationship between company index and market index affects the conditional variance (and conditional covariance), then conditional heteroscedasticity may be modelled with a function of the lagged error correction term. If shocks to the system that propagate on the first and the second moments change the volatility, then it is reasonable to study the behaviour of conditional variance as a function of short-run deviations (Lee, 1994). Given that short-run deviations from the long-run relationship between the company and market stock indices may affect the conditional variance and conditional covariance, then they will also influence the time-varying beta, as defined in equation 6.

The following bivariate GARCH(p,q)-X model may be used to represent the log difference of the company and the market indices:

$$\text{vech}(H_t) = C + \sum_{j=1}^{p} A_j \text{vech}(\varepsilon_{t-j})^2 + \sum_{j=1}^{q} B_j \text{vech}(H_{t-j}) + \sum_{j=1}^{k} D_j \text{vech}(z_{t-1})^2 \quad (9)$$

Once again, equations 3 and 4 (defined as before) also apply to the GARCH-X model. The squared error term (z_{t-1}) in the conditional variance and covariance equation (equation 9) measures the influences of the short-run deviations on conditional variance and covariance. The cointegration test between the log of the company stock index and the market index is conducted by means of the Engle-Granger (1987) test.[7]

As advocated by Lee (1994, p. 337), the square of the error-correction term (z) lagged once should be applied in the GARCH(1,1)-X model. The parameters D_{11} and D_{33} indicate the

---
[7] The following cointegration relationship is investigated by means of the Engle and Granger (1987) method:
$S_t = \eta + \gamma F_t + z_t$
where S_t and F_t are the log of the firm stock index and market price index, respectively. The residuals z_t are tested for unit root(s) to check for cointegration between S_t and F_t. The error correction term, which represents the short-run deviations from the long-run cointegrated relationship, has important predictive powers for the conditional mean of the cointegrated series (Engle and Yoo, 1987. Cointegration is found between the log of company index and market index for five firms. These results are available on request.

effects of the short-run deviations between the company stock index and the market stock index from a long-run cointegrated relationship on the conditional variance of the residuals of the log difference of the company and market indices, respectively. The parameter D_{22} shows the effect of the short-run deviations on the conditional covariance between the two variables. Significant parameters indicate that these terms have potential predictive power in modelling the conditional variance-covariance matrix of the returns. Therefore, last period's equilibrium error has significant impact on the adjustment process of the subsequent returns. If D_{33} and D_{22} are significant, then H_{12} (conditional covariance) and H_{22} (conditional variance of futures returns) are going to differ from the standard GARCH model H_{12} and H_{22}. For example, if D_{22} and D_{33} are positive, an increase in short-run deviations will increase H_{12} and H_{22}. In such a case, the GARCH-X time-varying beta will be different from the standard GARCH time-varying beta.

The methodology used to obtain the optimal forecast of the conditional variance of a time series from a GARCH model is the same as that used to obtain the optimal forecast of the conditional mean (Harris and Sollis 2003, p. 246)[8]. The basic univariate GARCH(p, q) is utilised to illustrate the forecast function for the conditional variance of the GARCH process due to its simplicity.

$$\sigma_t^2 = \alpha_0 + \sum_{i=1}^{q} \alpha_i u_{t-i}^2 + \sum_{j=1}^{p} \beta_j \sigma_{t-j}^2 \qquad (10)$$

Providing that all parameters are known and the sample size is T, taking conditional expectation the forecast function for the optimal h-step-ahead forecast of the conditional variance can be written as:

$$E(\sigma_{T+h}^2|\Omega_T) = \alpha_0 + \sum_{i=1}^{q} \alpha_i (u_{T+h-i}^2|\Omega_T) + \sum_{j=1}^{p} \beta_j (\sigma_{T+h-i}^2|\Omega_T) \qquad (11)$$

where Ω_T is the relevant information set. For $i \leq 0$, $E(u_{T+i}^2|\Omega_T) = u_{T+i}^2$ and $E(\sigma_{T+i}^2|\Omega_T) = \sigma_{T+i}^2$; for $i > 0$, $E(u_{T+i}^2|\Omega_T) = E(\sigma_{T+i}^2|\Omega_T)$; and for $i > 1$, $E(\sigma_{T+i}^2|\Omega_T)$ is obtained recursively. Consequently, the one-step-ahead forecast of the conditional variance is given by:

$$E(\sigma_{T+1}^2|\Omega_T) = \alpha_0 + \alpha_1 u_T^2 + \beta_1 \sigma_T^2 \qquad (12)$$

Although many GARCH specifications forecast the conditional variance in a similar way, the forecast function for some extensions of GARCH will be more difficult to derive. For instance, extra forecasts of the dummy variable I are necessary in the GARCH-GJR model. However, following the same framework, it is straightforward to generate forecasts of the

---

[8] Harris and Sollis (2003, p. 247) discuss the methodology in detail.

conditional variance and covariance using bivariate GARCH models, and thus the conditional beta.

4. KALMAN FILTER METHOD

In the engineering literature of the 1960s, an important notion called 'state space' was developed by control engineers to describe systems that vary through time. The general form of a state space model defines an observation (or measurement) equation and a transition (or state) equation, which together express the structure and dynamics of a system.

In a state space model, observation at time t is a linear combination of a set of variables, known as state variables, which compose the state vector at time t. If we denote the number of state variables by m and the $(m \times 1)$ vector by θ_t, the observation equation can be written as

$$y_t = z_t' \theta_t + u_t \qquad (13)$$

where z_t is assumed to be a known, the $(m \times 1)$ vector, and u_t is the observation error. The disturbance u_t is generally assumed to follow the normal distribution with zero mean, $u_t \sim N(0, \sigma_u^2)$. The set of state variables may be defined as the minimum set of information from present and past data such that the future value of a time series is completely determined by the present values of the state variables. This important property of the state vector is called the Markov property, which implies that the latest value of variables is sufficient to make predictions.

A state space model can be used to incorporate unobserved variables into, and estimate them along with, the observable model to impose a time-varying structure of the CAPM beta (Faff et al., 2000). Additionally, the structure of the time-varying beta can be explicitly modelled within the Kalman filter framework to follow any stochastic process. The Kalman filter recursively forecasts conditional betas from an initial set of priors, generating a series of conditional intercepts and beta coefficients for the CAPM.

The Kalman filter method estimates the conditional beta, using the following regression,

$$R_{it} = \alpha_t + \beta_{it} R_{Mt} + \varepsilon_t \qquad (14)$$

where R_{it} and R_{Mt} are the excess return on the individual share and the market portfolio at time t, and ε_t is the disturbance term. Equation (14) represents the observation equation of the state space model, which is similar to the CAPM model. However, the form of the transition equation depends on the form of stochastic processes that betas are assumed to follow. In other words, the transition equation can be flexible, such as using AR(1) or the random walk process. According to Faff et al. (2000), the random walk gives the best characterisation of the time-varying beta, while AR(1) and random coefficient forms of transition equations encounter the difficulty of convergence for some return series. Failure of convergence is indicative of a misspecification in the transition equation. Therefore, this paper considers the form of random walk, and thus the corresponding transition equation is

$$\beta_{it} = \beta_{it-1} + \eta_t \tag{15}$$

Equations (14) and (15) constitute a state space model. In addition, prior conditionals are necessary for using the Kalman filter to forecast the future value, which can be expressed by

$$\beta_0 \sim N(\beta_0, P_0) \tag{16}$$

The first two observations can be used to establish the prior condition. Based on the prior condition, the Kalman filter can recursively estimate the entire series of conditional beta.

5. DATA AND FORECASTING TIME-VARYING BETA SERIES

The data applied is weekly, ranging from January 1989 to December 2003. Twenty US firms are selected based on size (market capitalisation), industry and the product/service provided by the firm. Table 1 provides the details of the firms under study. The stock returns are created by taking the first difference of the log of the stock indices. The excess stock returns are created by subtracting the return on a risk-free asset from the stock returns. The risk-free asset applied is the US T-Bill Discount 3 Month. The proxy for market return is the return on index of S&P500.

To avoid the sample effect and overlapping issue, three forecast horizons are considered, including two one-year forecast horizons (2001 and 2003) and one two-year forecast horizon (2002 to 2003). All models are estimated for the periods 1989-2000, 1989-2001 and 1989-2002, and the estimated parameters are applied for forecasting over the forecast samples 2001, 2002-2003 and 2003.

It is important to point out that the lack of a benchmark is an inevitable weak point for studies on time-varying beta forecasts, since the beta value is unobservable in the real world. Although the point estimation of beta generated by the market model is a moderate proxy for the actual beta value, it is not an appropriate scale to measure a beta series forecasted with time variation. As a result, evaluation of forecast accuracy based on comparing conditional betas estimated and forecasted by the same approach cannot provide compelling evidence of the worth of the approach. To assess predictive performance, a logical extension is to examine returns out-of-sample. Recall the conditional CAPM equation

$$E(r_{i,t}|I_{t-1}) = \beta_{it-1} E(r_{m,t}|I_{t-1}) \tag{17}$$

With the out-of-sample forecasts of conditional betas, the out-of-sample forecasts of returns can be easily calculated by equation (17), in which the market return and the risk-free rate of return are actual returns observed. The relative accuracy of conditional beta forecasts can then be assessed by comparing the return forecasts with the actual returns. In this way, the issue of a missing benchmark can be settled.[9]

The methodology of forecasting time-varying betas will be carried out in several steps. In the first step, the actual beta series will be constructed by GARCH models and the Kalman filter approach, from 1989 to 2003. In the second step, the forecasting models will be used to forecast returns based on the estimated time-varying betas and be compared in terms of forecasting accuracy. In the third and last step, the empirical results of the performance of various models will be produced on the basis of hypothesis tests on whether the estimate is significantly different from the real value, which will provide evidence for comparative analysis of the merits of the different forecasting models.

Table 1. Company Profile Table

Name	Products	Industry	Market Capitalisation (m$)
American Electric Power	Electric power	Utilities	79.64
Alaska Air Group	Airline services	Transportation	725.18
Bank of America	Financial services	Financial	119503.30
Boeing	Aircraft, satellites, missile	Aerospace	33721.10
California Water Service	Water related services	Utilities	463.94
Delta Air Lines	Airline services	Transportation	1458.07
Ford Motor	Cars and trucks	Automotive	28163.04
General Electric	Engines, turbines, generators	Conglomerates	311755.30
Honeywell International	Aerospace equipments	Aerospace	28818.35
Microsoft	Software	Application software	295937.20
MGP Ingredients	Ingredients and distillery	Consumer Goods	120.49
New York Times	Media products	Publishing and newspapers	7078.13
Textron	Aircraft, vehicles, finance	Conglomerates	780.03
Utah Medical Products	Medical devices	Healthcare	120.66
Walt Disney	Entertainment products	Entertainment	47718.27
Wells Fargo & Company	Financial services	Financial	99643.50
Wendy's International	Restaurant services	Restaurant	4470.80
Florida Gaming	Jai-Alai games	Gaming Activities	12.23
Campbell Soup	Convenience food	Consumer Goods	11016.59
Bell Industries	Electronics	Wholesaler	21.50

---

[9] Brooks et al. (1998) provide a comparison in the context of the market model.

6. Measures of Forecast Accuracy

A group of measures derived from the forecast error are designed to evaluate *ex post* forecasts. To evaluate forecasts, different measures of forecast errors (MAE and MSE) are employed. Mean errors (ME) are employed to assess whether the models over or under-forecast return series. Among them, the most common overall accuracy measure is MSE (Diebold 2004, p. 298):

$$MSE = \frac{1}{n}\sum_{t=1}^{n} e_t^2 \tag{18}$$

$$MAE = \frac{1}{n}\sum_{t=1}^{n} |e_t| \tag{19}$$

$$ME = \frac{1}{n}\sum_{t=1}^{n} e \tag{20}$$

Where e is the forecast error defined as the difference between the actual value and the forecasted value.

The lower the forecast error measure, the better the forecasting performance. However, it does not necessarily mean that a lower MSE completely testifies superior forecasting ability, since the difference between the MSEs may be not significantly different from zero. Therefore, it is important to check whether any reductions in MSEs are statistically significant, rather than just compare the MSE of different forecasting models (Harris and Sollis 2003, p. 250).

Diebold and Mariano (1995) develop a test of equal forecast accuracy to test whether two sets of forecast errors, say e_{1t} and e_{2t}, have equal mean value. Using MSE as the measure, the null hypothesis of equal forecast accuracy can be represented as $E[d_t] = 0$, where $d_t = e_{1t}^2 - e_{2t}^2$. Supposing n, h-step-ahead forecasts have been generated, Diebold and Mariano (1995) suggest the mean of the difference between MSEs $\overline{d} = \frac{1}{n}\sum_{t=1}^{n} d_t$ has an approximate asymptotic variance of

$$Var(\overline{d}) \approx \frac{1}{n}\left[\gamma_0 + 2\sum_{k=1}^{h-1} \gamma_k\right] \tag{21}$$

where γ_k is the kth autocovariance of d_t, which can be estimated as:

$$\hat{\gamma}_k = \frac{1}{n} \sum_{t=k+1}^{n} (d_t - \bar{d})(d_{t-k} - \bar{d}) \qquad (22)$$

Therefore, the corresponding statistic for testing the equal forecast accuracy hypothesis is $S = \bar{d} / \sqrt{Var(\bar{d})}$, which has an asymptotic standard normal distribution. According to Diebold and Mariano (1995), results of Monte Carlo simulation experiments show that the performance of this statistic is good, even for small samples and when forecast errors are non-normally distributed. However, this test is found to be over-sized for small numbers of forecast observations and forecasts of two-steps ahead or greater.

Harvey et al. (1997) further develop the test for equal forecast accuracy by modifying Diebold and Mariano's (1995) approach. Since the estimator used by Diebold and Mariano (1995) is consistent but biased, Harvey et al. (1997) improve the finite sample performance of the Diebold and Mariano (1995) test by using an approximately unbiased estimator of the variance of $\bar{d}$. The modified test statistic is given by

$$S^* = \left[\frac{n+1-2h+n^{-1}h(h-1)}{n} \right]^{1/2} S \qquad (23)$$

Through Monte Carlo simulation experiments this modified statistic is found to perform much better than the original Diebold and Mariano at all forecast horizons and when the forecast errors are autocorrelated or have non-normal distribution. In this paper, we apply both the Diebold and Mariano test, and the modified Diebold and Mariano test. Both tests generate the same results, as daily forecasts have a sufficient amount of observations in each forecast sample.

7. GARCH AND KALMAN METHOD RESULTS

The GARCH model results obtained for all periods are quite standard for equity market data. Given their bulkiness, these results are not provided in order to save space but are available on request. The GARCH-X model is estimated only for nine companies: Alaska Air Group, Boeing, California Water Service, General Electric, Microsoft, MGP Ingredients, Utah Medical Products, Walt Disney and Florida Gaming. This is because cointegration between the log of the company stock index and the log of the market stock index is only found for these companies. The cointegration results are available on request. For the GARCH models, excepting the BEKK, the BHHH algorithm is used as the optimisation method to estimate the time-varying beta series. For the BEKK GARCH, the BFGS algorithm is applied.

The Kalman filter approach is the non-GARCH model applied in competition with GARCH for predicting the conditional beta. Once again, the BHHH algorithm is used as the optimisation method to estimate the twenty time-varying beta series. The Kalman filter results are also available on request.

The basic statistics indicate that the time-varying conditional betas estimated by means of the different GARCH models have positive and significant mean values. Most beta series show significant excess kurtosis. Hence, most conditional betas are leptokurtic. All beta series are rejected for normality with the Jarque-Bera statistics, usually at the 1% level. Compared to the results of GARCH models, betas generated by the Kalman filter approach show some different features. First, not all conditional betas can be calculated by means of the Kalman filter approach. Second, conditional betas have a wider range than those constructed by GARCH models. Third, skewness, kurtosis and Jarque-Bera statistics are more diversified. There are a few cases of symmetric distribution, mesokurtic, and a single case of normal distribution. These basic statistics of the estimated beta series are available on request.[10]

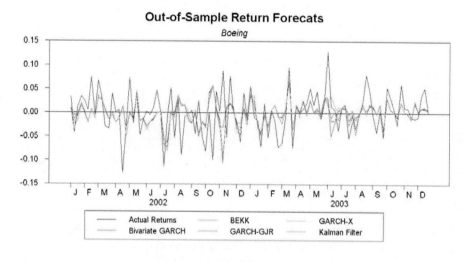

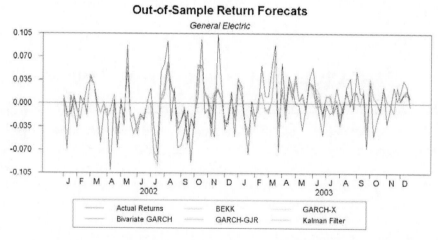

Figure 1.

---

[10] The augmented Dickey-Fuller test is applied to check for the stochastic structure of the beta series. All GARCH estimated beta series are found to have zero unit roots. Some of the beta estimated by means of the Kalman filter approach may contain one unit root. Therefore, conditional betas estimated by the Kalman filter show a different feature of dynamic structure from the ones generated by GARCH models. These results are also available on request.

8. FORECAST ERRORS BASED ON RETURN FORECASTS

As stated earlier, to avoid the sample effect and overlapping issue, three forecast horizons are considered, including two one-year (2001 and 2003) and one two-year (2002 to 2003). Also, as stated earlier, MAE, MSE and ME are the criteria applied to evaluate return forecasting performance. Given the bulkiness of these results only a summary is provided, tables of actual results are available on request. It is clear from the MAE and MSE statistics that the Kalman filter approach is the most accurate forecasting technique when forecasted returns are compared to actual returns. The Kalman filter outperforms GARCH class models in most forecasts over different forecast samples. All GARCH based models produce comparably accurate return forecasts. More precisely, BEKK and GJR are slightly superior to other GARCH models. Very little evidence of over or under prediction of return values is found during the two one-year forecasts but some evidence is found during the two-year forecast.

Figure 1 shows the return forecasted by the different methods and the actual return over the longer period (2002-2003) for two firms. All estimates seem to move together with the actual return, but because of the high frequency of the data it is difficult to say which method shows the closest correlation. Figures for other firms are available on request.

9. MODIFIED DIEBOLD AND MARIANO TESTS

As stated earlier, Harvey *et al.* (1997) propose a modified version that corrects for the tendency of the Diebold-Mariano statistic to be biased in small samples. Two criteria, including MSE and MAE derived from return forecasts, are employed to implement the modified Diebold-Mariano tests. Each time, the tests are conducted to detect superiority between two forecasting models, and thus there are ten groups of tests for five models. For each group there are a number of modified Diebold-Mariano tests for both MSE and MAE from return forecasts, between all applicable firms, and through three forecast samples.

Each modified Diebold-Mariano test generates two statistics, S_1 and S_2, based on two hypotheses:

1. H_0^1 : there is no statistical difference between two sets of forecast errors.

    H_1^1 : the first set of forecasting errors is significantly smaller than the second.

2. H_0^2 : there is no statistical difference between two sets of forecast errors.

    H_1^2 : the second set of forecasting errors is significantly smaller than the first.

It is clear that the sum of the *P* values of the two statistics (S_1 and S_2) is equal to unity. If we define the significance of the modified Diebold-Mariano statistics as at least 10% significance level of *t* distribution, adjusted statistics provide three possible answers to superiority between two rival models:

1. If S_1 is significant, then the first forecasting model outperforms the second.
2. If S_2 is significant, then the second forecasting model outperforms the first.
3. If neither S_1 nor S_2 is significant, then the two models produce equally accurate forecasts.

Tables 2 to 11 present the results of ten groups of modified Diebold-Mariano tests. Tables 2 to 5 provide a comparison between the Kalman filter approach and the four GARCH models. The Kalman filter is found to significantly outperform bivariate GARCH, BEKK GRACH and GJR GARCH models based on both the MSE and MAE (Tables 2 to 4).

Since GARCH-X can only be applied to nine firms a smaller group of forecast errors are available for Diebold-Mariano comparison tests between the GARCH-X and the Kalman filter models. Table 5 shows that for the majority of the firms the Kalman filter is superior to GARCH-X in the forecast samples of 2001 and 2002-2003. In the 2003 sample, only MAE indicates the superiority of Kalman filter in 25% of the cases; in contrast MSE shows equal accuracy of GARCH-X and Kalman filter in all cases.

Table 2. Percentage of Dominance of Kalman Filter over Bivariate GARCH

Hypothesis	2001 MSE	2001 MAE	2003 MSE	2003 MAE	2002-2003 MSE	2002-2003 MAE
Better	50	60	33.33	33.33	60	40
Worse	0	10	0	0	0	0
Equal Accuracy	50	30	66.67	66.67	40	60

Note: This table presents the proportion of firms that accept the three hypotheses. The statistic is the modified Diebold-Mariano test statistic, using MSE and MAE as the error criterion. Better means the former model dominate the later; while worse means the later model significantly outperforms the former. Equal accuracy indicates no significant difference between forecast errors. The significance is defined as at least 10% significance level of t distribution.

Table 3. Percentage of Dominance of Kalman Filter over BEKK GARCH

Hypothesis	2001 MSE	2001 MAE	2003 MSE	2003 MAE	2002-2003 MSE	2002-2003 MAE
Better	40	50	55.56	22.22	50	40
Worse	0	0	0	0	0	0
Equal Accuracy	60	50	44.44	77.78	50	60

Note: This table presents the proportion of firms that accept the three hypotheses. The statistic is the modified Diebold-Mariano test statistic, using MSE and MAE as the error criterion. Better means the former model dominate the later; while worse means the later model significantly outperforms the former. Equal accuracy indicates no significant difference between forecast errors. The significance is defined as at least 10% significance level of t distribution.

Table 4. Percentage of Dominance of Kalman Filter over GJR GARCH

Hypothesis	2001 MSE	2001 MAE	2003 MSE	2003 MAE	2002-2003 MSE	2002-2003 MAE
Better	60	60	33.33	33.33	50	60
Worse	0	0	0	0	0	0
Equal Accuracy	40	40	66.67	66.67	50	40

Note: This table presents the proportion of firms that accept the three hypotheses. The statistic is the modified Diebold-Mariano test statistic, using MSE and MAE as the error criterion. Better means the former model dominate the later; while worse means the later model significantly outperforms the former. Equal accuracy indicates no significant difference between forecast errors. The significance is defined as at least 10% significance level of t distribution.

Modified Diebold-Mariano tests are also applied amongst the GARCH models. Across different forecast horizons all tests based on MSE suggest that BEKK has enhanced predictive performance over the bivariate GARCH model (Table 6). Such enhancement is missing when MAE is used as the criterion; as both forecasting models are favoured by the same number of firms. In general, equal accuracy is supported by more than 60% of firms.

Table 5. Percentage of Dominance of Kalman Filter over GARCH-X

Hypothesis	2001 MSE	2001 MAE	2003 MSE	2003 MAE	2002-2003 MSE	2002-2003 MAE
Better	60	40	0	25	75	50
Worse	0	0	0	0	0	0
Equal Accuracy	40	60	100	75	25	50

Note: This table presents the proportion of firms that accept the three hypotheses. The statistic is the modified Diebold-Mariano test statistic, using MSE and MAE as the error criterion. Better means the former model dominate the later; while worse means the later model significantly outperforms the former. Equal accuracy indicates no significant difference between forecast errors. The significance is defined as at least 10% significance level of t distribution.

Table 6. Percentage of Dominance of Bivariate GARCH over BEKK GARCH

Hypothesis	2001 MSE	2001 MAE	2003 MSE	2003 MAE	2002-2003 MSE	2002-2003 MAE
Better	10	15	15	15	15	20
Worse	15	15	20	15	20	20
Equal Accuracy	75	70	65	70	65	60

Note: This table presents the proportion of firms that accept the three hypotheses. The statistic is the modified Diebold-Mariano test statistic, using MSE and MAE as the error criterion. Better means the former model dominate the later; while worse means the later model significantly outperforms the former. Equal accuracy indicates no significant difference between forecast errors. The significance is defined as at least 10% significance level of t distribution.

In Table 7, the Diebold-Mariano tests provide evidence that the bivariate GARCH outperforms GJR by having a higher percentage of dominance across three forecast samples in terms of both MSE and MAE. On the other hand, equal accuracy is supported by more than half of the firms; thus forecasting performance of these two models is rather close.

Bivariate GARCH is slightly superior to GARCH-X with a few more firms showing significantly smaller forecast errors (Table 8). However, there is no significant difference between MSE and MAE in at least two thirds of the firms. Therefore, bivariate GARCH and GARCH-X have comparable forecasting ability in most cases.

Table 7. Percentage of Dominance of Bivariate GARCH over GJR GARCH

Hypothesis	2001		2003		2002-2003	
	MSE	MAE	MSE	MAE	MSE	MAE
Better	20	25	10	15	40	35
Worse	5	5	5	5	5	5
Equal Accuracy	75	70	85	80	55	60

Note: This table presents the proportion of firms that accept the three hypotheses. The statistic is the modified Diebold-Mariano test statistic, using MSE and MAE as the error criterion. Better means the former model dominate the later; while worse means the later model significantly outperforms the former. Equal accuracy indicates no significant difference between forecast errors. The significance is defined as at least 10% significance level of t distribution.

Table 8. Percentage of Dominance of Bivariate GARCH over GARCH-X

Hypothesis	2001		2003		2002-2003	
	MSE	MAE	MSE	MAE	MSE	MAE
Better	22.22	22.22	11.11	11.11	11.11	11.11
Worse	11.11	11.11	11.11	0	0	0
Equal Accuracy	66.67	66.67	77.78	88.89	88.89	88.89

Note: This table presents the proportion of firms that accept the three hypotheses. The statistic is the modified Diebold-Mariano test statistic, using MSE and MAE as the error criterion. Better means the former model dominate the later; while worse means the later model significantly outperforms the former. Equal accuracy indicates no significant difference between forecast errors. The significance is defined as at least 10% significance level of t distribution.

Results of Diebold-Mariano tests between BEKK GARCH and GARCH-GJR are reported in Table 9. It shows more firms accepting that the BEKK has significantly smaller forecast errors than GJR. But the superiority of BEKK is not evident since at least half of the firms indicate equal accuracy through different forecast samples.

In the 2001 sample, BEKK dominates GARCH-X in a few more cases (Table 10). In contrast, both MSE and MAE provide evidence for the dominance of GARCH-X in the 2003 sample. Both models produce equally accurate forecasts in the 2002-2003 sample.

Table 9. Percentage of Dominance of BEKK GARCH over GJR GARCH.

Hypothesis	2001		2003		2002-2003	
	MSE	MAE	MSE	MAE	MSE	MAE
Better	15	25	25	20	35	15
Worse	10	15	10	10	15	15
Equal Accuracy	75	60	65	70	50	70

Note: This table presents the proportion of firms that accept the three hypotheses. The statistic is the modified Diebold-Mariano test statistic, using MSE and MAE as the error criterion. Better means the former model dominate the later; while worse means the later model significantly outperforms the former. Equal accuracy indicates no significant difference between forecast errors. The significance is defined as at least 10% significance level of t distribution.

Table 10. Percentage of Dominance of BEKK GARCH over GARCH-X.

Hypothesis	2001		2003		2002-2003	
	MSE	MAE	MSE	MAE	MSE	MAE
Better	22.22	44.44	0	0	11.11	11.11
Worse	11.11	11.11	22.22	22.22	11.11	11.11
Equal Accuracy	66.67	45.55	77.78	77.78	77.78	77.78

Note: This table presents the proportion of firms that accept the three hypotheses. The statistic is the modified Diebold-Mariano test statistic, using MSE and MAE as the error criterion. Better means the former model dominate the later; while worse means the later model significantly outperforms the former. Equal accuracy indicates no significant difference between forecast errors. The significance is defined as at least 10% significance level of t distribution.

Table 11. Percentage of Dominance of GJR GARCH over GARCH-X.

Hypothesis	2001		2003		2002-2003	
	MSE	MAE	MSE	MAE	MSE	MAE
Better	22.22	11.11	0	0	0	0
Worse	22.22	33.33	22.22	11.11	11.11	0
Equal Accuracy	55.56	55.56	77.78	88.89	88.89	100

Note: This table presents the proportion of firms that accept the three hypotheses. The statistic is the modified Diebold-Mariano test statistic, using MSE and MAE as the error criterion. Better means the former model dominate the later; while worse means the later model significantly outperforms the former. Equal accuracy indicates no significant difference between forecast errors. The significance is defined as at least 10% significance level of t distribution.

Table 11 reports the results from the Diebold-Mariano tests between GJR and GARCH-X forecasting models. In general, GARCH-X is found to have dominance over GJR to some extent, especially in the forecast horizon of 2003. However, overall forecasting performance of the two models is fairly similar, since more than half of the firms indicate equal accuracy.

In summary, the Diebold-Mariano comparison tests imply that the Kalman filter is the most accurate forecasting model, dominating GARCH type models in terms of return forecasts. Forecast accuracy of the GARCH type models is not considerably distinctive, since most firms provide evidence of equal accuracy among GARCH models. However, for some firms there does seem to be indications of the relative superiority of particular GARCH models over the others. In general, BEKK is the best specification with significant smaller forecast errors, followed by bivariate GARCH. GARCH-X models are a little inferior to bivariate GARCH; while GJR GARCH produces the most inaccurate out-of-sample forecasts.

CONCLUSION

This paper empirically estimates the daily time-varying beta and attempts to forecast the estimated weekly betas of twenty US firms. Since the beta (systematic risk) is the only risk that investors should be concerned about, prediction of the beta value helps investors by making their investment decisions easier. The value of beta can also be used by market participants to measure the performance of fund managers through the Treynor ratio. For corporate financial managers, forecasts of the conditional beta benefit them both in capital structure decisions and also in investment appraisals. This paper empirically investigates the forecasting ability of four different GARCH models: standard bivariate GARCH, bivariate BEKK, bivariate GARCH-GJR, and the bivariate GARCH-X. The paper also studies the forecasting ability of the non-GARCH method the Kalman filter approach. The GARCH models apply the conditional variance information to construct the conditional beta series. The Kalman approach recursively estimates the beta series from an initial set of priors, generating a series of conditional alphas and betas in the market model.

The tests are carried out in two steps. In the first step, the actual beta series are constructed by GARCH models and the Kalman filter approach from 1989 to 2003. In the second step, the forecasting models are used to forecast returns based on the estimated time-varying betas and be compared in terms of forecasting accuracy. To avoid the sample effect, three forecast horizons are considered, including two one-year forecasts, 2001 and 2003, and one two-year horizon from 2002 to 2003. Two sets of forecasts are made and the different methods applied are compared.

In the third and last step, the empirical results of the performance of various models are produced on the basis of hypothesis tests as to whether the estimate is significantly different from the real value, which will provide evidence for comparative analysis of the merits of the different forecasting models. Various measures of forecast errors are calculated on the basis of beta forecasts to assess the relative superiority of the alternative models. In order to evaluate the level of forecast errors between conditional beta forecasts and actual values, mean absolute errors (MAE), mean squared errors (MSE), and mean errors (ME) are applied.

Forecast errors based on return forecasts are employed to evaluate the out-of-sample forecasting ability of both GARCH and non-GARCH models. Measures of forecast errors overwhelmingly support the Kalman filter approach. The last comparison technique used is the modified Diebold-Mariano test. This test is conducted to detect superiority between two forecasting models at a time. The results again find evidence in favour of the Kalman filter approach, relative to GARCH models. This result is similar to Brooks et al. (1998) and Faff et al. (2000). The BEKK GARCH appears to have somewhat more accurate forecasts than the other bivariate GARCH models. However, the domination of BEKK over other GARCH models is not considerably evident.

As CAPM betas are widely used by market participants and academic researchers for a variety of purposes, empirical evidence from this paper can be helpful for those who use the beta for their decision making or research development in US stock markets. GARCH models are found to be more suitable to estimate rather than to forecast the time-varying beta. Both bivariate GARCH and BEKK are appropriate models to describe the dynamic process of conditional betas. However, despite their theoretical advantage, GARCH-X and GJR are less competent in modally US weekly beta series. To forecast the time-varying beta, the Kalman filter is a better method than GARCH models. However, if the purpose of the beta forecast is not directly related to investment in the stock market, BEKK can be a good choice, since it effectively captures the time variation features of systematic risk and produces moderately accurate and consistent forecasts of systematic risk. Thus, BEKK can be excellent to establish measures for risk management purpose. If forecasted beta is to be used for decision making in stock markets, then the Kalman filter is a better choice than the GARCH models, since it is superior to GARCH models in terms of return forecasts. Results presented in this paper advocate further research in this field, applying different markets, time periods and methods.

REFERENCES

Akgiray, V. (1989). Conditional Heteroscedasticity in Time Series of Stock Returns: Evidence and Forecast, *Journal of Business*, *Vol. 62*, pp. 55-80.

Alexander, C. (2001). *Market Models: A Guide to Financial Data Analysis*, Chichester: Wiley.

Baillie, R. T. & Myers, R. J. (1991). Bivariate GARCH Estimation of the Optimal Commodity Future Hedge, *Journal of Applied Econometrics*, *Vol. 6*, pp. 109-124.

Black, F. (1976). Studies of Stock Market Volatility Changes, *Proceedings of the American Statistical Association, Business and Economics Statistics Section*, pp. 177-181.

Black, A., Fraser, P. & Power, D. (1992). UK Unit Trust Performance 1980-1989: A Passive Time-varying Approach, *Journal of Banking and Finance, Vol. 16*, pp. 1015-1033.

Berndt, E., Hall B., Hall R. & Hausman J. (1974). Estimation and Inference in Nonlinear Structural Models, *Annals of Economic and Social Measurement*, *Vol. 3*, pp. 653-665.

Bodurtha, J. & Mark, N. (1991). Testing the CAPM with Time-Varying Risk and Returns, *Journal of Finance, Vol. 46*, pp. 1485-1505.

Bollerslev, T. (1988). On the Correlation Structure for the Generalized Autoregressive Conditional Heteroscedastic Process, *Journal of Time Series Analysis*, *Vol. 9*, pp. 121-131.

Bollerslev, T., Chou, R. & Kroner, K. (1992). ARCH Modeling in Finance, *Journal of Econometrics*, *Vol. 52*, pp. 5-59.

Bollerslev, T., Engle, R. & Nelson, D. (1994). ARCH Models, In Engle, R. and McFadden, D. (Eds.), *Handbook of Econometrics, vol. 4*, Elsevier Science, New York, pp. 2960-3038.

Bollerslev, T., Engle, R. F. & Wooldridge, J. M. (1988). A Capital Asset Pricing Model with Time-Varying Covariances, *The Journal of Political Economy*, Vol. *96*, pp. 116-131.

Bos, T. & Newbold, P. (1984). An Empirical Investigation of the Possibility of Stochastic Systematic Risk in the Market Model, *Journal of Business*, Vol. *57*, pp. 35-41.

Brailsford, T. J. & Faff, R. W. (1996). An Evaluation of Volatility Forecasting Techniques, *Journal of Banking and Finance*, Vol. *20*, pp. 419-438.

Braun, P. A., Nelson, D. B. & Sunier, A. M. (1995). Good News, Bad News, Volatility, and Betas, *Journal of Finance*, Vol. *50*, pp. 1575-1603.

Brooks, C. (2002). *Introductory Econometrics for Finance*, Cambridge: Cambridge University Press.

Brooks, R. D., Faff, R.W. & McKenzie, M. D. (1998). Time-Varying Beta Risk of Australian Industry Portfolios: A Comparison of Modelling Techniques, *Australian Journal of Management*, Vol. *23*, pp. 1-22.

Broyden, C. G. (1965). A Class of Methods for Solving Nonlinear Simultaneous Equations, *Mathematics of Computation*, Vol. *19*, pp. 577-93.

Christie, A. (1982). The Stochastic Behavior of Common Stock Variances: Value, Leverage and Interest Rate Effects, *Journal of Financial Economics*, Vol. *10*, pp. 407-432.

Day, T. E. & Lewis, C. M. (1992). Stock Market Volatility and the Information Content of Stock Index Options, *Journal of Econometrics*, Vol. *52*, pp. 267-287.

Diebold, F. X. (2004) *Elements of Forecasting*, Third Edition, Ohio: Thomson South-Western.

Diebold, F. X. & Mariano, R. S. (1995). Comparing Predictive Accuracy, *Journal of Business and Economic Statistics*, Vol., pp. 253-263.

Dimson, E. & Marsh, P. (1990). Volatility Forecasting without Data-Snooping, *Journal of Banking and Finance*, Vol. *44*, pp. 399-421.

Engle, C. & Rodrigues A. (1989). Tests of International CAPM with Time-Varying Covariances, *Journal of Applied Econometrics*, Vol. *4*, pp. 119-138.

Engle, R. & Ng, V. (1993). Measuring and Testing the Impact of News on Volatility, *Journal of Finance*, Vol. 48, pp. 1749-1778.

Engle. R. & Granger, C. (1987). Cointegration and Error Correction: Representation, Estimation and Testing, *Econometrica*, Vol. 55, pp. 251-276.

Engle, R. F. & Kroner, K. F. (1995). Multivariate Simultaneous GARCH, *Econometric Theory*, Vol. *11*, pp. 122-150.

Engle, R. F. & Yoo, B. S. (1987) Forecasting and testing in Cointegrated Systems, *Journal of Econometrics*, Vol. *35*, pp. 143-159.

Fabozzi, F. & Francis, J. (1978). Beta as a Random Coefficient, *Journal of Financial and Quantitative Analysis*, Vol. *13*, pp. 101-116.

Faff, R. W., Hillier, D. & Hillier, J. (2000). Time Varying Beta Risk: An Analysis of Alternative Modelling Techniques, *Journal of Business Finance and Accounting*, Vol. *27*, pp. 523-554.

Fletcher, R. & Powell, M. J. D. (1963). A Rapidly Convergent Descent Method for Minimisation, *Computer Journal*, Vol. *6*, pp. 163-68.

Fornari, F. & Mele, A. (1996). Modeling the Changing Asymmetry of Conditional Variances, *Economics Letters*, Vol. *50*, pp. 197-203.

Franses, P. H. & Van Dijk, D. (1996). Forecasting Stock Market Volatility Using Non-Linear GARCH Models, *Journal of Forecasting*, Vol. *15*, pp.229-235.

Giannopoulos, K. (1995). Estimating the Time-Varying Components of International Stock Markets Risk, *European Journal of Finance*, Vol. *1*, pp. 129– 164.

Glosten, L., Jagannathan, R. & Runkle, D. (1993). On the Relation between the Expected Value and the Volatility of the Nominal Excess Return on Stocks, *Journal of Finance*, Vol. *48*, pp. 1779-1801.

Gonzales-Rivera G. (1996). Time-Varying Risk The Case of American Computer Industry, *Journal of Empirical Finance*, Vol. *2*, pp. 333-342.

Hansen, L. & Richard, S. (1987). The Role of Conditioning Information in Deducing Testable Restriction Implied by Dynamic Asset Pricing Models, *Econometrica*, Vol. *55,* pp. 587-614.

Harris, R. & Sollis, R. (2003). *Applied Time Series Modelling and Forecasting*, New York: John Wiley.

Harvey, D., Leybourne, S. J. & Newbold, P. (1997). Testing the Equality of Prediction Mean Squared Errors, *International Journal of Forecasting*, Vol. *13*, pp. 281–291.

Hentschel, L. (1995). All in the Family: Nesting Symmetric and Asymmetric GARCH Models, *Journal of Financial Economics*, Vol. *39*, pp. 71-104.

Klemkosky, R. & Martin, J. (1975). The Adjustment of Beta Forecasts, *Journal of Finance*, Vol. 30, pp. 1123-1128.

Koutmos, G., Lee, U. & Theodossiou, P. (1994). Time-Varying Betas and Volatility Persistence in International Stock Markets, *Journal of Economics and Business*, Vol. *46*, pp. 101-112.

Lee, T. H. (1994). Spread and Volatility in Spot and Forward Exchange Rates, *Journal of International Money and Finance*, Vol. *13*, pp. 375-383.

Lintner, J. (1965). The Valuation of Risk Assets and the Selection of Risky Investments in Stock Portfolios and Capital Budgets, *Review of Economics and Statistics*, Vol. *47*, pp. 13-37.

Markowitz, H. (1952). Portfolio Selection, *Journal of Finance*, Vol. *7*, pp. 77-91.

Meade, N. (2002). A Comparison of the Accuracy of Short Term Foreign Exchange Forecasting Methods, *International Journal of Forecasting*, Vol.*18*, pp. 67-83.

Muth, J. (1961). Rational Expectation and the Theory of Price Movements, *Econometrica*, Vol. *29*, pp. 1-23.

Ng, L. (1991). Tests of the CAPM with Time-Varying Covariances: A Multivariate GARCH Approach, *Journal of Finance*, Vol. *46*, pp. 1507-1521.

Pagan, A. & Schwert, G. W. (1990). Alternative Models for Conditional Stock Volatilities, *Journal of Econometrics*, Vol. *46*, pp. 267-290.

Rosenberg, B. & Guy, J. (1976a). Prediction of the Beta from Investment Fundamentals. Part 1, *Financial Analysts Journal*, Vol. *32*, pp. 60-72.

Rosenberg, B. & Guy, J. (1976b). Prediction of the Beta from Investment Fundamentals. Part 2, *Financial Analysts Journal*, Vol. *32*, pp.62-70.

Schwert, G. W. & Seguin, P. J. (1990). Heteroscedasticity in Stock Returns, *Journal of Finance*, vol. *4*, pp. 1129–55.

Sharpe, W. F. (1964). Capital Asset Price: A Theory of Market Equilibrium under Conditions of Risk, *Journal of Finance*, Vol. *19*, pp. 425–442.

Tse, Y. K. (2000). A Test for Constant Correlations in a Multivariate GARCH Model, *Journal of Econometrics*, Vol. *98*, pp. 107-127.

Wahab, M. (1995). Conditional Dynamics and Optimal Spreading in the Precious Metals Futures Markets, *Journal of Futures Markets*, Vol. *15*, pp. 131-166.
Well, C. (1994). Variable Betas on the Stockholm Exchange 1971-1989, *Applied Economics*, Vol. 4, pp. 75-92.
West, K. D. & Cho, D. (1995). The Predictive Ability of Several Models of Exchange Rate Volatility, *Journal of Econometrics*, Vol. *69*, pp. 367-391.
Yun, J. (2002). Forecasting Volatility in the New Zealand Stock Market, *Applied Financial Economics*, Vol. *12*, pp. 193-202.

In: Kalman Filtering
Editor: Joaquín M. Gomez

ISBN: 978-1-61761-462-0
© 2011 Nova Science Publishers, Inc.

Chapter 5

ENSEMBLE FORECASTING THROUGH EVOLUTIONARY COMPUTING AND DATA ASSIMILATION: APPLICATION TO ENVIRONMENTAL SCIENCES

M. Kashif Gill[1], Mark Wigmosta, Andre Coleman and Lance Vail[2]

[1]WindLogics, Inc, St. Paul, MN, USA
[2]Pacific Northwest National Laboratory, Richland, WA, USA

ABSTRACT

A distributed modeling system for short-term to seasonal streamflow forecasts with the ability to utilize daily remotely-sensed snow cover products and real-time streamflow and meteorology measurements is presented herein. The modeling framework employs the state-of-the-art data assimilation and evolutionary computing strategies to accurately forecast environmental variables i.e., streamflow. Spatial variability in watershed characteristics and meteorology is represented using a raster-based computational grid. Snow accumulation and melt, simplified soil water movement, and evapotranspiration are simulated in each computational unit. The model is run at a daily time-step with surface runoff and subsurface flow aggregated at the watershed scale. The model is parameterized using a multi-objective evolutionary computing scheme using Swarm Intelligence. This approach allows the model to be updated with spatial snow water equivalent from National Operational Hydrologic Remote Sensing Center's (NOHRSC) Snow Data Assimilation (SNODAS) and observed streamflow using an ensemble Kalman-based data assimilation strategy that accounts for uncertainty in weather forecasts, model parameters, and observations used for updating. The daily model inflow forecasts for the Dworshak Reservoir in north-central Idaho are compared to observations. The April-July volumetric forecasts issued by the U.S. Army Corps of Engineers (USACE) for Water Years 2000 – 2007 are also compared with model

---

[*] Corresponding author: E-mail address: nakamori@edu.kagoshima-u.ac.jp, TEL: +81(Japan)-(0)99-285-7866, FACSIMILE: +81(Japan)-(0)99-285-7735

forecasts. October 1 and March 1 volumetric forecasts are comparable to those issued by the USACE's regression based method. An improvement in March 1 forecasts is shown by pruning the initial ensemble set based on their similarity with the observed meteorology. The short-term (one-, three-, and seven-day) forecasts using Kalman-assimilation of streamflow show excellent agreement with observations. The scheme shows great potential for the use of data assimilation in modeling streamflow and other environmental variables.

Keywords: Data Assimilation, Streamflow, Forecasting, Distributed Hydrologic Model, Snow, Parameter estimation, Remote Sensing

1. INTRODUCTION

Approximately 70-80% of the water supply in the Western U.S. results from mountain snowmelt where water resource managers require accurate and timely predictions of water supplies. These predictions are important to balance often conflicting needs between hydropower production, flood control, municipal and agricultural water use, and environmental constraints. Inaccurate streamflow forecasts can have severe implications on water allocations and can result in losses of property, agricultural and hydropower production, and environmental stability. Reliable forecasts on the other hand, can be used to mitigate losses not only in extreme years (drought and floods) but also in the years which are not categorized as significant hydrological events. The purpose of this chapter is to demonstrate an ensemble streamflow forecasting system using data assimilation of streamflow and spatial snowpack that considers explicitly the uncertainty in meteorological forecasts, model structure and parameters, and the observations used for model updating.

Seasonal streamflow forecasting of snowmelt runoff which is based on statistical methods, dates back to the 1900's and still dominate in practice over more recent dynamic model-based methods [*McGuire et al.*, 2006]. The Natural Resource Conservation Service (NRCS) and the National Oceanic and Atmospheric Administration (NOAA) produce seasonal volumetric forecasts at over 750 locations in the western United States. These forecasts are based on principal components regression techniques using current snow water equivalent (SWE), fall and spring precipitation, base flow, and climate indices [*Garen*, 1992; *Pagano et al.*, 2004]. The skill of these forecasts prior to January is essentially that of climatology [*Lettenmaier and Garen*, 1979]. Forecast skill generally improves between January and April as the snow accumulation season progresses. Regions with wet winters and dry springs show greater improvement than those with dry winters and wet springs, while mixed rain-snow basins in the Pacific Northwest generally do not show significant improvement (Pagano et al., 2004). Forecast accuracy begins to decline through late spring and early summer as streamflow is less dependent on winter snowpack, until finally, there is little predictive skill by August [*McGuire et al.*, 2006].

The National Weather Service (NWS) introduced an approach for ensemble streamflow prediction (ESP) in the 1970's [*Twedt et al.*, 1977]. Probabilistic streamflow forecasts are produced firstly, by running the hydrologic model up to the forecast date with observed meteorology to obtain an estimate of initial conditions. The second step is then forcing the model during the forecast period by re-sampling from a pool of historically observed meteorological sequences. While the ESP method has the advantage of capturing the entire range of climatic variability in the historic record, in most implementations, the model parameters are fixed, thus model parameter error is not accounted for [*McGuire et al.*, 2006]. Furthermore, the underlying hydrologic model structure makes it difficult to update the model state by assimilating point or spatial observations.

Recent and evolving developments in remote-sensing of parameters such as precipitation, soil moisture, snowpack, vegetation cover, and surface topography are yielding spatial and temporal data that are driving a revolution in hydrologic science [*NRC*, 2001]. Due to their lumped design, many operational forecasting methodologies and lumped conceptual hydrologic models are poorly suited to take advantage of current spatial data products (such as NEXRAD rainfall and MODIS or SNODAS snowpack data), or real-time point measurements (such as streamflow or in-situ meteorology).

To be of maximum benefit, streamflow forecasts should consider explicitly the uncertainty in meteorological forecasts, model structure and parameters, and the observations used for model updating. This has been an active area of research in the last decade and has resulted in enormous improvements in terms of our understanding and the development of new techniques. Various approaches have been designed to address the issues of parameter and model uncertainty [*Thiemann et al.*, 2001; *Vrugt et al.*, 2002; *Vrugt et al.*, 2003; *Kaheil et al.*, 2006]. In *Vrugt et al.*, [2002], a parameter estimation method based on the localization of information (PIMLI) approach was proposed. The PIMLI approach combines the strengths of the generalized sensitivity analysis (GSA) method [*Spear and Hornberger*, 1980], the Bayesian recursive estimation (BARE) algorithm [*Thiemann et al.*, 2001], and the Metropolis algorithm [*Metropolis et al.*, 1953].

Recent focus has been laid on the use of data assimilation schemes to deal with uncertainty, making them very suitable for use in hydrologic forecasting. The commonly employed data assimilation approach is known as sequential data assimilation (SDA). The SDA approach advances one step at a time and once new observations become available, it uses those to update the system state. Sequential data assimilation explicitly takes into account input uncertainty, model uncertainty, and output uncertainty [*Weerts and Serafy*, 2006]. This has resulted in a series of approaches within the hydrologic forecasting community. *Vrugt et al.*, [2005, 2006] presented an approach for uncertainty characterization through the use of simultaneous optimization and data assimilation (SODA) for application in hydrologic models. *Weerts and Serafy* [2006] used particle filtering and ensemble Kalman filtering for flood forecasting and showed that Ensemble Kalman filter (EnKF) performed better than the particle filters employed in the study. The reason for better performance of EnKF is attributed to better handling of uncertainties and less sensitivity to the misspecifications within the model. *Moradkhani et al.* [2005a] presented an approach for dual-state parameter estimation using EnKF along with kernel smoothing of parameters. The results are shown for streamflow forecasting using a conceptual rainfall-runoff model. In another effort a sequential Monte Carlo/Particle filtering approach is used for the characterization and estimation of parameters and uncertainty [*Moradkhani et al.*, 2005b].

Gill et al., [2007] applied a hybrid approach combining the strengths of Statistical Learning Theory (SLT) and data assimilation (DA) via the use of Support Vector Machine (SVM) and ensemble Kalman filter for estimation and forecasting of soil moisture. Ajami et al., [2007] presents an approach called the Integrated Bayesian Uncertainty Estimator (IBUNE) which is a combination of Markov chain Monte Carlo (MSMC) and Bayesian

Model Averaging (BMA), to explicitly account for the major uncertainties of hydrologic rainfall-runoff predictions. The BMA approach is an attempt to replace the traditional one-model approach, otherwise referred to as "One-method syndrome" [*Wood and Lettenmaier*, 2006]. In another research effort, *Vrugt and Robinson* [2007] compared BMA with EnKF for probabilistic streamflow forecasting and found that EnKF outperformed BMA. *Liu and Gupta* [2007] developed a thorough overview of methods for treatment of uncertainty within hydrologic prediction models.

This study focuses on development of a spatially-distributed ensemble hydrologic forecasting methodology employing a state-of-the-art sequential data assimilation method, EnKF, using ground-based observations of streamflow, and satellite/model-based observations of SWE. Some of the recent efforts in this regard, in particular those using snow assimilation, are discussed in the following.

Andreadis and Lettenmaier [2006] used EnKF for snow data assimilation using snow cover area (SCA) from the Moderate Resolution Imaging Spectroradiometer (MODIS) satellite-based sensor and reported a modest improvement in streamflow estimation over the non-updated results. Similarly, *Slater and Clark* [2006] used an ensemble square root Kalman filter for updating model estimates of SWE for streamflow forecasting. *Clark et al.* [2006] used SCA information for EnKF updating and reported minor improvements in streamflow estimates. Similarly, *McGuire et al.* [2006] used MODIS SCA in a rule-based approach to update SWE distribution within the watershed model. For example, if the model shows snow in a pixel whereas the satellite does not, then SWE was either totally removed or was redistributed to locations where both the model and satellite agree. It is also important to note that all these approaches [by *Andreadis and Lettenmaier*, 2006; *Clark et al.*, 2006; *McGuire et al.*, 2006] do not use SWE; instead rely on SCA satellite observations. The current study on the other hand directly employs SWE from the National Operational Hydrologic Remote Sensing Center's (NOHRSC) Snow Data Assimilation (SNODAS) products. Furthermore, the current study employs a methodology for EnKF data assimilation of both SWE and streamflow observations, unlike previous studies where only one or the other is used.

The remainder of this chapter is organized as follows. The model description, parameter estimation, and the EnKF are presented in section 2. The case study and the data description are presented in section 3. The experimental design for data assimilation along with the results and discussion are presented in section 4, and lastly, the conclusions are presented in section 5.

2. MODELING FRAMEWORK

The current chapter provides a method to address uncertainty using an ensemble forecasting approach that also involves the assimilation of observations as they become available. A key component for improved forecasting is the ability to utilize multiple spatial and temporal data products in an integrated manner to 1) force the model, and 2) update critical state variables, such as the spatial distribution of SWE and soil moisture. This approach required the development of a spatially distributed model with the ability to utilize multiple data sources and provide sufficient process representation and structure allowing for spatial updating of critical state variables. Primary inputs to the model include precipitation, mean air temperature, elevation, and land cover. The model is run over a spatially-based computational grid that is structured to align with the pixel resolution of the remote-sensing snow products used for updating. An ensemble of model output is produced using an input ensemble consisting of past traces of meteorological forcing as a surrogate for future conditions. The approach also accounts for model parameter uncertainty through a sampling strategy within the calibrated parameter space. The model is calibrated using an evolutionary computation scheme called "Particle Swarm Optimization" (PSO) [*Eberhart and Kennedy*, 1995; *Gill et al.*, 2006]. With this approach, the ensemble of model output generated using a meteorological ensemble and model parameters sampled from the feasible parameter space, is combined with the streamflow and SWE observations to update the model state using EnKF data assimilation

2.1. Distributed Hydrological Model

2.1.1. Process representation

The Distributed Hydrologic Model (DHM) is developed to estimate unregulated streamflow at the outlet of the drainage basin. Spatial variability in basin characteristics and meteorology is represented using a raster-based computational mesh (Figure) at a spatial scale consistent with the remote-sensing product(s) in use (e.g., ~ 781-m for SNODAS or 500-m for MODIS). Snow accumulation and melt, evapotranspiration, and simplified vertical soil water movement are simulated in each computational unit (cell). Downward percolation out of the soil rooting zone from each cell is summed for all computational units at each time step and input to the subsurface saturated zone, which produces streamflow at the basin outlet. Any surface runoff produced is also aggregated and assumed to leave the basin during the time step it was generated. The DHM does not allow explicit cell-to-cell water flow routing; rather surface runoff and subsurface flow are aggregated at the basin scale. Uncertainty in weather forecasts is addressed explicitly by using ensembles of possible future meteorological realizations to drive the model. This approach allows the model to be updated with spatial snow cover and measured streamflow using an ensemble Kalman-type data assimilation strategy as described in Section 2.2.

Snow accumulation and melt

All storages are expressed as a depth per unit area and fluxes as depth per unit area per day. The snow accumulation and melt process in each grid cell is represented through a simple mass balance approach based on SWE:

$$SWE^{t+\Delta t} = SWE^{t} + P_s - M \qquad (1)$$

where P_s is the depth of snowfall, M is the depth of snowmelt, and t and $t + \Delta t$ refer to the beginning and end of the time step, respectively. The phase of precipitation is based on current air temperature (T_a) and a threshold rain-snow temperature (T_{rs}):

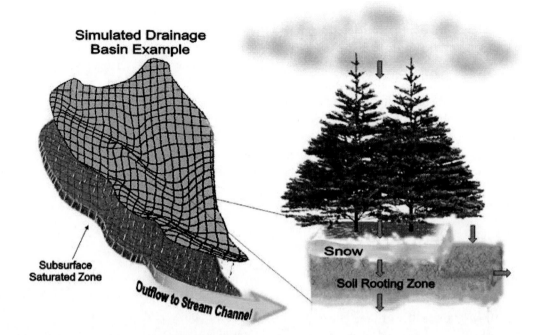

Figure 1. Distributed Hydrologic Model.

$$\begin{aligned} P_r = P;\ P_s = 0 & \quad for \quad T_a \geq T_{rs} \\ P_r = 0;\ P_s = P & \quad for \quad T_a < T_{rs} \end{aligned} \quad (2)$$

where P is the depth of precipitation and P_r is the depth of rainfall. A temperature index method is used to model snowmelt:

$$\begin{aligned} M = mF(T_a - T_b) & \quad for \quad T_a > T_b \\ M = 0 & \quad for \quad T_a \leq T_b \end{aligned} \quad (3)$$

where mF is the snowmelt factor and T_b is the base temperature for melt to occur.

Rooting zone soil moisture

The rooting zone in each grid cell receives rainfall and snowmelt from above and loses water through evapotranspiration and percolation to the subsurface saturated zone below. Mass balance within each cell is given by:

$$S_{rz}^{t+\Delta t} = S_{rz}^{t} + P_r + M - ET - Q_{rz} \quad (4)$$

where S_{rz} is the depth of water storage in the rooting zone, ET is the depth of evapotranspiration, and Q_{rz} is the depth of water discharged to the lower saturated zone.

Evapotranspiration is calculated from the current water storage, a threshold storage, and potential evapotranspiration (*PET*) demand:

$$ET = PET \quad \text{for} \quad S_{rz} > S_c$$
$$ET = PET\left(\frac{S_{rz}}{S_c}\right) \quad \text{for} \quad S_{rz} \leq S_c \tag{5}$$

where S_c is the water content above which evapotranspiration occurs at the potential rate. The Hargreaves equation [*ASCE*, 1996] is used to calculate daily PET based on solar radiation and grid cell air temperature.

The rate of water discharged to the saturated zone from each rooting zone grid cell (q_v) is a non-linear function of water content:

$$q_v = K_{s\_rz}\left(\frac{S_{rz}}{S_{rz\_max}}\right)^{n\_rz} \tag{6}$$

where K_{s_rz} is the rooting zone soil saturated hydraulic conductivity, S_{rz_max} is the maximum water holding capacity of the rooting zone, and n_rz is an exponent. The discharge volume over the time step (Q_{rz}) is given by:

$$Q_{rz} = 1/2\left(q_v(S_{rz}^t) + q_v(\hat{S}_{rz})\right)\Delta t \tag{7}$$

where $\hat{S}_{rz}$ is an updated storage that includes precipitation and snowmelt given by:
$$\hat{S}_{rz} = S_{rz}^t + M + P_r$$

Overland flow is generated in grid cells when the rooting zone water content exceeds its maximum holding capacity:

$$Q_o = S_{rz}^{t+\Delta t} - S_{rz\_max} \quad \text{for} \quad S_{rz}^{t+\Delta t} > S_{rz\_max}$$
$$Q_o = 0 \quad \text{for} \quad S_{rz}^{t+\Delta t} \leq S_{rz\_max} \tag{8}$$

where Q_o is the depth of overland flow. Overland flow from any grid cell is assumed to contribute directly to basin outflow during the time step it was generated. When this occurs, the excess water is removed and the cell rooting zone soil moisture is set equal to S_{rz_max}.

Subsurface saturated zone

The subsurface saturated zone receives percolation from each rooting zone grid cell and discharges water to the channel system. Mass balance within this zone is given by:

$$S_{sat}^{t+\Delta t} = S_{sat}^{t} + \sum Q_{rz} - Q_{sat} \tag{9}$$

where $\sum Q_{rz}$ is the total volume of water received from all of the rooting zone cells (expressed as a depth per unit area), and Q_{sat} is the discharge depth per unit area from the saturated zone. The rate of water discharged from the saturated zone to the channel system (q_{sat}) is a non-linear function of water content:

$$q_{sat} = q_s \left(\frac{S_{sat}}{S_o}\right)^{n\_sat} \tag{10}$$

where q_s is the discharge coefficient, S_o is a nominal storage capacity, and n_sat is an exponent. The discharge volume over the time step is given by:

$$Q_{sat} = 1/2 \left(q_{sat}\left(S_{sat}^{t}\right) + q_{sat}\left(\hat{S}_{sat}\right)\right)\Delta t \tag{11}$$

where $\hat{S}_{sat}$ is an updated storage that includes discharge from all rooting zones given by:
$\hat{S}_{sat} = S_{sat}^{t} + \sum Q_{rz}$.

2.1.2. PSO Calibration and parameter uncertainty

There are nine hydrologic model parameters (see Table 1) that needs to be determined through calibration. These parameters consist of three parameters for snow routine namely: base temperature T_b, rain-snow threshold temperature T_{RS}, melt factor mF; four parameters from rooting zone soil moisture accounting routine namely: root zone hydraulic conductivity $K_{s\text{-}rz}$, root zone exponent n_rz, root zone maximum storage $Smax$, root zone critical storage S_c; and two parameters from lower zone saturated zone accounting routine namely: saturated zone discharge coefficient q_s, and saturated zone discharge coefficient n_sat. The Particle Swarm Optimization (PSO) [*Eberhart and Kennedy*, 1995; *Kennedy and Eberhart*, 1995; *Eberhart et al.*, 1996] is used for the calibration of model parameters. PSO is an evolutionary computing technique based on a social-psychological metaphor of individuals living in a social world. It is inspired from the behavior individuals in bird flocks or fish schools as they hunt for food and other resources competing yet interacting with each other for the good of the swarm (flock or school). PSO has gained in popularity due to its simplistic design and effective and efficient convergence to the optimum. In the current approach, PSO is evolved using root mean square error between simulated and observed hydrograph as the objective function. The internal parameters of PSO are fixed and have been adopted from *Gill et al.* [2006].

Table 1. Hydrologic parameters and the ranges

Parameters	Range
Tbase (Tb)	-2 - +2 °C
SnowRainTemp (T_{RS})	-2 - +2 °C
Melt Factor (mF)	0.1 – 5 mm/day/°C
Root zone hydraulic conductivity (K_{s-rz})	0.1 – 4 mm/d
Root zone exponent (n_rz)	0.1 – 4
Root zone maximum storage (Smax)	400 – 1200 mm
Root zone critical storage (S_c)	200 – 1200 mm
Saturated zone discharge coefficient (q_s)	0.1 – 4 mm/d
Saturated zone discharge exponent (n_sat)	0.1 – 4

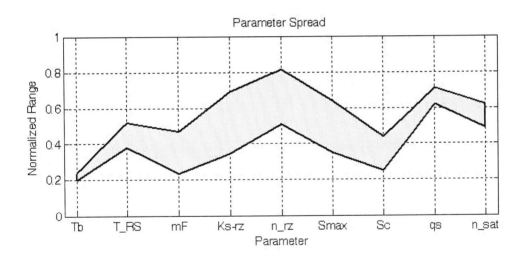

Figure 2. Feasible parameter space.

Model parameter uncertainty is represented using an approach that employs multiple runs of PSO with random initializations of parameters from the parameter region specified through Table 1. The approach is based on the principle of "equifinality" addressed in detail by *Beven* [1993] and *Beven and Freer* [2001] stating there can be multiple parameter sets providing equally feasible solutions against the objective function. Each PSO run is completed using 5,000 function evaluations, giving a so-called "optimum parameter set". The approach thence, provides a feasible region within the parameter space. The feasible parameter space/region obtained through PSO is shown in Figure . The parameter ranges are normalized between the maximum and the minimum for plotting and interpret the parameter sensitivity, determinability, and interaction. The parameters with a wider spread have higher uncertainty and lesser determinability compared to the ones with a tighter spread having lesser uncertainty and higher determinability.

2.2. Data Assimilation via Kalman Filtering

The classical Kalman filter (KF) was introduced by R. E. Kalman in 1960 [*Kalman, 1960*], mainly for application in the communication and control of random signals. The classical Kalman Filter deals with state estimation for linear processes by updating model predictions using observational measurements. The assimilation is performed at each time step when new observations become available. The state progression in a hydrologic model can be expressed as follows:

$$\mathbf{x}_{t+1} = \mathbf{F}(\mathbf{x}_t, \xi_t, \theta) + \mathbf{w}_t \tag{12}$$

where $\mathbf{F}(.)$ is a non-linear function representing the hydrologic model that relates state $\mathbf{X}_t$ at time t to state $\mathbf{X}_{t+1}$ at time $t+1$; ξ_t represents the input forcing to the hydrologic model, and $\theta \in \Theta \subseteq \Re^k$ represents the parameter set. The term $\mathbf{W}_t(\sim N(0,\mathbf{Q}))$ represents the process noise, and $\mathbf{Q}$ is the process noise covariance. Similarly, an observation equation can be suggested:

$$\mathbf{y}_{t+1} = \mathbf{H}(\mathbf{x}_{t+1}) + \mathbf{v}_{t+1} \tag{13}$$

where y_{t+1} is the observation at time step 't+1' and $\mathbf{H}(.)$ is the measurement operator that maps the state-space into the measurement or output space with a noise component $\mathbf{V}_{t+1}(\sim N(0,\mathbf{R}))$ and $\mathbf{R}$ is observation noise covariance. Now at each time step when observation becomes available, an update analysis equation can be written as follows:

$$\mathbf{x}_{t+1} = \mathbf{x}_{t+1}^- + \mathbf{K}(\mathbf{y}_{t+1} - \mathbf{H}(\mathbf{x}_{t+1}^-)) \tag{14}$$

The difference term in Eq. 17 between measurement and model predictions $(\mathbf{y}_{t+1} - \mathbf{H}(\mathbf{x}_{t+1}^-))$ is called the residual or measurement *innovation*. If the residual term is zero, the measurement and predictions are in complete agreement. $\mathbf{K}$ in Eq. 14 is the gain-term (also known as Kalman gain or weighting matrix), which employs the error covariance calculated from the ensemble of model states. It is given as:

$$\mathbf{K} = \mathbf{P}_{t+1}^- \mathbf{H}^T (\mathbf{H} \mathbf{P}_{t+1}^- \mathbf{H}^T + \mathbf{R})^{-1} \tag{15}$$

The Kalman gain, K in Eq. 19 defines the weighting to be applied to the actual measurements. If the measurement error covariance $\mathbf{R}$ approaches zero, the actual measurement $\mathbf{y}_{t+1}$ is trusted more, and the prediction is trusted less. Similarly, as the a priori estimate error covariance $\mathbf{P}_{t+1}^-$ approaches zero, the actual measurement D is given less weight and the predictions are given more weight. It should be noted that the mean of the ensemble of state estimation is used to compute the variances for prediction and updated estimation.

2.2.1. Ensemble Kalman filter

The Ensemble Kalman filter method by *Evensen* [1994] has an advantage over traditional filtering techniques as it uses a limited number of model states and results in faster convergence. The EnKF extends KF to deal with the non-linearities in data assimilation and ensemble generation problems, and is a very popular tool in weather, ocean, and hydrologic prediction modeling. A comprehensive literature review on data assimilation and filtering techniques can be found in *Drecourt* [2003].

The EnKF moves forward in a sequential manner. To begin, since the initial system state is typically not well known, the initial ensemble of the system state should be generated. Theoretically, this population should correctly reflect the error statistics of model state estimation, but practically, a modest misspecification of the initial state ensemble does not influence the result over time [*Evensen*, 2002]. The notation presented here is similar to that used in most of the EnKF literature [*Evensen*, 1994, 2003; *Reichle et al.*, 2002; *Hamill*, 2006; *Vrugt et al.*, 2006]. EnKF approximates the state estimates using an ensemble of model states, assuming that a large enough sample of system states is available to represent a specific probability distribution function (pdf) and is used for estimation of the statistical moments of the state conditioned on pdf. Below are the various steps within EnKF framework to solve equations 12-15.

Firstly, define the prior and posterior states ensembles. Let $\mathbf{X}_t^-$ represents the prior state estimate ensemble matrix $\{\mathbf{x}_{t,1}^-, \mathbf{x}_{t,2}^- \ldots \mathbf{x}_{t,n}^-\}$ at time t having size $m \times n$; $\mathbf{X}_t$ stands for the posterior state estimate ensemble $\{\mathbf{x}_{t,1}, \mathbf{x}_{t,2} \ldots \mathbf{x}_{t,n}\}$ at time t having size $m \times n$; $\mathbf{X}_{t,1}^-$ represents a (single member of ensemble) vector of m model states at time t, and n is the ensemble size. The ensemble mean, $\overline{\mathbf{X}}_t^-$, being considered as the best estimate, is used to calculate the model error.

$$\overline{\mathbf{x}}_t^- = 1/n \sum_{i=1}^{n} \mathbf{x}_{t,i}^- \qquad (16)$$

Two key components of EnKF are the prediction and updating steps. These steps are also defined as predictor and corrector steps, respectively. The predictor is responsible for projecting the state and error covariance estimates forward in time to obtain the *a priori* estimates for the next time step. The corrector incorporates a new measurement into the *a priori* estimate to obtain an improved *a posteriori* estimate for the same time. The algorithm is similar to the predictor-corrector method popular in solving numerical problems [*Welch and Bishop*, 2002].

Prediction

The prior state estimate is calculated from the posterior estimation in the previous time step. Based on this estimate, the state prior mean and covariance can be calculated:

$$\mathbf{X}^-_{t+1} = \mathbf{F}(\mathbf{X}_t, \xi_t, \theta) + \mathbf{W}_t \tag{17}$$

$$\mathbf{P}^-_{t+1} = E[(\mathbf{X}^-_{t+1} - \overline{\mathbf{X}}^-_{t+1})(\mathbf{X}^-_{t+1} - \overline{\mathbf{X}}^-_{t+1})^T] \tag{18}$$

where P^-_{t+1} represents the $m \times m$ prior estimate of covariance; $\overline{\mathbf{X}}^-_{t+1} = \{\overline{\mathbf{x}}^-_{t+1}, \overline{\mathbf{x}}^-_{t+1} \ldots \overline{\mathbf{x}}^-_{t+1}\}$ represents $m \times n$ state ensemble mean matrix with all the n columns same as $\overline{\mathbf{X}}^-_{t+1}$; and E is the expectation operator.

Updating

The observations are treated as a random variable. To do so, a sample of observations is generated from a distribution with the mean equal to the observation mean and the variance equal to the observation variance **R**. This distribution reflects prior knowledge of the user about observation error statistics. Using **D** to stand for the observation sample matrix obtained after adding perturbations using **R**, the equations used are:

$$\mathbf{X}_{t+1} = \mathbf{X}^-_{t+1} + \mathbf{P}^-_{t+1}\mathbf{H}^T(\mathbf{H}\mathbf{P}^-_{t+1}\mathbf{H}^T + \mathbf{R})^{-1}[\mathbf{D} - \mathbf{H}(\mathbf{X}^-_{t+1})] \tag{19}$$

$$\mathbf{P}_{t+1} = E[(\mathbf{X}_{t+1} - \overline{\mathbf{X}}_{t+1})(\mathbf{X}_{t+1} - \overline{\mathbf{X}}_{t+1})^T] \tag{20}$$

The product term $\mathbf{P}^-_{t+1}\mathbf{H}^T(\mathbf{H}\mathbf{P}^-_{t+1}\mathbf{H}^T + \mathbf{R})^{-1}$ in Eq. 19 is the approximation of Kalman gain term in equation 15.

2.3. Model Setup and Preprocessing

The ensemble of meteorological forecasts is generated by sampling historical record based on a similarity index known as *Hausdorff norm*, of current conditions with previous years. The Hausdorff norm [*Chavent and Lechevallier*, 2002, *Jesorky et al.*, 2001] is a metric between two point sets defined as the maximum distance of a set to the nearest point in the other set (See appendix for more details). All available meteorological data for the given region of interest are first processed for missing and erroneous values. The geographic coordinates of each meteorology station are used to extract elevation values from digital elevation models (DEM), which are then used to generate elevation-based lapse rates for precipitation and temperature.

In preparation for streamflow forecasting, the nine parameters of the hydrologic model are calibrated from historical data using the PSO algorithm. The PSO is run 50 times, each with 5,000 function evaluations and starting with different initial conditions, thus giving an ensemble of parameter set of size 50. This ensemble set defines the boundaries for the so-called "feasible parameter region", $\Theta \in \Re^m$, where m is the dimension of parameter set (see Figure 33).

2.4. Streamflow Forecasting

A three-step process is used to generate forecasts and update the model (Figure 3).

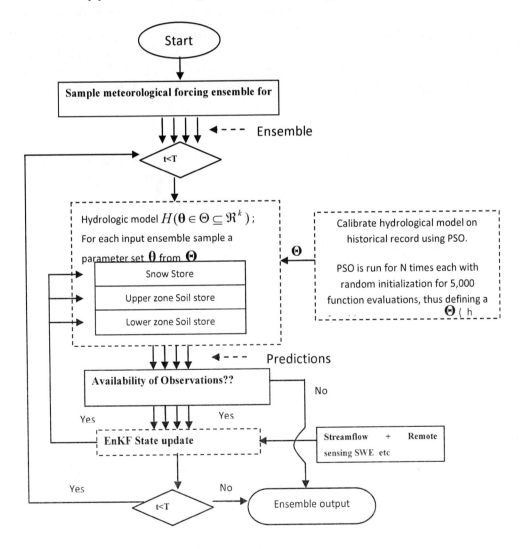

Figure 3. Ensemble Forecasting Flowchart.

Input ensemble generation: Input members of the meteorology ensemble for a particular year 'i', are generated by finding the statistical neighbors with year indices 'q' of the previous year 'i-1' from the historical data using the *Hausdorff norm* [*Chavent and Lechevallier*, 2002, *Jesorky et al.*, 2001] similarity measure index. The term 'q' represents indices for the selected years based on the desired ensemble size, which is chosen to be 50 in the present study. Once 'q' statistically similar neighbors to the previous year 'i-1' have been picked, the ensemble members for the current year 'i' are chosen to be the years immediately following 'q', which have indices 'q+1' in the sequence of data.

Prediction: Once the input ensembles to the year 'i' are sampled from the historical data, the next step is to make a prediction for the year 'i'. The predictions are made using the input ensembles with parameter uncertainty accounted for by sampling parameters from the feasible region, Θ. Each ensemble member of meteorological input is run with a parameter set drawn from the feasible region, Θ, which is defined by 50 PSO optimized runs, thus giving an ensemble of model sates. It is assumed that the process noise term $\mathbf{W}_t$ is embedded in the ensemble set thus produced. This way, the current approach not only accounts for uncertainty in the inputs and parameters, but also reduces the computational burden required to run individual ensemble members. The model error is specified through the covariance between the ensemble members of states. The observation noise covariance R is specified to be 0.1% (10% error) considering the fact that these observation come from the inverse flow routing.

Update: The daily streamflow observations and SNODAS SWE estimates are assimilated with the model output ensemble via EnKF to update model states. SWE is updated directly, model simulated SWE in each grid cell is replaced by the corresponding Kalman filter estimated value, (SWE_{kf}), that is model $SWE^{+\Delta t} = SWE_{kf}$ in Eq. 1.

A mass balance approach is used to update soil moisture states in the saturated zone, and if necessary, the rooting zone based on the difference between model and Kalman Filter estimated discharge:

$$\Delta S_{sat} = (Q_{sat} - Q_{kf})\Delta t \qquad (12)$$

where ΔS_{sat} is the change in saturated zone storage, Q_{kf} is the Kalman filter estimated stream discharge, and Q_{sat} is the model simulated discharge given through Eq. 11. When, $\Delta S_{sat} < S_{sat}^{t+\Delta t}$, the saturated zone storage is updated so, $S_{sat}^{t+\Delta t} = S^{t+\Delta t} - \Delta S_{sat}$. When, $\Delta S_{sat} > S_{sat}^{t+\Delta t}$, $\Delta S_{sat} = \Delta S_{sat} - S_{sat}^{t+\Delta t}$ and $S_{sat}^{t+\Delta t}$ is set equal to zero. In this case, to conserve mass the rooting zone storages are reduced by:

$$S_{rz}^{t+\Delta t} = S_{rz}^{t+\Delta t} - \Delta S_{sat}\left(\frac{Q_{rz}}{\sum Q_{rz}}\right) \qquad (13)$$

3. DWORSHAK CASE STUDY

The streamflow forecasting methodology is applied to the 6,325 km^2 Dworshak watershed in north-central Idaho State (Figure 4) and compared to the U.S. Army Corps of Engineers (USACE) volumetric forecasts and to daily inflows estimated from observations. The USACE uses the NRCS approach based on principal component regression technique using current SWE, fall and spring precipitation, base flow, and climate indices to produce seasonal forecasts at the Dworshak reservoir [http://www.nwd-wc.usace.army.mil /report/dwrf/]. The Dworshak watershed is located within the greater Snake and Columbia River watersheds and is characterized by mountainous and rugged terrain with elevations ranging from 300 – 2400-m. This area exhibits a dominant land cover of coniferous forest and hydrography is generally an east to west flowing system with high-gradient streams. The watershed is influenced by orographic precipitation events sourced from Pacific-based moisture where the long-term mean annual precipitation ranges from 700-mm in the lower elevations to 1650-mm in the upper reaches of the basin [*NRCS*, 2007; *WRCC*, 2007].

3.1. Data Description

A series of spatial and spatio-temporal datasets is assembled into a spatially-distributed computational grid with the ability to provide observation data to the model in near real-time. Inputs required to run the model include precipitation, surface air temperature, elevation, and land cover. The elevation data in the Dworshak watershed is established using USGS 10-meter digital elevation models with anomalous data detection being run through a 3x3 kernel filter. Land cover data for the Dworshak watershed has a fairly uniform and temporally-stable coniferous forest cover with some rock outcrops and open-water areas. A high-resolution 30-meter land-cover dataset from the Idaho GAP Analysis project [*Redmond*, 1996; *Homer*, 1998] is used for model input. Real-time instrumentation sensors in the watershed provide streamflow, precipitation, temperature, and snow water equivalent data into the model as well as providing historic information for the establishment of temperature and precipitation lapse rates (see Figure).

Observed streamflow used for model updating refers to daily inflow to the Dworshak Reservoir estimated by the USACE using a mass balance approach called inverse flow routing based primarily on measured changes in pool elevation and the amount of water released from the dam. The potential error associated with back-calculating inflow via mass balance is expressed using an ensemble of "observed" inflows. Meteorological data is provided at the eastern and high-end of the Dworshak watershed, the NRCS SNOTEL station, Hoodoo Basin. The SNOTEL station provides precipitation, maximum and minimum temperature, and snow water equivalent measurements at an elevation of 1844-m. Monthly precipitation and temperature elevation lapse rates are established using data from five SNOTEL stations, two COOP and one AgriMet weather station, all located within the basin (see Figure). Temperature and precipitation lapse rates are used to distribute meteorological inputs to all model grid cells using an elevation difference approach. Finally, SNODAS SWE data is collected and processed daily for the model snow state updating [*NOHRSC*, 2007].

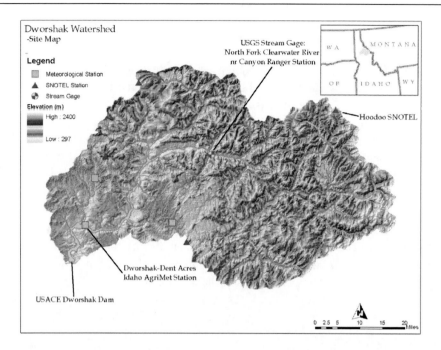

Figure 4. Site map of the Dworshak watershed showing instrumentation sites.

4. EXPERIMENTAL DESIGN

Short-term and seasonal forecasts are generated with and without observational data updating. In phase one, the meteorological forecast ensemble for a particular year is used to make daily forecasts for the entire year, starting on October 1st. These are considered as ensemble streamflow forecasts without any update via EnKF. The volumetric seasonal forecasts for April-July, issued by USACE are compared with ones obtained from the model.

In the second phase, the model is updated with observed streamflow, with and without SWE updating, from October 1st till March 1st and no update from March 1st till the end of the water year. The volumetric forecasts for April-July for this case are referred to as March 1st forecasts and are compared to the March 1st forecasts issued by USACE for the same period of interest (April-July). A pick and choose method described in the following is also tested to seek improved seasonal forecasts. The SNODAS SWE estimates are available at the 781x781m cells and are used directly to update model estimated snow sates. The soil moisture states are updated with the streamflow observations provided by the USACE at the Dworshak dam.

Short-term (one-, three-, and seven-day) forecasts are generated with streamflow updating till the current time steps. The forecasts are made such that the model is updated at current time step 'j', is then run for the desired forecast in advance ('j+3' for the three-day case), and then moves onto the 'j+1' for another update, and is run for another three days, and so on.

4.1. Results and Discussion

The model calibration results using PSO are shown in Figure for all the nine parameters. It can be noticed that a feasible parameter region is defined within the entire parameter space, which is then used to sample parameters for each run of ensemble members.

Long-term volumetric forecasts without SWE updating

Long-term volumetric forecasts are obtained using model simulations forced with historical climate traces as a surrogate for the future climate based on the similarity measure as explained previously. The volumetric forecasts are estimated for the melt season of April-July for the water years 2000-2007. October 1st forecasts show a slightly better correlation coefficient (0.2) than the USACE (0.002), but the results are generally poor. In the second phase, an EnKF scheme was used to update model states using daily streamflow observations till March 1st and no update was done for the rest of the year. The March 1st R^2 is 0.66 for the current method compared to 0.64 for USACE forecasts (Figure 5 (left)). It can be noticed that the forecast estimates from USACE and the current method show reasonable results for the March 1st.

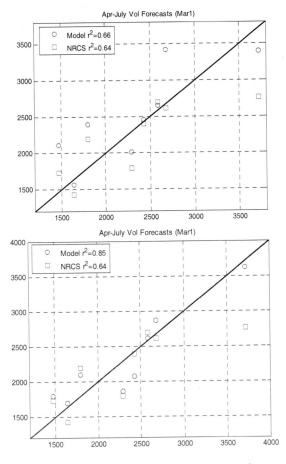

Figure 5. Streamflow volume forecasts for April-July issued on March 1st without (left) and with (right) refinement.

In a bid to improve March 1st forecasts, a pick and choose methodology was used that selects those members of the initial meteorological ensemble that show maximum similarity with meteorology observed between October and February of the current water year. A roulette wheel selection is done so that the members with highest fitness have a higher chance of being picked compared to those with poorer fitness. After pruning the initially picked ensemble set, a new ensemble set is obtained which is then used to produce March 1st forecasts (Figure -right). These predictions are found to be better than the ones without pruning, increasing the R^2 from 0.66 to 0.85.

Short-term model forecasts without SWE updating

The short-term forecasts are obtained by updating model ensembles with observed streamflow until the current time step and then simulating the ensembles without any update for the desired forecast period (e.g., one-, three-, and seven-day). The short-term forecasts for one-, three-, and seven-days in WY 2006 are shown in Figure 6 to Figure8. The gray-colored region represents the ensemble spread obtained after EnKF updates, whereas the observations are shown by the solid continuous line. The empirical cumulative probability distributions (CDF) are also shown for each of the one-, three-, and seven-day forecasts. Results for the one-day forecasts in Figure show good agreement with the observations. It is also found that the updated model spread brackets the observations. Similar results are found for three-day forecasts shown in Figure. The results for the seven-day forecasts modestly deteriorate in terms of overestimation and increased thicknesses of the ensemble spread, but are still acceptable (Figure). It is also found that the low flows are accurately modeled, whereas the major peak in the hydrograph is bracketed, but generally overestimated in each case. The magnitude of overestimation for the peaks is higher for the seven-day forecast compared to the one- and three-day forecasts as can also be noticed in the CDF plots. Notice, in Figure, the seven-day forecast CDF has a wider thickness of ensemble spread compared to the one- and three-day forecasts (Figure and Figure). It is also noticed that the model ensemble spread gets larger as the magnitude of the streamflow increases. In order to further evaluate one-, three-, and seven-day forecasts, goodness-of-fit measures are also estimated for each of the forecasts against observed flows. The results for root mean square error (RMSE), Nash-Sutcliffe efficiency coefficient (NS), bias, mean absolute error (MAE), mean square error (MSE), and coefficient of determination (R^2) are summarized in Table 2 and Figure. The goodness-of-fit results show a good fit between the forecasts and the observations. It is further noticed that the values deteriorate as the forecast lead time increase.

Table 2. Forecast goodness-of-fit

	k-day Forecasts		
	one-day	three-day	seven-day
RMSE	0.48	0.58	0.97
NS	0.95	0.93	0.82
Bias	-0.13	-0.11	-0.31
MAE	0.29	0.34	0.59
MSE	0.23	0.33	0.95
r^2	0.97	0.95	0.89

Ensemble Forecasting through Evolutionary Computing and Data Assimilation

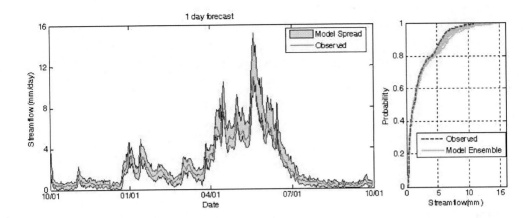

Figure 6. Streamflow forecasts 1-day.

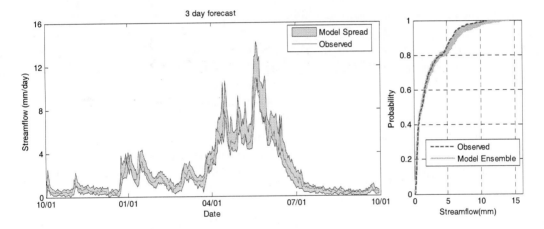

Figure 7. Streamflow forecasts 3-day.

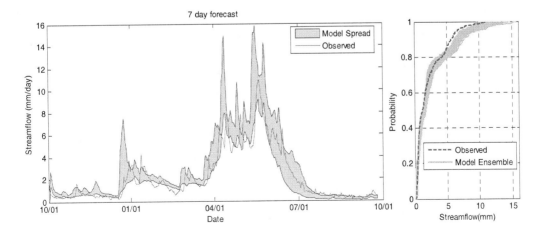

Figure 8. Streamflow forecasts 7-day.

Forecast results with SWE updating

The seasonal and short-term forecasts are also analyzed after updating with SNODAS SWE. The SWE assimilation is shown in Figure9 for the grid cell containing the Hoodoo SNOTEL station, showing the non-updated model simulated SWE ensemble, the SNODAS observation, and the updated model mean using ensemble Kalman filter. It can be noticed that the model spread (gray region) becomes thicker as the melt season starts. This is the result of higher uncertainty in the snow model parameter, melt factor (m_f), and presents a perfect case showing the combined model and parameter uncertainty. The model predictions are assimilated through EnKF and an updated SWE is produced. The results in Figure show spatial coverage of SWE within the entire basin after an update with SNODAS SWE on May 1, 2006.

The use of SNODAS SWE to update snow states within the model was found to improve three-day forecasts, but degrade the one- and seven-day forecasts when compared to using streamflow data assimilation alone. Hydrograph peaks were highly overestimated in the one- and three-days forecasts with SNODAS SWE updating. The March 1st seasonal forecasts were also degraded with assimilation of SWE data.

A number of factors may have contributed to these results, including model error, issues with the SNODAS data, and the fact that the SNODAS SWE data were not used in model calibration to derive the feasible parameter space. We are in the process of comparing spatial and temporal model simulated and SNODAS SWE against SNOTEL SWE and higher resolution MODIS snow cover. If the SNODAS data show better agreement, this would point toward the use of SNODAS (and or MODIS) in both model calibration and ongoing data assimilation. If model simulated SWE shows a closer match, this may indicate some spatial limitations to the SNODAS product, as the model is forced to conserve both the grid cell snow water balance and the basin wide water balance, while SNODAS is only concerned with the former.

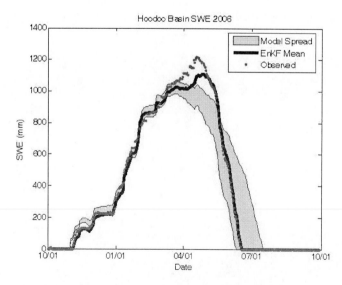

Figure 9. SWE estimates for WY 2006 showing the model ensemble spread without updating (gray), SNODAS values (red), and the updated SWE (mean) using EnKF (black).

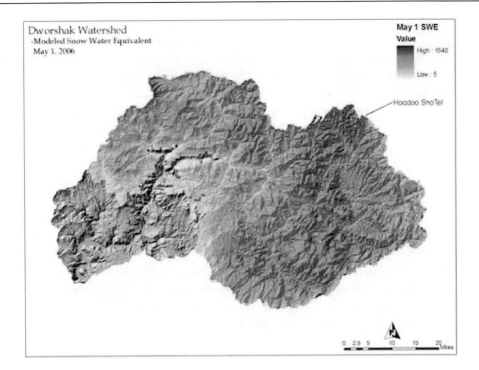

Figure 10. Updated SWE for May 1, 2006.

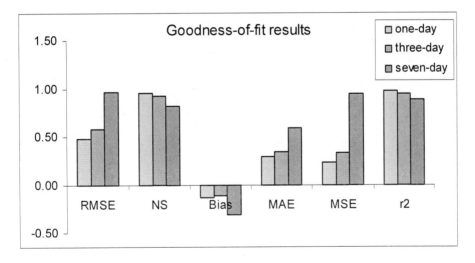

Figure 11. Forecast goodness-of-fit.

CONCLUSION

The current research presents an ensemble daily streamflow forecasting approach using a spatially-distributed hydrologic model with data assimilation via an Ensemble Kalman filter. The method is used for short-term to seasonal streamflow forecasts and was demonstrated in the snow-dominated Dworshak watershed in north-central Idaho. The forecast volumes for the snowmelt season (April-July) were compared to USACE forecasts and observations in water years 2000-2007. October 1 forecasts were made without streamflow updating and contain minimal information, with an R^2 of 0.2 (versus 0.002 for USACE). In the second phase, the model is updated with observed streamflow from October until the beginning of March and then run without updates for the remainder of the water year, resulting in an R^2 of 0.66 versus 0.64 for USACE forecasts. In a bid to improve March 1st forecasts, a pruning methodology was used that selects those members of the initial meteorological ensemble that show maximum similarity with meteorology observed between October and February of the current water year. This method produced an R^2 of 0.85.

The chapter also presents the results for short-term (one-, three-, and seven-day) forecasts employing streamflow updates. The one- and three-day forecasts show good agreement with the observations (R^2 = 0.97 and 0.95, with NS = 0.95 and 0.93, respectively), where as the seven-day forecast resulted in slight deterioration (R^2 = 0.89 and NS = 0.82). The research also employed the use of SNODAS SWE for updating snow state within the model that produced partial improvement in some cases, but generally degraded model performance. A systematic analysis of these results is the subject of ongoing research, as is the use of short-term meteorological forecasts from National Oceanic and Atmospheric Administration's (NOAA) NCEP and NDFD forecasting models.

APPENDIX

Hausdorff norm: Let $A = \{a_1, \ldots a_m\}$ and $B = \{b_1, \ldots b_n\}$ are two point sets. Then Haudorff norm $H(A,B)$, between the two sets is defined as:

$$H(A, B) = \max(h(A, B), h(B, A))$$

Where, $h(A, B) = \max_{a \in A} \{\min_{b \in B} \|a - b\|\}$

ACKNOWLEDGMENTS

The author wishes to thank the Pacific Northwest Regional Collaboratory (PNWRC - www.pnwrc.org), Pacific Northwest National Laboratory's Energy Infrastructure Operations Center, and NASA Solutions Network for their support in this work.

REFERENCES

Ajami, N. K., Q. Duan, & S. Sorooshian, (2007). An integrated hydrologic Bayesian multimodel combination framework: Confronting input, parameter, and model structural uncertainty in hydrologic prediction, *Water Resour. Res.*, *43*, W01403, doi: 10.1029/2005WR004745.

Andreadis K. M. & Lettenmaier, D. P. (2006). Assimilating remotely-sensed snow observations into a macroscale hydrology model. *Adv. Water Resour.*, *29*, 872–886.

ASCE, (1996). *Hydrology Handbook*, ASCE Manuals and Reports on Engineering Practice No. 28, American Society of Civil Engineers, New York, New York.

Beven, K. (1993). Prophecy, reality and uncertainty in distributed hydrological modeling, *Adv. Water Resour.*, *16*, 41–51.

Beven, K. & Freer, J. (2001). Equifinality, data assimilation, and data uncertainty estimation in mechanistic modelling of complex environmental systems using the GLUE methodology, *J. Hydrol.*, *249*, 11– 29.

Chavent, M. & Lechevallier, Y. (2002). Dynamical clustering of interval data: Optimization of an adequacy criterion based on hausdorff distance, In *Classification, Clustering, and Data Analysis* (K. Jajuga, A. Sokolowski and H.-H. Bock, eds.). 53–60. Sringer Verlag, Berlin.

Clark, M. P., Slater, A. G., Barrett, A. P., Hay, L. E., McCabe, G. J., Rajagopalan, B. & Leavesley, G. H. (2006). Assimilation of snow covered area information into hydrologic and land-surface models. *Adv. Water Resour.*, *29*, 1209–1221.

Drecourt, J. P. (2003). Kalman Filtering in Hydrological Modeling, DAIHM Technical Report, *DHI Water & Environment*, http://projects.dhi.dk/daihm/Files/KFlitreview.pdf, accessed July 20, 2003.

Eberhart, R. C. & Kennedy, J. (1995). A new optimizer using particle swarm theory, in Proceedings of the Sixth International Symposium on *Micro Machine and Human Science*, 1995. MHS '95, pp. 39– 43, doi:10.1109/MHS.1995.494215, IEEE Press, Piscataway, N. J.

Eberhart, R. C., Dobbins, R. W. & Simpson, P. (1996). *Computational Intelligence PC Tools*, Elsevier, New York.

Evensen, G. (1994). Sequential Data Assimilation With a Nonlinear Quasi-Geostrophic Model Using Monte-Carlo Methods to Forecast Error Statistics. *Journal of Geophysical Research*, *99*, 10143-10162.

Evensen, G. (2002). Sequential Data Assimilation for Nonlinear Dynamics: The Ensemble Kalman Filter. In: *Ocean Forecasting: Conceptual Basis and Applications*, N. Pinardi, and J.D. Woods (Editors). Springer-Verlag, Berlin, Germany, 101-120.

Evensen, G., (2003). The Ensemble Kalman Filter: Theoretical Formulation and Practical Implementation. *Ocean Dynamics*, *53(4)*, 343-367.

Garen, D. C. (1992). Improved techniques in regression-based stream volume forecasting, *J. Water Resour. Plan. Manage.*, *118(6)*, 654-670.

Gill, M. K., Kaheil, Y. H., Khalil, A., McKee, M. & Bastidas, L. (2006). Multiobjective particle swarm optimization for parameter estimation in hydrology, *Water Resour. Res.*, *42*, W07417, doi:10.1029/2005WR004528.

Gill, M. K., Kemblowski, M. W. & McKee, M. (2007). Soil Moisture Data Assimilation Using Support Vector Machines and Ensemble Kalman Filter. *Journal of the American Water Resources Association (JAWRA) 43(4)*, 1004-1015. DOI: 10.1111/j.1752-1688.2007.00082.x.

Hamill, T. M. (2006). Ensemble-based atmospheric data assimilation. In: Palmer TN, Hagedorn R, editors. *Predictability of weather and climate*. Cambridge Press.

Homer, C. G. (1998). Idaho and western Wyoming land cover classification report and metadata. Remote Sensing / GIS Laboratories, Department of Geography and Earth Resources, Utah State University. Contract Number 1422-D910-A3-0210.

Jesorsky, O., Kirchberg, K. & Frischholz, R. (2001). Robust face detection using the Hausdorff distance," in 3rd Int. Conf. on *Audio- and Video-Based Biometric Person Authentication*, pp. 90-95, Halmstad, Sweden, June 2001.

Kaheil, Y. H., Gill, M. K., McKee, M. & Bastidas, L. (2006). A new Bayesian recursive technique for parameter estimation, *Water Resour. Res., 42*, W08423, doi:10.1029/2005WR004529.

Kalman, R. E. (1960). A New Approach to linear Filtering and Prediction problem, *Transactions of the ASME-Journal of Basic Engineering., 82*(series D), 35-45.

Kennedy, J. & Eberhart, R. C. (1995). Particle swarm optimization, in Proceedings of IEEE International Conference on *Neural Networks, IV, vol. 4*, 1942– 1948, doi:10.1109/ICNN.1995.488968, IEEE Press, Piscataway, N. J.

Lettenmaier, D. P. & Garen, D. C. (1979). *Evaluation of streamflow forecasting methods*, Proc. Western Snow Conference, 48-55.

Liu, Y. & Gupta, H. V. (2007). Uncertainty in hydrologic modeling: Toward an integrated data assimilation framework, *Water Resour. Res., 43*, W07401, doi:10.1029/2006WR005756.

McGuire, M., Wood, A. W., Hamlet, A. F. & Lettenmaier, D. P. (2006). Use of satellite data for streamflow and reservoir storage forecasts in the Snake River Basin, *J. Water Resour. Plan. Manage., 132(2)*, 97-110.

Metropolis, N., Rosenbluth, A. W., Rosenbluth, M. N., Teller, A. H. & Teller, E. (1953). Equations of state calculations by fast computing machines, *J. Chem. Phys., 21*, 1087–1091.

Moradkhani, H., Sorooshian, S., Gupta, H. V. & Hauser, P. R. (2005a). Dual state-parameter estimation of hydrological models using ensemble Kalman filter, *Adv. Water Resour., 28*, 135– 147.

Moradkhani, H., Hsu, K., Gupta, H. V. & Sorooshian, S. (2005b). Uncertainty assessment of hydrologic model states and parameters: Sequential data assimilation using particle filter, *Water Resour. Res., 41*, W05012, doi:10.1029/2004WR003604.

National Research Council, (2001). *"Grand challenges in environment sciences."* National Academy Press, Washington, D.C.

NOHRSC (2007). Snow Data Assimilation System (SNODAS) data products at NSIDC, National Operational Hydrologic Remote Sensing Center. Boulder, CO: National Snow and Ice Data Center. Digital media.

NRCS (2007). SNOTEL Site Information and Reports for Hoodoo Basin, Natural Resources Conservation Service, ren, D. & Sorooshian, S. (2004). Evaluation of official western U.S. seasonal water supply outlooks, 1922-2002, *J. Hydrometeorology, Vol. 5*, 896-909.

Redmond, R. L., Ma, Z., Tady, T. P., Winne, J., Schumacher, J., Troutwine, J. & Holloway, S. W. (1996). *Mapping existing vegetation and land cover across western Montana and northern Idaho*, Wildlife Spatial Analysis Laboratory, University of Montana, Final Report Contract Number, 53-0343-4-000012.

Reichle, R. H., McLaughlin, D. B. & Entekhabi, D. (2002). Hydrologic data assimilation with the ensemble Kalman filter. *Mon. Wea. Rev., 130*, 103–114.

Slater, A. G. & Clark, M. P. (2006). Snow Data Assimilation via an Ensemble Kalman Filter. *Journal of Hydrometeorology, 7(3)*, 478:493.

Spear, R. C. & Hornberger, G. M. (1980). Eutrophication in the Peel Inlet, II, Identification of critical uncertainties via generalized sensitivity analysis, *Water Res., 14*, 43-49.

Twedt, T. M., Jr. Schaake, J. C. & Peck, E. L. (1977). National Weather Service extended streamflow prediction, Proc., *Western Snow Conf.*, 52-57.

Thiemann, M., Trosset, M., Gupta, H. & Sorooshian, S. (2001). Bayesian recursive parameter estimation for hydrologic models, *Water Resour. Res., 37(10),* 2521–2535.

Vrugt, J. A., Bouten, W., Gupta, H. V. & Sorooshian, S. (2002). Toward improved identifiability of hydrologic model parameters: The information content of experimental data, *Water Resour. Res., 38(12)*, 1312, doi:10.1029/2001WR001118.

Vrugt, J. A., Gupta, H. V., Bouten, W. & Sorooshian, S. (2003). A Shuffled Complex Evolution Metropolis algorithm for optimization and uncertainty assessment of hydrologic model parameters, *Water Resour. Res., 39(8).* 1201, doi:10.1029/2002WR001642.

Vrugt, J. A., Diks, C. G. H., Gupta, H. V., Bouten, W. & Verstraten, J. M. (2005). Improved treatment of uncertainty in hydrologic modeling: Combining the strengths of global optimization and data assimilation, *Water Resour. Res., 41*, W01017, doi:10.1029/2004WR003059.

Vrugt, J. A., Gupta, H., Nualláin, B. Ó. & Bouten, W. (2006). Real-Time Data Assimilation for Operational Ensemble Streamflow Forecasting. *Journal of Hydrometeorology, 7(3)*, 548.

Vrugt, J. A. & Robinson, B. A. (2007). Treatment of uncertainty using ensemble methods: Comparison of sequential data assimilation and Bayesian model averaging, *Water Resour. Res., 43*, W01411, doi:10.1029/2005WR004838.

WRCC, (2007). *Idaho Annual Precipitation Summary*, Western Regional Climate Center, http://www.wrcc.dri.edu/htmlfiles/id/id.ppt.ext.html, accessed on September 3.

Weerts, A. H. & El Serafy, G. Y. H. (2006). Particle filtering and ensemble Kalman filtering for state updating with hydrological conceptual rainfall-runoff models, *Water Resour. Res., 42*, W09403, doi:10.1029/2005WR004093.

Welch, G. & Bishop, G. (2002). An Introduction to the Kalman Filter. Technical Report 95-041, University of North Carolina at Chapel Hill, Department of Computer Science: http://www.cs.unc.edu/~welch/media, *accessed January, 13*, 2002.

Wigmosta, M. S., Vail, L. W. & Lettenmaier, D. P. (1994). A Distributed Hydrology-Vegetation Model for Complex Terrain, *Water Resources Research, 30(6)*, 1665-1679.

Wigmosta, M. S., Nijssen, B. & Storck, P. (2002). The Distributed Hydrology Soil Vegetation Model, In *Mathematical Models of Small Watershed Hydrology and Applications,* V.P. Singh, D.K. Frevert, eds., Water Resource Publications, Littleton, CO., p. 7-42.

Wood, A. W. & Lettenmaier, D. P. (2006). A Test Bed for New Seasonal Hydrologic Forecasting Approaches in the Western United States. *Bull. Amer. Meteor. Soc., 87*, 1699–1712.

In: Kalman Filtering
Editor: Joaquín M. Gomez

ISBN: 978-1-61761-462-0
© 2011 Nova Science Publishers, Inc.

Chapter 6

ATTITUDE DETERMINATION USING KALMAN FILTERING FOR LOW EARTH ORBIT MICROSATELLITES

Si Mohammed and Mohammed Arezki
Centre of Space Techniques, Space Mechanic Division, Arzew, Algeria

ABSTRACT

As part of the attitude determination and control system on board microsatellites, the kalman filter is implemented to generate an estimate attitude upon sensor observation.

This chapter presents an overview and analysis of some attitude estimators based on Kalman filtering for application on low Earth orbit microsatellites. Various steps of the design and in orbit implementation of the estimators are described. These estimators are used during nominal attitude of the microsatellite. One type of attitude estimator is based on small Euler angles. For this mode two estimator versions have been used, the full estimator and its reduced version. This estimator is useful for on board computation where energy resources are very limited. Another type of attitude estimator is based on the quaternion parameter. The advantage of the quaternion parameter is the none singularity. For this mode two estimator versions have been used, the six state and the seven state estimators.

Numerical analysis and in orbit test results clearly indicate that the reduced version is better than the full version for specifically attitude determination and control system configuration of a microsatellite. Also, the results prove that the six state estimator is better than the seven state estimator in terms of computational demand.

1. INTRODUCTION

Various sensors can be employed to measure the attitude of an orbiting body to a varying degree of accuracy. For example in Alsat-1 first Algerian microsatellite in orbit, three 3-axis flux gate magnetometers are used to measure the geomagnetic field vector in microsatellite's body coordinates and when compared to the expected direction obtained from a model of the geomagnetic field (e.g. IGRF), full attitude and angular rate of the microsatellite can be estimated [1]. For a calibrated magnetometer, during of low solar activity periods, the attitude angles can be estimated with an accuracy of less than 1 degree per axis. Four 2-axis (azimuth and elevation) analog sun sensors are used to determine the position of the sun relative to the microsatellite body and when the sun position is known relative to the microsatellite position by using orbital models (sun and microsatellite), full attitude can be estimated. The four sensors cover the full 360 degrees azimuth range and a 120 degrees elevation range. Each axis has a ± 60 degrees range and can measure the sun vector with an accuracy of 0.3 degree. More accurate attitude measurements can be acquired by using a horizon and sun sensors or a star sensor.

Unfortunately the most accurate sensors, due to a limited field of view, can only give useful measurements during a fraction of the orbit, whereas the least accurate sensors with an increased measurement range and longer availability of useful data during an orbit, have mostly increased levels of measurement noise. To maximize the attitude determination accuracy and to have continuous availability of attitude information, some form of onboard sate estimation will be needed.

The application of Kalman filters in the aerospace field has gone a long way since the original work by R.E. Kalman on prediction theory in 1960 [2], [3]. Lefferts et. al. [4] reviews the Kalman filtering methods and their development during the sixties and seventies when applied to the Appolo program and subsequent spacecraft attitude determination problems. The kalman filter is also used extensively in orbit determination problems [5], [6].

The Kalman filter is implemented to determinate the attitude for many reasons compared to other filtering techniques. Again and regarding other techniques, data generated must be cyclically processed and discarded rather being stored due to the limitation on board the microsatellite. From this point of view, the Kalman filter becomes ideal for processing large amount of data. It is an optimal, recursive, data processing algorithm that generate an estimate attitude upon sensor observation. Although the Kalman filter assumes a linear system model, it was found in many applications with non linear system. There are two general methods that can be used for linearizing the non linear discrete time Kalman filter. When the filter algorithm is applied to such a linear representation of a non linear system, it is called a linearised kalman filter. This method linearise about some nominal trajectory in state space that does not depend on the measurement data. However, sometimes the linear approximation leads to diverge. The satisfactory can be obtained by using the extended Kalman filter (EKF). This method linearise about the Kalman filter's estimated trajectory that is continually updated with the states estimates resulting from measurements. Because the attitude motion of the microsatellite is nonlinear, the extended Kalman filter will be used for optimal attitude estimation. The basic theoretical concepts and the mathematical models of the Kalman filter are described in [7], [8].

The attitude of the microsatellite can be defined by Euler angles (roll, pitch and yaw) [9]. Unfortunately, the kinematic equations involve non linear and computationally expensive trigonometric functions and the angles become undefined for some rotations. Despite theses difficulties, the Euler angles representation still used for attitude determination operation on board microsatellites.

Under the assumption of small Euler angles, the resulting attitude estimator using Kalman filter becomes fairly concise. This attitude estimator is extremely useful on board where computational resources are very limited. This attitude estimator is dedicated for a spinning and a gravity gradient stabilised microsatellite. Specifically an attitude determination and control system axially symmetric for a microsatellite, the pitch angle dynamics is decoupled from the attitude dynamic equation. The dimension of the state vector to be estimated is 6 including the angular rate terms and which requires manipulating 6X6 matrices. This is not ideal both from a program size as well as processing time point of view. We take advantage of the decoupled nature of the attitude dynamic equation and design the pitch and roll/yaw estimator separately. The attitude estimation will be done by, so called, the small libration attitude estimator – reduced version. This attitude estimator is implemented on board Alsat-1 [10] and has been studied in [11].

A better representation of the attitude that has no singularities and no trigonometric functions in the attitude matix, will be the Euler symmetric parameters or quaternion [9]. The development of the kalman filter for the quaternion representation was motived by the requirement of real time autonomous attitude determination for attitude control [4].

Since the first application of the quaternion representation to the spacecraft attitude determination by Ickes [12], it has become the most popular means through which we represent the kinematics of orbiting bodies. The quaternion representation has been used for attitude estimation problems [4], [13] - [21].

The attitude estimator using Kalman filter most commonly implemented for the attitude determination is the 7 state quaternion version. This is due, in part, to the relative ease of 7 state mathematics in comparison with other attitude estimators that have been utilized for microsatellite operations. In addition, the 7 state quaternion version has well-established reputation as an effective and reliable estimation technique. The state vector is comprised of the four-element quaternion attitude vector combined with the three-element body rates vector, with respect to the inertial frame.

However, this estimator necessarily involves 7x7 matrices, increasing the computational demand upon the onboard computer of the microsatellite. Hence, any means of significantly reducing the computational intensity of the attitude estimator while not sacrificing operability is highly desirable. The 6 state quaternion version can be found in the following papers [4,18,21]. It was implemented on board Alsat-1 as an attempt to realize this objective.

For a three axis active gravity gradient stabilized microsatellite, such as Alsat-1, the simplified Euler attitude equation is normally used to design the attitude determination and control system (ADCS) algorithm. The main objective of such a microsatellite is to gain the nadir attitude pointing and basically no attitude slew maneuver is required. However for experimental reasons, the six state and seven state quaternion attitude estimators were implemented during the Alsat-1 lifetime. This chapter presents the design implementation of these attitude estimators using Kalman filter by taking into consideration the advantage of the experiments achieved on board Alsat-1. These attitude estimators are used after the commissioning phase and during the nominal operation of the microsatellite.

Alsat-1 is 3 axis stabilized microsatellite, using a pitch momentum wheel and yaw reaction wheel, with dual redundant 3 axis magnetorquer. A gravity-gradient boom is employed to provide a high degree of platform stability. Two vector magnetometers and four dual axis sun sensors are carried to determine the attitude [1].

The organization of this chapter proceeds as follows. Section 2 and 3 present a brief review of the microsatellite attitude kinematic and dynamic equations, followed by a summary of the extended Kalman filter algorithm. The various steps of the design of the attitude estimator based on small Euler angles and on the quaternion including the related versions are discussed in section 4. Section 5 reports the simulation results. Section 6 describes the implementation of the attitude estimators and the in orbit results during nadir attitude pointing. Finally, the conclusion of this work is presented in section 7 in order to demonstrate the usefulness of the attitude estimators.

2. Preliminaries

In this section, a brief review of the dynamic and kinematic equations of motion for a three-axis active gravity gradient stabilised microsatellite is shown.

2.1. Dynamic Equations

The dynamic of the microsatellite in the inertial space is governed by Euler's equations of motion. With the added influence of the gravity gradient boom and reaction wheel angular momentum, the equation in vector form can be expressed as [9]

$$\mathbf{I}\dot{\boldsymbol{\omega}}_{BY} = \mathbf{N}_{GG} + \mathbf{N}_D + \mathbf{N}_M - \boldsymbol{\omega}_{BY} \times (\mathbf{I}\boldsymbol{\omega}_{BY} + \mathbf{h}) - \dot{\mathbf{h}} \qquad (1)$$

where,

$\boldsymbol{\omega}_{BY} = \begin{bmatrix} \omega_x & \omega_y & \omega_z \end{bmatrix}^T$: inertially referenced body angular rate vector;

$\mathbf{I}$: moment of inertia (MOI) tensor of microsatellite;

$\mathbf{h} = \begin{bmatrix} h_x & h_y & h_z \end{bmatrix}^T$: wheel angular momentum vector;

$$\mathbf{N}_{GG} = 3\omega_0^2 \left[\mathbf{u} \times \mathbf{I}\mathbf{u}\right] \qquad (2)$$

$\mathbf{N}_{GG}$: gravity gradient torque vector,

$\mathbf{u}$: unit zenith vector expressed in the body axis coordinates;

$\mathbf{N}_D = \begin{bmatrix} N_{dx} & N_{dy} & N_{dz} \end{bmatrix}^T$: external disturbance torque vector such as aerodynamic torque and solar radiation pressure torque ;

$\mathbf{N}_M = \begin{bmatrix} N_{mx} & N_{my} & N_{mz} \end{bmatrix}^T$: applied torque vector by 3-axis magnetorquers.

The Euler attitude angle satisfies the following equation

$$\dot{\phi} = \omega_{ox} \cos\psi - \omega_{oy} \sin\psi \qquad (3.a)$$

$$\dot{\theta} = (\omega_{ox} \sin\psi + \omega_{oy} \cos\psi)\sec\phi \qquad (3.b)$$

$$\dot{\psi} = \omega_{oz} + (\omega_{ox} \sin\psi + \omega_{oy} \cos\psi)\tan\phi \qquad (3.c)$$

where ϕ, θ and ψ are roll, pitch and yaw respectively and $\boldsymbol{\omega}_{LO} = [\omega_{ox} \; \omega_{oy} \; \omega_{oz}]^T$ is an orbit reference body angular velocity vector. This vector can be derived by

$$\boldsymbol{\omega}_{LO} = \boldsymbol{\omega}_{BY} - \mathbf{A}\boldsymbol{\omega}_0 \qquad (4)$$

where $\boldsymbol{\omega}_0 = [0 \; -\omega_0 \; 0]^T$ is an orbital rate vector and $\mathbf{A}$ as the direction cosine matrix (DCM) given by

$$\mathbf{A} = \begin{bmatrix} c\psi c\theta + s\psi s\phi s\theta & s\psi c\phi & -c\psi s\theta + s\psi s\phi c\theta \\ -s\psi c\theta + c\psi s\phi s\theta & c\psi c\phi & s\psi s\theta + c\psi s\phi c\theta \\ c\phi s\theta & -s\phi & c\phi c\theta \end{bmatrix} \qquad (5)$$

where,
 ϕ: roll angle;
 θ: pitch angle;
 ψ: yaw angle;
 c: cosine function;
 s: sine function.

Using attitude matrix, Equation (4) becomes

$$\omega_{ox} = \omega_x + \omega_0 \cos\phi \sin\psi \qquad (6.a)$$

$$\omega_{oy} = \omega_y + \omega_0 \cos\phi \cos\psi \qquad (6.b)$$

$$\omega_{oz} = \omega_z - \omega_0 \sin\phi \qquad (6.c)$$

Substituting Equation (4) into Equation (3) yields

$$\dot{\phi} = \omega_x \cos\psi - \omega_y \sin\psi \qquad (7.a)$$

$$\dot{\theta} = (\omega_x \sin\psi + \omega_y \cos\psi)\sec\phi + \omega_0 \qquad (7.b)$$

$$\dot{\psi} = \omega_z + (\omega_x \sin\psi + \omega_y \cos\psi)\tan\phi \qquad (7.c)$$

Note that Euler equation has a singularity when the roll angle ϕ equals 90 degrees.

2.2. Kinematic Equations

If the Euler attitude angle singularity mentioned above may become significant, then quaternion parameter to describe the attitude motion may be another alternative. The kinematic of the microsatellite will be modelled making use of quaternion parameter. The parametrization of the quaternion vector is done as [9]

$$q_1 \equiv e_{xo} \cos(\frac{1}{2}\Phi) \quad q_2 \equiv e_{yo} \cos(\frac{1}{2}\Phi) \quad q_3 \equiv e_{zo} \cos(\frac{1}{2}\Phi) \quad q_4 \equiv \cos(\frac{1}{2}\Phi) \tag{8}$$

where,

$\mathbf{e} = \begin{bmatrix} e_{xo} & e_{yo} & e_{zo} \end{bmatrix}^T$: Euler vector in orbit referenced coordinate;

Φ: rotation angle around the Euler vector.

The equation of the quaternion is introduced by the following vector set of differential equations [9]

$$\dot{\mathbf{q}} = \frac{1}{2}\Omega \mathbf{q} = \frac{1}{2}\Lambda(\mathbf{q})\omega_{LO} \tag{9}$$

where,

$$\Omega = \begin{bmatrix} 0 & \omega_{oz} & -\omega_{oy} & \omega_{ox} \\ -\omega_{oz} & 0 & \omega_{ox} & \omega_{oy} \\ \omega_{oy} & -\omega_{ox} & 0 & \omega_{oz} \\ -\omega_{ox} & -\omega_{oy} & -\omega_{oz} & 0 \end{bmatrix} \tag{10}$$

$$\Lambda(\mathbf{q}) = \begin{bmatrix} q_4 & -q_3 & q_2 \\ q_3 & q_4 & -q_1 \\ -q_2 & q_1 & q_4 \\ -q_1 & -q_2 & -q_3 \end{bmatrix} \tag{11}$$

The direction cosine matrix **A** can be written in quaternion form as follows

$$\mathbf{A} = \begin{bmatrix} q_1^2 - q_2^2 - q_3^2 + q_4^2 & 2(q_1q_2 + q_3q_4) & 2(q_1q_3 - q_2q_4) \\ 2(q_1q_2 - q_3q_4) & -q_1^2 + q_2^2 - q_3^2 + q_4^2 & 2(q_2q_3 + q_1q_4) \\ 2(q_1q_3 + q_2q_4) & 2(q_2q_3 - q_1q_4) & -q_1^2 - q_2^2 + q_3^2 + q_4^2 \end{bmatrix} \tag{12}$$

3. KALMAN FILTERING

We review in this section the principal equations for the extended kalman filter [7], [8] in order to introduce the necessary notation.

The estimation of the Kalman filter operates basically in two cycles, propagation and correction. During the first cycle, the filter propagates the system's state using a model to predict the dynamic's state. The correction cycle inputs state data (measurements) and uses these values of observations to correct the difference between propagated and measured microsatellite state data. However, during the correction phase noise and imperfection are being added to the measurements themselves. To overcome this, during the correction phase it is envisaged that the Kalman filter balances data from propagation and the state model using optimization theory. This process will yield to correct estimate of the system's state. Due to the iteration process, the corrected estimate of the system's state is used as an initial condition for the Kalman filter propagation cycle.

3.1. Propagation Cycle

State propagation

Given the initial condition on the state vector $\hat{\mathbf{X}}$, its propagation from t_k to t_{k+1} in the absence of measurements is computed numerically unless there is an analytical solution and is obtained as

$$\begin{cases} \hat{\mathbf{X}}_{k+1}^- = \hat{\mathbf{X}}_k^+ + \int_{t_k}^{t_{k+1}} \dot{\mathbf{X}}(\mathbf{X}, t) dt \\ \mathbf{X}(t_k) = \hat{\mathbf{X}}_k^+ \end{cases} \quad (13)$$

Covariance Propagation

Given the initial condition on the error covariance matrix, its propagation in the absence of measurements is obtained as

$$\mathbf{P}_{k+1}^- = \mathbf{\Phi}_k \mathbf{P}_k^+ \mathbf{\Phi}_k^T + \mathbf{Q}_k \quad (14)$$

where the process noise matrix $\mathbf{Q}_k$ is defined by [8]

$$\mathbf{Q}_k = \int_{t_k}^{t_{k+1}} \int_{t_k}^{t_{k+1}} \mathbf{\Phi}(t_{k+1}, u) E\left[\mathbf{w}(u)\mathbf{w}(v)^T\right] \mathbf{\Phi}(t_{k+1}, v)^T du dv \quad (15)$$

The state transition matrix satisfies the following differential equation

$$\frac{d\Phi(t,\tau)}{dt} = \mathbf{F}(t)\Phi(t,\tau) \qquad (16)$$

During the interval of $[t_k, t_{k+1}]$, it can be assumed the matrix $\mathbf{F}$ is a constant and neglecting terms which are higher than the first order, so the solution of the Equation (16) yields

$$\Phi = e^{\mathbf{F}\Delta t} = \mathbf{I} + \sum_{n \geq 1} \frac{(\mathbf{F}\Delta t)^n}{n!} \approx \mathbf{I} + \mathbf{F}\Delta t \qquad (17)$$

where $\Delta t = t_{k+1} - t_k$.

3.2. Correction Cycle

Kalman Gain

The Kalman gain is obtained as

$$\mathbf{K}_{k+1} = \mathbf{P}_{k+1}^{-} \mathbf{H}_{k+1}^{T} \left[\mathbf{H}_{k+1} \mathbf{P}_{k+1}^{-} \mathbf{H}_{k+1}^{T} + \mathbf{R}_{k+1} \right]^{-1} \qquad (18)$$

where the observation matrix $\mathbf{H}_{k+1}$ is given by

$$\mathbf{H}_{k+1} = \left. \frac{\partial (\text{obs})}{\partial \mathbf{X}} \right|_{\hat{\mathbf{X}} = \hat{\mathbf{X}}_{k+1}^{-}} \qquad (19)$$

The observation noise variance matrix $\mathbf{R}$ is assumed to be a constant diagonal matrix.

State Update

The state update immediately following the measurement is given by

$$\hat{\mathbf{X}}_{k+1}^{+} = \hat{\mathbf{X}}_{k+1}^{-} + \mathbf{K}_{k+1}(\mathbf{Z}_{k+1} - \hat{\mathbf{Z}}_{k+1}^{-}) \qquad (20)$$

where $\mathbf{Z}_{k+1}$ is actual observation given by sensors.

Covariance Update

The error covariance matrix immediately following the measurement is updated as

$$\mathbf{P}_{k+1}^{+} = \left[\mathbf{I} - \mathbf{K}_{k+1} \mathbf{H}_{k+1} \right] \mathbf{P}_{k+1}^{-} \qquad (21)$$

3.3. Note on the Process Noise Covariance Matrix Q

The process noise covariance will compensate the uncertainty of the model used in the system equation. If $\Delta t = t_{k+1} - t_k$, then Equation (15) becomes

$$Q_k = \int_0^{\Delta t}\int_0^{\Delta t} \Phi(\Delta t, u) E\left[w(u)w(v)^T\right] \Phi(\Delta t, v)^T du dv \qquad (22)$$

For Alsat-1 attitude determination, it is assumed that the Euler angles and angular velocity term have process noise, then the **E** matrix is given by

$$E\left[w(u)w(v)^T\right] = \begin{bmatrix} O & O \\ O & \begin{matrix} \sigma_x^2 \delta(u-v) & 0 & 0 \\ 0 & \sigma_y^2 \delta(u-v) & 0 \\ 0 & 0 & \sigma_z^2 \delta(u-v) \end{matrix} \end{bmatrix} \qquad (23)$$

where
  δ: Dirac delta function,
  O: zero matrix.

Substituting the approximation $\Phi \approx I + F\Delta t$ and Equation (23) into Equation (22), after integration the **Q** matrix yields

$$Q = Q_1 \Delta t + Q_2 \frac{(\Delta t)^2}{2} + Q_3 \frac{(\Delta t)^3}{3} \qquad (24)$$

where

$$Q_1 = \begin{bmatrix} O & O \\ O & S \end{bmatrix} \qquad (25)$$

$$Q_2 = \begin{bmatrix} O & F_{12}S \\ SF_{12}^T & SF_{22}^T + F_{22}S \end{bmatrix} \qquad (26)$$

$$Q_3 = \begin{bmatrix} F_{12}SF_{12}^T & F_{12}SF_{22}^T \\ F_{22}SF_{12}^T & F_{22}SF_{22}^T \end{bmatrix} \qquad (27)$$

The **S** matrix is obtained as

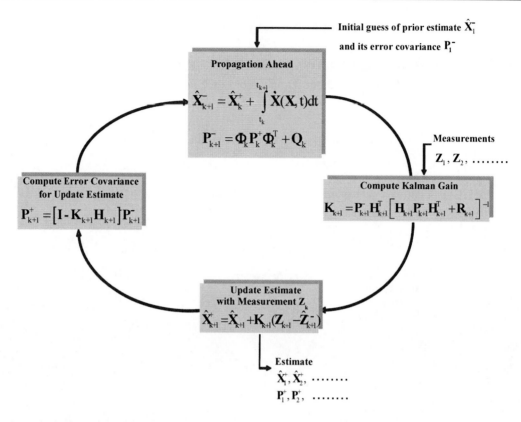

Figure 1. Kalman Filter Loop Process.

$$S = \begin{bmatrix} \sigma_x^2 & 0 & 0 \\ 0 & \sigma_y^2 & 0 \\ 0 & 0 & \sigma_z^2 \end{bmatrix} \quad (28)$$

Note that in practice, the process noise covariance matrix Q is assumed diagonal. The Kalman filter loop process is depicted in Figure 1

4. ATTITUDE ESTIMATOR

Once the microsatellite attitude becomes predictable after the initial attitude acquisition, it is required to start the attitude state estimation process. We have to design an attitude estimator including the actuators dynamics under the restriction of the processing power of the on board computer. The attitude estimator is assumed to be presented by small Euler angles and the quaternion parameter including the related versions of the estimators.

In this section, the various steps of the design and the implementation of the kalman filter are examined for these attitude estimators. In each case, explicit expressions of the system equation, transition matrix and observation equation are developed.

This work is suitable for a 3 axis stabilised microsatellite, equipped with three (reaction/momentum) wheels and three magnetorquers. The attitude sensors considered here are the sun sensor and the magnetometer.

4.1. Attitude Estimator – Small Euler Angles

4.1.1. Full version

The state vector to be estimated is 6 dimensional such that

$$\mathbf{X} = \begin{bmatrix} \phi & \theta & \psi & \dot{\phi} & \dot{\theta} & \dot{\psi} \end{bmatrix}^T \tag{29}$$

4.1.1.1. System equation

Under the assumption of small Euler angles, the first order approximated attitude equation (note that the second order terms have already been truncated) is given by

$$\ddot{\phi} = (4\omega_0^2 \alpha + \omega_0 \frac{h_y}{I_x})\phi - \frac{h_z}{I_x}\dot{\theta} + (\omega_0 + \alpha\omega_0 + \frac{h_y}{I_x})\psi - \frac{\dot{h}_x}{I_x} + \omega_0 \frac{h_z}{I_x} + \frac{N_{mx}}{I_x} + w_x \tag{30.a}$$

$$\ddot{\theta} = -\omega_0 \frac{h_x}{I_y}\phi - 3\omega_0^2 \beta \theta - \omega_0 \frac{h_z}{I_y}\psi + \frac{h_z}{I_y}\dot{\phi} - \frac{h_x}{I_y}\dot{\psi} - \frac{\dot{h}_y}{I_y} + \frac{N_{my}}{I_y} + w_y \tag{30.b}$$

$$\ddot{\psi} = (-\omega_0^2 \gamma + \omega_0 \frac{h_y}{I_z})\psi + \frac{h_x}{I_z}\dot{\theta} + (-\omega_0 - \frac{h_y}{I_z} + \omega_0 \gamma)\dot{\phi} - \frac{\dot{h}_z}{I_z} + \frac{N_{mz}}{I_z} + w_z \tag{30.c}$$

where,

$\mathbf{w} = [w_x \ w_y \ w_z]^T$: zero mean system noise vector;
$\mathbf{N} = [N_{mx} \ N_{my} \ N_{mz}]^T$: applied magnetorquer control firing;
$\mathbf{I} = \text{diag}[I_x \ I_y \ I_z]^T$: moment of inertia tensor of the microsatellite;
$\mathbf{h} = [h_x \ h_y \ h_z]^T$: wheel angular momentum vector;
ω : orbital rate.

$$\alpha = \frac{I_z - I_y}{I_x}, \ \beta = \frac{I_x - I_z}{I_y}, \ \gamma = \frac{I_y - I_x}{I_z}$$

4.1.1.2. State Transition Matrix

The state transition matrix is defined by

$$\Phi \approx \mathbf{I}_{6\times 6} + \frac{\partial(\dot{\phi} \ \dot{\theta} \ \dot{\psi} \ d\dot{\phi} \ d\dot{\theta} \ d\dot{\psi})}{\partial(\phi \ \theta \ \psi \ \dot{\phi} \ \dot{\theta} \ \dot{\psi})} \Delta t = \mathbf{I}_{6\times 6} + \begin{bmatrix} \mathbf{O}_{3\times 3} & \mathbf{I}_{3\times 3} \\ \mathbf{F}_{21} & \mathbf{F}_{22} \end{bmatrix} \Delta t \tag{31}$$

The sub-matrices $\mathbf{F}_{21}$ and $\mathbf{F}_{22}$ are given as follows

$$\mathbf{F}_{21} = \begin{bmatrix} 4\omega_0^2\alpha + \omega_0 \dfrac{h_y}{I_x} & 0 & 0 \\ -\omega_0 \dfrac{h_x}{I_y} & -3\omega_0^2\beta & -\omega_0 \dfrac{h_z}{I_y} \\ 0 & 0 & -\omega_0^2\gamma + \omega_0 \dfrac{h_y}{I_z} \end{bmatrix} \qquad (32)$$

$$\mathbf{F}_{21} = \begin{bmatrix} 0 & -\dfrac{h_z}{I_x} & \omega_0(1+\alpha) + \dfrac{h_y}{I_x} \\ \dfrac{h_z}{I_y} & 0 & -\dfrac{h_x}{I_y} \\ -\omega_0 - \dfrac{h_y}{I_z} + \omega_0\gamma & \dfrac{h_x}{I_z} & 0 \end{bmatrix} \qquad (33)$$

4.1.1.3. Observation Equation

The sensor observation model is as follows

$$\mathbf{Z}_{BY} = \mathbf{A}\mathbf{Z}_{LO} + \mathbf{n} \qquad (34)$$

where $\mathbf{Z}_{BY} = [Z_x^{BY}\ Z_y^{BY}\ Z_z^{BY}]^T$, $\mathbf{Z}_{LO} = [Z_x^{LO}\ Z_y^{LO}\ Z_z^{LO}]^T$ are normalised sensor vector with respect to the body axis and local orbit coordinate system and $\mathbf{n} = [n_x\ n_y\ n_z]^T$ is a zero mean measurement noise vector.

Therefore the observation matrix becomes

$$\mathbf{H} = \frac{\partial \mathbf{Z}_{BY}}{\partial \mathbf{X}} = \begin{bmatrix} \dfrac{\partial \mathbf{Z}_{BY}}{\partial(\phi\ \theta\ \psi)} & \dfrac{\partial \mathbf{Z}_{BY}}{\partial(\dot\phi\ \dot\theta\ \dot\psi)} \end{bmatrix} \qquad (35)$$

$$\mathbf{H} = \begin{bmatrix} \dfrac{\partial \mathbf{A}}{\partial \phi}\mathbf{Z}_{LO} & \dfrac{\partial \mathbf{A}}{\partial \theta}\mathbf{Z}_{LO} & \dfrac{\partial \mathbf{A}}{\partial \psi}\mathbf{Z}_{LO} & \mathbf{0}_{3\times 3} \end{bmatrix} \qquad (36)$$

where,

$\mathbf{Z}_{BY}$: body referenced measurements, directly from onboard sensors;

Z_{LO} : orbit referenced measurements, from orbit model prediction;
O_{3x3} : 3x3 zero matrix.

The partial derivatives of the attitude matrix with respect to the Euler angles are introduced in Appendix A.

4.1.2. Reduced version

For this case, we used the Alsat-1 which is an axially symmetric microsatellite with Y momentum/Z reaction wheels and the off diagonal products of inertia elements in the MOI tensor will be zero. The deployed boom along the Z-axis also increases the MOI elements I_x and I_y to a much larger and equal value. This value is called the transverse MOI, I_T.

The attitude estimator implemented on Alsat-1 after the initial attitude acquisition (boom deployment) [22] and during the nominal attitude is given by [10,11]

$$\ddot{\phi} = (4\omega_0^2 \frac{I_z}{I_T} - 1 + \frac{h_y \omega_0}{I_T})\phi - \frac{h_z}{I_T}\dot{\theta} + (\frac{I_z}{I_T}\omega_0 + \frac{h_y}{I_T})\dot{\psi} + \frac{h_z \omega_0}{I_T} + \frac{N_{mx}}{I_T} + w_x \quad (37.a)$$

$$\ddot{\theta} = 3\omega_0^2(\frac{I_z}{I_T} - 1)\theta - \frac{h_z \omega_0}{I_T}\psi + \frac{h_z}{I_T}\dot{\phi} - \frac{\dot{h}_y}{I_T} + \frac{N_{my}}{I_T} + w_y \quad (37.b)$$

$$\ddot{\psi} = (-\frac{h_y}{I_z} - \omega_0)\dot{\phi} + \frac{h_y \omega_0}{I_z}\psi - \frac{\dot{h}_z}{I_z} + \frac{N_{mz}}{I_z} + w_z \quad (37.c)$$

Under the assumptions mentioned above, the analysis of Equations (37.a) to (37.c) shows that the pitch dynamic is decoupled from the roll/yaw dynamic, so the respective estimators can be designed separately.

The state vector of the roll/yaw estimator is defined by

$$\mathbf{X} = \begin{bmatrix} \phi & \psi & \dot{\phi} & \dot{\psi} \end{bmatrix}^T \quad (38)$$

and the state vector of the pitch estimator is defined by

$$\mathbf{X} = \begin{bmatrix} \theta & \dot{\theta} \end{bmatrix}^T \quad (39)$$

The various steps of the design and the implementation of the reduced version and comparison with its full version are described in detail in [11].

4.2. Attitude Estimator – Quaternion

4.2.1. Attitude estimator – 7 state quaternion version

The state vector to be estimated is 7 dimensional such that

$$\mathbf{X} = \begin{bmatrix} \mathbf{q} & \mathbf{\omega}_{BY} \end{bmatrix}^T = \begin{bmatrix} q_1 & q_2 & q_3 & q_4 & \omega_x & \omega_y & \omega_z \end{bmatrix}^T \tag{40}$$

4.2.1.1. System equation

From Equations (1), (2) and (9), the system equation becomes

$$\begin{bmatrix} \dot{q}_1 \\ \dot{q}_2 \\ \dot{q}_3 \\ \dot{q}_4 \end{bmatrix} = \frac{1}{2} \begin{bmatrix} 0 & \omega_{oz} & -\omega_{oy} & \omega_{ox} \\ -\omega_{oz} & 0 & \omega_{ox} & \omega_{oy} \\ \omega_{oy} & -\omega_{ox} & 0 & \omega_{oz} \\ -\omega_{ox} & -\omega_{oy} & -\omega_{oz} & 0 \end{bmatrix} \begin{bmatrix} q_1 \\ q_2 \\ q_3 \\ q_4 \end{bmatrix} \tag{41}$$

$$\dot{\omega}_x = \frac{N_x^{MT}}{I_x} + 3\omega_0^2 \alpha a_{23} a_{33} - \alpha \omega_y \omega_z - \frac{h_z}{I_x} \omega_y + \frac{h_y}{I_x} \omega_z - \frac{\dot{h}_x}{I_x} + w_x \tag{42.a}$$

$$\dot{\omega}_x = \frac{N_y^{MT}}{I_y} + 3\omega_0^2 \beta a_{13} a_{33} - \beta \omega_x \omega_z - \frac{h_x}{I_y} \omega_z + \frac{h_z}{I_y} \omega_x - \frac{\dot{h}_y}{I_y} + w_y \tag{42.b}$$

$$\dot{\omega}_z = \frac{N_z^{MT}}{I_z} + 3\omega_0^2 \gamma a_{13} a_{23} - \gamma \omega_x \omega_y - \frac{h_y}{I_z} \omega_x + \frac{h_x}{I_z} \omega_y - \frac{\dot{h}_z}{I_z} + w_z \tag{42.c}$$

where,

$\mathbf{N}^{MT} = \begin{bmatrix} N_x^{MT} & N_y^{MT} & N_z^{MT} \end{bmatrix}^T$: applied torque vector by 3-axis magnetorquers;
$\mathbf{w} = [w_x \ w_y \ w_z]^T$: zero mean system noise vector;
a_{ij} : (i,j) component of the DCM matrix.

$$\alpha = \frac{I_z - I_y}{I_x}, \ \beta = \frac{I_x - I_z}{I_y}, \ \gamma = \frac{I_y - I_x}{I_z}$$

4.2.1.2. State transition matrix

The state transition matrix is defined by

$$\Phi \approx \mathbf{I}_{7\times7} + \begin{bmatrix} \dfrac{\partial \dot{q}}{\partial q} & \dfrac{\partial \dot{q}}{\partial \omega} \\ \dfrac{\partial \dot{\omega}}{\partial q} & \dfrac{\partial \dot{\omega}}{\partial \omega} \end{bmatrix} \Delta t = \mathbf{I}_{7\times7} + \begin{bmatrix} \mathbf{F}_{11} & \mathbf{F}_{12} \\ \mathbf{F}_{21} & \mathbf{F}_{22} \end{bmatrix} \Delta t \quad (43)$$

The sub-matrices $\mathbf{F}_{ij}$ are given as follows [23]

$$\mathbf{F}_{11} = \begin{bmatrix} q_3 q_1 \omega_0 & q_1 q_4 \omega_0 + \dfrac{1}{2}\omega_{oz} & (1-q_1^2)\omega_0 - \dfrac{1}{2}\omega_{oy} & -q_1 q_2 \omega_0 + \dfrac{1}{2}\omega_{ox} \\ q_3 q_2 \omega_0 - \dfrac{1}{2}\omega_{oz} & q_2 q_4 \omega_0 & -q_1 q_2 \omega_0 + \dfrac{1}{2}\omega_{oz} & (1-q_2^2)\omega_0 + \dfrac{1}{2}\omega_{oy} \\ (q_3^2-1)\omega_0 + \dfrac{1}{2}\omega_{oy} & q_3 q_4 \omega_0 - \dfrac{1}{2}\omega_{oz} & -q_1 q_3 \omega_0 & -q_2 q_3 \omega_0 + \dfrac{1}{2}\omega_{oz} \\ q_3 q_4 \omega_0 - \dfrac{1}{2}\omega_{oz} & (q_4^2-1)\omega_0 - \dfrac{1}{2}\omega_{oy} & -q_1 q_4 \omega_0 - \dfrac{1}{2}\omega_{oz} & -q_2 q_4 \omega_0 \end{bmatrix} \quad (44.a)$$

$$\mathbf{F}_{12} = \dfrac{1}{2} \begin{bmatrix} q_4 & -q_3 & q_2 \\ q_3 & q_4 & -q_1 \\ -q_2 & q_1 & q_4 \\ -q_1 & -q_2 & -q_3 \end{bmatrix} \quad (44.b)$$

$$\mathbf{F}_{21} = \begin{bmatrix} 6\omega_0^2\alpha(-a_{23}q_1 + a_{33}q_4) & 6\omega_0^2\alpha(-a_{23}q_2 + a_{33}q_3) \\ 6\omega_0^2\beta(+a_{33}q_3 - a_{13}q_1) & 6\omega_0^2\beta(-a_{33}q_4 - a_{13}q_2) \\ 6\omega_0^2\gamma(+a_{13}q_4 + a_{23}q_3) & 6\omega_0^2\gamma(+a_{13}q_3 - a_{23}q_4) \\ 6\omega_0^2\alpha(+a_{23}q_3 + a_{33}q_2) & 6\omega_0^2\alpha(+a_{23}q_4 + a_{33}q_1) \\ 6\omega_0^2\beta(+a_{33}q_1 + a_{13}q_3) & 6\omega_0^2\beta(-a_{33}q_2 + a_{13}q_4) \\ 6\omega_0^2\gamma(+a_{13}q_2 + a_{23}q_1) & 6\omega_0^2\gamma(+a_{13}q_1 - a_{23}q_2) \end{bmatrix} \quad (44.c)$$

$$\mathbf{F}_{22} = \begin{bmatrix} 0 & -\alpha\omega_z - \dfrac{h_z}{I_x} & -\alpha\omega_y + \dfrac{h_y}{I_x} \\ -\beta\omega_z + \dfrac{h_z}{I_y} & 0 & -\beta\omega_x \\ -\gamma\omega_y - \dfrac{h_y}{I_z} & -\gamma\omega_x & 0 \end{bmatrix} \quad (44.d)$$

4.2.1.3. Observation Equation

The sensor observation model is same as Equation (34)
Therefore the observation matrix becomes

$$\mathbf{H} = \frac{\partial \mathbf{Z}_{BY}}{\partial \mathbf{X}} = \left[\frac{\partial \mathbf{Z}_{BY}}{\partial (q_1 \, q_2 \, q_3 \, q_4)} \quad \frac{\partial \mathbf{Z}_{BY}}{\partial (\omega_x \, \omega_y \, \omega_z)} \right] \quad (45)$$

$$\mathbf{H} = \left[\frac{\partial \mathbf{A}}{\partial q_1} \mathbf{Z}_{LO} \quad \frac{\partial \mathbf{A}}{\partial q_2} \mathbf{Z}_{LO} \quad \frac{\partial \mathbf{A}}{\partial q_3} \mathbf{Z}_{LO} \quad \frac{\partial \mathbf{A}}{\partial q_4} \mathbf{Z}_{LO} \quad \mathbf{0}_{3 \times 3} \right] \quad (46)$$

The partial derivatives of the attitude matrix with respect to the quaternion are introduced in Appendix B.

The process noise covariance matrix $\mathbf{Q}$ implemented on board Alsat-1 is given by

$$\mathbf{Q} \approx \begin{bmatrix} \frac{\sigma^2}{12}\left(\frac{q_4^2}{I_x^2}+\frac{q_3^2}{I_y^2}+\frac{q_2^2}{I_z^2}\right)\Delta T^3 & 0 & 0 & 0 & 0 & 0 & 0 \\ 0 & \frac{\sigma^2}{12}\left(\frac{q_3^2}{I_x^2}+\frac{q_4^2}{I_y^2}+\frac{q_1^2}{I_z^2}\right)\Delta T^3 & 0 & 0 & 0 & 0 & 0 \\ 0 & 0 & \frac{\sigma^2}{12}\left(\frac{q_2^2}{I_x^2}+\frac{q_1^2}{I_y^2}+\frac{q_4^2}{I_z^2}\right)\Delta T^3 & 0 & 0 & 0 & 0 \\ 0 & 0 & 0 & \frac{\sigma^2}{12}\left(\frac{q_1^2}{I_x^2}+\frac{q_2^2}{I_y^2}+\frac{q_3^2}{I_z^2}\right)\Delta T^3 & 0 & 0 & 0 \\ 0 & 0 & 0 & 0 & \frac{\sigma^2}{I_x^2}\Delta T & 0 & 0 \\ 0 & 0 & 0 & 0 & 0 & \frac{\sigma^2}{I_y^2}\Delta T & 0 \\ 0 & 0 & 0 & 0 & 0 & 0 & \frac{\sigma^2}{I_z^2}\Delta T \end{bmatrix} \quad (47)$$

4.2.2. Attitude estimator – 6 state quaternion version

The state vector to be estimated is similar to (40), but the estimator will only used to compute a 6 element differential vector, i.e. one element less than the 7 state quaternion version. The reason for this reduction in state is : the attitude quaternion has been linearised in a special way. Instead of expressing the predicted quaternion in terms of an estimated value plus a differential, it is expressed in terms of a differential quaternion times the estimated quaternion using quaternion multiplication.

$$\mathbf{q} = \hat{\mathbf{q}} \otimes \begin{bmatrix} \delta \mathbf{q} \\ 1 \end{bmatrix} = \begin{bmatrix} 1 & \delta q_3 & -\delta q_2 & \delta q_1 \\ -\delta q_3 & 1 & \delta q_1 & \delta q_2 \\ \delta q_2 & -\delta q_1 & 1 & \delta q_3 \\ -\delta q_1 & -\delta q_2 & -\delta q_3 & 1 \end{bmatrix} \begin{bmatrix} \hat{q}_1 \\ \hat{q}_2 \\ \hat{q}_3 \\ \hat{q}_4 \end{bmatrix} \tag{48}$$

The differential quaternion has just three unknowns, the fourth is not needed because the differential Euler angles is small and no attitude singularity can occur.

The predicted body rates are expressed in terms of an estimated value plus a differential as follows

$$\omega = \hat{\omega} + \delta \omega \tag{49}$$

Note that the cross matrix definition is utilized extensively throughout this process. The cross matrix is defined specifically as

$$\mathbf{C}(\mathbf{a}) = \begin{bmatrix} 0 & -a_3 & a_2 \\ a_3 & 0 & -a_1 \\ -a_2 & a_1 & 0 \end{bmatrix} \tag{50}$$

The auxiliary state vector to be estimated is 6 dimensional such that

$$\mathbf{X} = \begin{bmatrix} \delta q_1 & \delta q_2 & \delta q_3 & \delta \omega_x & \delta \omega_y & \delta \omega_z \end{bmatrix}^T$$

4.2.2.1. System Equation

The full non linear model of the microsatellite can be obtained from the Euler Equation (1) and quaternion Equation (9). The linearised differential system equation to be used is given by

$$\delta \dot{\mathbf{q}} = -\mathbf{C}(\omega) \delta \mathbf{q} + \frac{1}{2} \delta \omega \tag{51}$$

$$\mathbf{I} \delta \dot{\omega} = (\delta \mathbf{N}_{GG} + \delta \mathbf{N}_{MT}) \delta \mathbf{q} - (\mathbf{C}(\omega) \mathbf{I} - \mathbf{C}(\omega \mathbf{I} + \mathbf{h})) \delta \omega \tag{52}$$

The two matrices $\delta \mathbf{N}_{GG}$ and $\delta \mathbf{N}_{MT}$ are obtained as follows

$$\delta \mathbf{N}_{GG} = 6 \omega_0^2 (\mathbf{C}(\omega) \mathbf{I} - \mathbf{C}(\omega \mathbf{I})) \mathbf{C}(\mathbf{u}) \tag{53}$$

$$\delta \mathbf{N}_{MT} = 2 \mathbf{C}(\mathbf{m}) \mathbf{C}(\mathbf{B}_{BY}) \tag{54}$$

where,

$$\mathbf{u} = -\mathbf{A}(\mathbf{q})\begin{bmatrix} 0 & 0 & 1 \end{bmatrix}^T$$

$$\mathbf{B}_{BY} = \mathbf{A}(\mathbf{q})\mathbf{B}_{LO}$$

$\mathbf{m} = [m_x\ m_y\ m_z]^T$: magnetic torquer of the magnetorquer.

Through the results above, the system equation for the 6 state quaternion version is assumed as

$$\begin{bmatrix} \delta\dot{\mathbf{q}} \\ \delta\dot{\boldsymbol{\omega}} \end{bmatrix} = \begin{bmatrix} -\mathbf{C}(\boldsymbol{\omega}) & \frac{1}{2}\mathbf{I}_{3x3} \\ \mathbf{I}^{-1}(\delta\mathbf{N}_{GG} + \delta\mathbf{N}_{MT}) & -\mathbf{I}^{-1}(\mathbf{C}(\boldsymbol{\omega})\mathbf{I} - \mathbf{C}(\boldsymbol{\omega}\mathbf{I} + \mathbf{h})) \end{bmatrix} \begin{bmatrix} \delta\mathbf{q} \\ \delta\boldsymbol{\omega} \end{bmatrix} + \begin{bmatrix} 0 \\ \mathbf{w} \end{bmatrix} \quad (55)$$

where,

w : zero mean system noise vector.

4.2.2.2. State Transition Matrix

The state transition matrix is defined by

$$\Phi \approx \mathbf{I}_{6x6} + \begin{bmatrix} \dfrac{\partial(\delta\dot{\mathbf{q}})}{\partial(\delta\mathbf{q})} & \dfrac{\partial(\delta\dot{\mathbf{q}})}{\partial(\delta\boldsymbol{\omega})} \\ \dfrac{\partial(\delta\dot{\boldsymbol{\omega}})}{\partial(\delta\mathbf{q})} & \dfrac{\partial(\delta\dot{\boldsymbol{\omega}})}{\partial(\delta\boldsymbol{\omega})} \end{bmatrix} \cdot \Delta t \quad (56)$$

The sub-matrices $\mathbf{F}_{ij}$ can be obtained directly from the system equation as follows

$$\Phi \approx \mathbf{I}_{6x6} + \begin{bmatrix} -\mathbf{C}(\boldsymbol{\omega}) & \frac{1}{2}\mathbf{I}_{3x3} \\ \mathbf{I}^{-1}(\delta\mathbf{N}_{GG} + \delta\mathbf{N}_{MT}) & -\mathbf{I}^{-1}(\mathbf{C}(\boldsymbol{\omega})\mathbf{I} - \mathbf{C}(\boldsymbol{\omega}\mathbf{I} + \mathbf{h})) \end{bmatrix} \cdot \Delta t \quad (57)$$

The sub-matrices $\mathbf{F}_{ij}$ are introduced in Appendix C.

4.2.2.3. Observation Equation

Same sensor observation model is used defined in Equation (34)
Therefore the observation matrix becomes

$$\mathbf{H} = \frac{\partial \mathbf{Z}_{BY}}{\partial \mathbf{X}} = \begin{bmatrix} \dfrac{\partial \mathbf{Z}_{BY}}{\partial(\delta q_1\ \delta q_2\ \delta q_3)} & \dfrac{\partial \mathbf{Z}_{BY}}{\partial(\delta\omega_x\ \delta\omega_y\ \delta\omega_z)} \end{bmatrix} \quad (58)$$

$$\mathbf{H} = \left[\frac{\partial \mathbf{A}}{\partial (\delta q_1)} \mathbf{Z}_{LO} \quad \frac{\partial \mathbf{A}}{\partial (\delta q_2)} \mathbf{Z}_{LO} \quad \frac{\partial \mathbf{A}}{\partial (\delta q_3)} \mathbf{Z}_{LO} \quad \mathbf{O}_{3x3} \right] \quad (59)$$

$$\mathbf{H} = [2\mathbf{C}(\mathbf{AZ}_{LO}) \quad \mathbf{O}_{3x3}] \quad (60)$$

The process noise covariance matrix **Q** implemented on board Alsat-1 is given by

$$\mathbf{Q} \approx \begin{bmatrix} \frac{1}{12}\frac{\sigma^2}{I_x^2}\Delta T^3 & 0 & 0 & 0 & 0 & 0 \\ 0 & \frac{1}{12}\frac{\sigma^2}{I_y^2}\Delta T^3 & 0 & 0 & 0 & 0 \\ 0 & 0 & \frac{1}{12}\frac{\sigma^2}{I_z^2}\Delta T^3 & 0 & 0 & 0 \\ 0 & 0 & 0 & \frac{\sigma^2}{I_x^2}\Delta T & 0 & 0 \\ 0 & 0 & 0 & 0 & \frac{\sigma^2}{I_y^2}\Delta T & 0 \\ 0 & 0 & 0 & 0 & 0 & \frac{\sigma^2}{I_z^2}\Delta T \end{bmatrix} \quad (61)$$

Table 1 demonstrates the sheer magnitude of the reduction in the number of matrix elements during a sampling time of 10 seconds.

The results below indicate clearly that the reduction in size of the 6 state quaternion version is a great benefit comparing to the 7 state quaternion version in terms of computational demand. Note that the reduction in the number of matrix elements should correlate with the processing speed.

Table 1. Estimator Element Comparison

	7 State	Elements	6 State	Elements
X	(7x1)	7	(7x1)	7
Y			(6x1)	6
P	(7x7)	49	(6x6)	36
Φ	(7x7)	49	(6x6)	36
Q	(7x7)	49	(6x6)	36
H	(3x7)	21	(3x6)	18
K	(7x3)	21	(6x3)	18
	Total for 10 sec	196	Total	157
	Total for one orbit (100 mins)	19600	Total	157000

5. SIMULATIONS RESULTS

The simulation results presented in this section were obtained with an attitude determination simulator for low earth orbit microsatellite using C code, MATLAB and SIMULINK.

The attitude estimators were run for two orbits, with the initial conditions detailed below (see Tables 2 - 8) which are inputs for the Alsat-1 ADCS flight code. The simulator incorporates the magnetometer and sun sensor as sensors for attitude determination. The gravity gradient torque and the aerodynamic torque were the disturbance modeled.

Table 2. Orbit characteristics

Orbit	Circular
Inclination [degree]	98
Altitude [km]	686
sampling period [sec]	10

Table 3. Inertia tensor (Microsatellite configuration I)

I_{xx} [kgm^2]	152.9
I_{xy} [kgm^2]	0.0
I_{xz} [kgm^2]	0.0
I_{yx} [kgm^2]	-0.25
I_{yy} [kgm^2]	152.5
I_{yz} [kgm^2]	0.0005
I_{zx} [kgm^2]	0.1
I_{zy} [kgm^2]	0.0
I_{zz} [kgm^2]	4.91

Table 4. Measurement error variance (Measurement noise covariance matrix R)

Magnetometer measurement error variance in X/Y/Z axis [microTesla]^2	$(0.3)^2$
Sun sensor measurement error variance in X/Y/Z axis [degree]^2	$(0.1)^2$

Table 5. Initial uncertainty of Euler angles and rates (Initial covariance matrix P) Small Euler angles estimator

Initial roll angle [degree]^2	10^2
Initial pitch angle [degree]^2	10^2
Initial yaw angle [degree]^2	10^2
Initial angular rate [degree/sec]^2	$(0.01)^2$
Initial angular rate degree/sec]^2	$(0.01)^2$
Initial angular rate [degree/sec]^2	$(0.01)^2$

Table 6. Process noise intensity of Euler angles and rates (System noise covariance matrix Q) Small Euler angles estimator

Intensity for φ term σ_ϕ^2 [degree]^2	$(10^{-5})^2$
Intensity for θ term σ_θ^2 [degree]^2	$(10^{-5})^2$
Intensity for ψ term σ_ψ^2 [degree]^2	$(10^{-5})^2$
Intensity for $\dot\phi$ term $\sigma_{\dot\phi}^2$ [degree/sec]^2	$(10^{-3})^2$
Intensity for $\dot\theta$ term $\sigma_{\dot\theta}^2$ [degree/sec]^2	$(10^{-3})^2$
Intensity for $\dot\psi$ term $\sigma_{\dot\psi}^2$ [degree/sec]^2	$(10^{-3})^2$

Table 7. Initial uncertainty of quaternion and angular rates (initial covariance matrix P) Quaternion estimator

Initial q_1	$(0.3)^2$
Initial q_2	$(0.3)^2$
Initial q_3	$(0.3)^2$
Initial q_4	$(0.3)^2$
Initial angular rate ω_x (°/s)^2	$(0.01)^2$
Initial angular rate ω_y (°/s)^2	$(0.01)^2$
Initial angular rate ω_z (°/s)^2	$(0.01)^2$

Table 8. Process noise intensity of angular rates (system noise covariance matrix Q) Quaternion estimator

Intensity for ω_x term $\sigma_{\omega x}^2$ (°/s)^2	$(10^{-5})^2$
Intensity for ω_y term $\sigma_{\omega y}^2$ (°/s)^2	$(10^{-5})^2$
Intensity for ω_z term $\sigma_{\omega z}^2$ (°/s)^2	$(10^{-5})^2$

Due to a lack of obvious physical interpretation of the quaternion, the Euler angles will normally be used to present the attitude during simulation tests.

To evaluate the performance of the attitude estimators in terms of errors, given by the RMS, an analysis and a comparison are performed.

5.1. Attitude Estimator – Small Euler Angles

The attitude error estimation of the full estimator and its reduced version is shown below in figures 2 to 5. The estimator takes in the order of 10 minutes to converge [11].

Tabulated error magnitudes (see Table 9) for the full estimator and its reduced version are presented below the graphs.

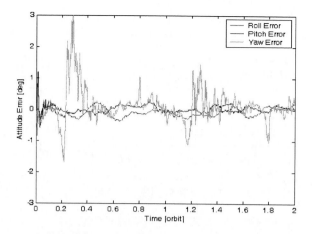

Figure 2. Attitude error Small Euler angles - Full version.

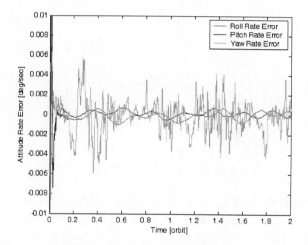

Figure 3. Attitude rate error Small Euler angles - Full version.

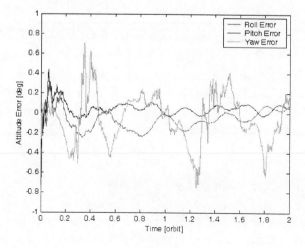

Figure 4. Attitude error Small Euler angles - Reduced version.

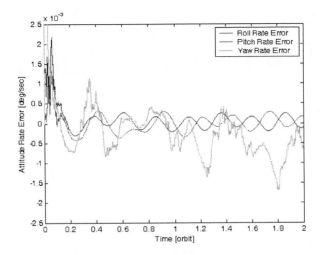

Figure 5. Attitude rate error Small Euler angles - Reduced version.

Table 9. Attitude Error Estimation Comparison

	RMS Full Estimator	RMS Reduced Estimator
Roll [deg]	0.15	0.08
Pitch [deg]	0.07	0.07
Yaw [deg]	0.14	0.13
Roll Rate [milli-deg/sec]	0.25	0.07
Pitch Rate [milli-deg/sec]	0.14	0.08
Yaw Rate [milli-deg/sec]	1.08	0.64
	Magnitude of Error	Magnitude of Error
Angles [deg]	0.22	0.16
Rate [milli-deg/sec]	1.12	0.65

The magnitude of the RMS error results indicates that the angular error is approximately 0.22 deg and 0.16 deg for the full estimator and its reduced version respectively and the rate error is about 1.12 milli-deg/sec and 0.65 milli-deg/sec for the full estimator and its reduced version respectively.

The error estimation of the reduced version is better than the full version. This takes the advantage in the computation time reduction to execute the reduced version.

5.2. Attitude Estimator – Quaternion

The attitude error estimation of the 7 state estimator and its reduced version are presented below in figures 6 to 9. The estimator converge after 5 minutes.

Tabulated error magnitudes (see Table 10) for the 7 state estimator and its reduced version are presented below the graphs.

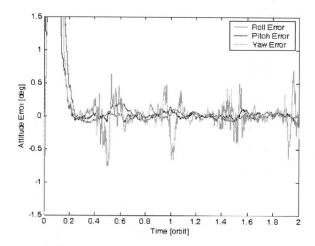

Figure 6. Attitude error Quaternion - 6 state.

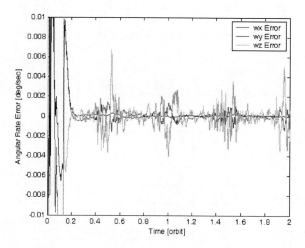

Figure 7. Attitude rate error Quaternion - 6 state.

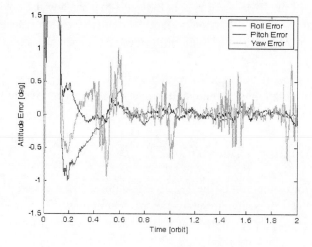

Figure 8. Attitude error Quaternion - 7 state.

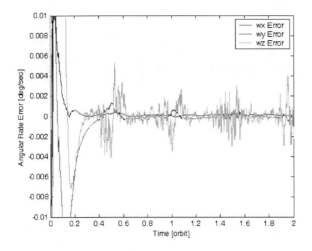

Figure 9 Attitude rate error Quaternion - 7 state.

Table 10. Attitude Error Estimation Comparison

	RMS Quaternion - 7 state	RMS Quaternion - 6 state
Roll [deg]	0.09	0.09
Pitch [deg]	0.03	0.04
Yaw [deg]	0.06	0.06
ω_x [milli-deg/sec]	0.08	0.09
ω_y [milli-deg/sec]	0.15	0.16
ω_z [milli-deg/sec]	0.32	0.36
	Magnitude of Error	Magnitude of Error
Angles [deg]	**0.11**	**0.11**
Rate [milli-deg/sec]	**0.37**	**0.41**

The magnitude of the RMS error results indicates that the angular error is approximately 0.11 deg for both estimators and the rate error is about 0.41 milli-deg/sec and 0.37 milli-deg/sec for the 6 state estimator and its full version respectively.

Note that the magnitude of the RMS error of the 6 state estimator and its full version are almost similar.

6. IN ORBIT RESULTS

Filtering problems are generally related to the control and tasks procedures automatisation. In general, the objective is not to know the state of the system but to control its evolution. If we consider the example of Alsat-1which is an Earth observation microsatellite, controlling its attitude to nadir attitude pointing for imaging is more important than knowing its estimated attitude.

This section describes the in orbit results of the Alsat-1 attitude determination and control system for nadir attitude pointing control (imaging mode) using the attitude estimators described previously in section 4.

The Y-wheel pitch and Z-wheel yaw attitude controllers implemented on Alsat-1 for the nadir attitude pointing control (imaging mode) are [24,25]

$$\left.\begin{array}{l} N_{y-wheel} = K_{d\theta}\dot{\theta} + K_{p\theta}(\theta - \theta_{ref}) \\ \Omega_{y-wheel-cmd} = \int N_{y-wheel} dt / I_{wheel} \end{array}\right\} \quad (62)$$

$$\left.\begin{array}{l} N_{z-wheel} = K_{d\psi}\dot{\psi} + K_{p\psi}(\psi - \psi_{ref}) \\ \Omega_{z-wheel-cmd} = \int N_{z-wheel} dt / I_{wheel} \end{array}\right\} \quad (63)$$

The saturation limits are $|N_{y-wheel}| \leq 3 \times 10^{-3}$ Nm and $500\text{RPM} \leq |\Omega_{y-wheel-cmd}| \leq 3000\text{RPM}$.

where,

Kp : proportional gain;
Kd : derivative gain;
θ_{ref} : reference pitch angle (nominally zero for the nadir pointing);
ψ_{ref} : reference yaw angle (nominally zero for the nadir pointing);
N_{wheel} : torque wheel required;
I_{wheel} : wheel inertia;
Ω_{wheel} : wheel command.

To maintain the wheel momentum at a certain reference level and to damp any nutation in roll and yaw, a magnetorquer cross-product control law is used

$$\mathbf{M} = \frac{\mathbf{e} \times \mathbf{B}}{\|\mathbf{B}\|} \quad (64)$$

The error vector for the magnetorquer cross-product control algorithm including Y and Z wheel is given by [24,25]

$$\mathbf{e} = \begin{bmatrix} K_{dx}\dfrac{\omega_{ox}}{\omega_0} \\ K_y(h_y - h_{y-ref}) \\ K_z(h_z - h_{z-ref}) \end{bmatrix} \quad (65)$$

where,

M : magnetorquer switch-on time;
B : magnetometer measured magnetic field vector;
ω_0 : orbit angular rate in rad/s;
ω_{0x} : X orbit referenced angular rate of the microsatellite in rad/s;
K_{dx} : derivative gain;
K : momentum maintenance gain constant;
h_{y-ref} : reference pitch wheel momentum;
h_y : pitch wheel momentum measurement in Nms;
h_{z-ref} : reference yaw wheel momentum;
h_z : yaw wheel momentum measurement in Nms.

The gains we have used are tabulated below.
The nominal Y-wheel speed command of –1500 RPM has been chosen.

Table 11. Controller gains implemented on board Alsat-1 for the attitude nadir pointing control

$K_{d\theta}$	$K_{p\theta}$	$K_{d\psi}$	$K_{p\psi}$	K_y	K_z	K_{dx}
2.0	8.25×10-3	5×10-3	0.25	50	50	8

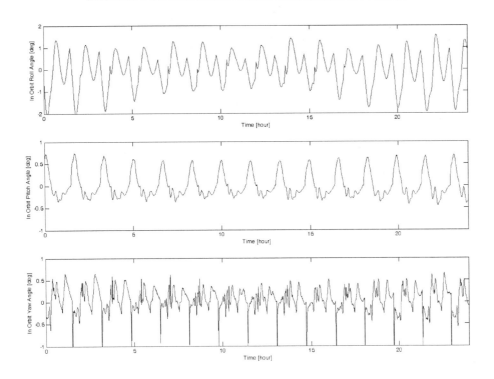

Figure 10. In orbit Alsat-1 attitude during imaging mode Small Euler angles estimator – Reduced version.

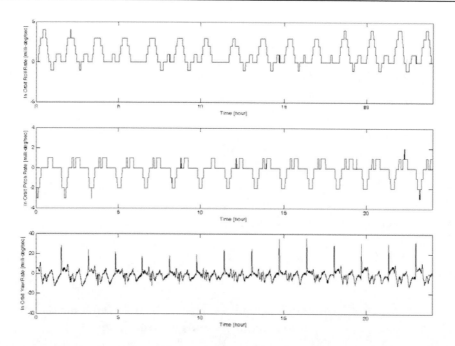

Figure 11. In orbit Alsat-1 attitude rate during imaging mode Small Euler angles estimator – Reduced version.

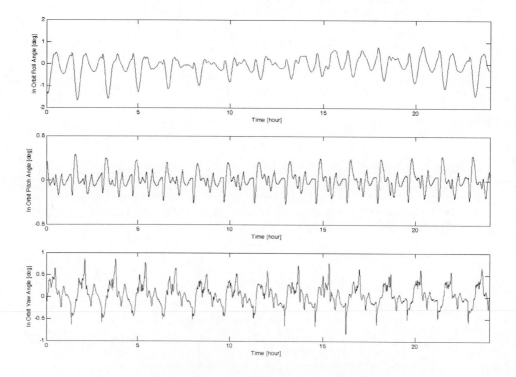

Figure 12. In orbit Alsat-1 attitude during imaging mode Quaternion estimator – 7 state.

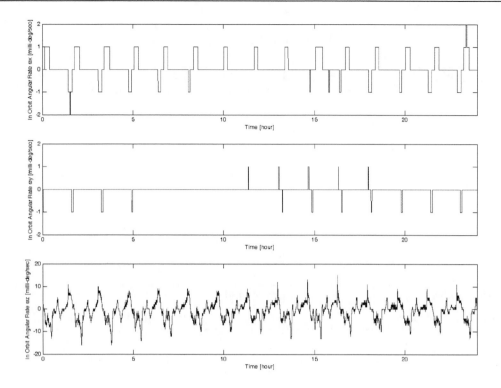

Figure 13. In orbit Alsat-1 attitude rate during imaging mode Quaternion estimator – 7 state.

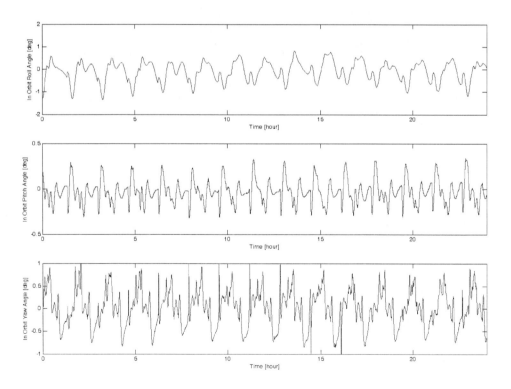

Figure 14. In orbit Alsat-1 attitude during imaging mode Quaternion estimator – 6 state.

Table 12. Alsat-1 attitude requirements, in orbit attitude on Alsat-1 using different estimators

Attitude	Alsat-1 requirements	Small Euler angles Reduced version	7 State Quaternion Version	6 State Quaternion Version
Roll [deg]	≤ 1	0.26	0.08	0.10
Pitch [deg]	≤ 1	0.05	0.02	0.03
Yaw [deg]	≤ 0.5	0.16	0.08	0.06
ω_x [milli-deg/sec]	≤ 5	0.06	0.04	0.02
ω_y [milli-deg/sec]	≤ 5	0.48	0.03	0.06
ω_z [milli-deg/sec]	≤ 5	2.82	1.44	1.33
	Magnitude			
Angles [deg]	**1.5**	**0.31**	**0.12**	**0.12**
Rate [milli-deg/sec]	**8.66**	**2.86**	**1.44**	**1.33**

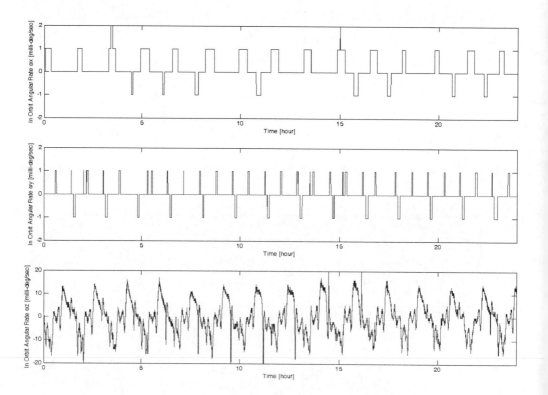

Figure 15. In orbit Alsat-1 attitude rate during imaging mode Quaternion estimator – 6 state.

The in orbit attitude profiles are shown in figure 10 to figure 15. These results were obtained directly from the on board ADCS algorithm implemented on board Alsat-1. Standard deviation of the attitude using different attitude estimators are tabulated in Table 12.

The in orbit results shown in table 12 meet the Alsat-1 requirements listed on the above table. Notice that the attitude and attitude rate results using small Euler angles estimator are 5 and 3 times better than their requirements respectively. Although the in orbit results using quaternion estimator are quite satisfactory against the requirements, attitude and attitude rate results are 12 and 6 times better than their requirements respectively. In addition, note that the magnitude of the attitude of the 6 state quaternion version and the 7 state quaternion version are almost identical.

Conclusion

The primary focus of this chapter is to detail the design implementation of the attitude estimators based on Kalman filtering for application on low Earth orbit microsatellites, particularly, for an optimal application on Alsat-1. These attitude estimators are used after the initial attitude acquisition of the microsatellite in order to obtain the attitude knowledge and also during the nominal operation task such as the imaging mode. In orbit test results clearly indicate that all attitude estimators meet the attitude determination and control system requirements during nominal operation. The reduced version of small Euler angles of an axially symmetric configuration and a specifically attitude control system of a microsatellite, such as Alsat-1, is better than the full version in term of accuracy and computational demand. Although, the implemented 6 state quaternion version on board Alsat-1 to date is better than the 7 state quaternion version in terms of computational demand. However, the accuracy is almost similar. According to these results, the 6 state is a more logical choice for the Alsat-1 attitude determination and control system.

To conclude, we have demonstrated successful operation of low cost attitude estimators to be used on a low earth orbit microsatellite.

Appendix A: Partial Derivative of the Attitude Matix

The partial derivatives of the attitude matrix **A** with respect to the Euler angles are given by

$$\frac{\partial \mathbf{A}}{\partial \phi} = \begin{bmatrix} s\psi s\theta c\phi & -s\psi s\phi & s\psi c\theta c\phi \\ c\psi s\theta c\phi & -c\psi s\phi & c\psi c\theta c\phi \\ -s\theta s\phi & -c\phi & -c\theta s\phi \end{bmatrix} \qquad (A.1)$$

$$\frac{\partial \mathbf{A}}{\partial \theta} = \begin{bmatrix} -c\psi s\theta + s\psi c\theta s\phi & 0 & -c\psi c\theta - s\psi s\theta s\phi \\ s\psi s\theta + c\psi c\theta s\phi & 0 & s\psi c\theta - c\psi s\theta s\phi \\ c\theta c\phi & 0 & -s\theta c\phi \end{bmatrix} \quad (A.2)$$

$$\frac{\partial \mathbf{A}}{\partial \psi} = \begin{bmatrix} -s\psi c\theta + c\psi s\theta s\phi & c\psi c\phi & s\psi s\theta + c\psi c\theta s\phi \\ -c\psi c\theta - s\psi s\theta s\phi & -s\psi c\phi & c\psi s\theta - s\psi c\theta s\phi \\ 0 & 0 & 0 \end{bmatrix} \quad (A.3)$$

APPENDIX B: PARTIAL DERIVATIVE OF THE ATTITUDE MATIX

The partial derivatives of the attitude matrix **A** with respect to the quaternion are given by

$$\frac{\partial \mathbf{A}}{\partial q_1} = 2 \begin{bmatrix} q_1 & q_2 & q_3 \\ q_2 & -q_1 & q_4 \\ q_3 & -q_4 & -q_1 \end{bmatrix} \quad (B.1)$$

$$\frac{\partial \mathbf{A}}{\partial q_2} = 2 \begin{bmatrix} -q_2 & q_1 & -q_4 \\ q_1 & q_2 & q_3 \\ q_4 & q_3 & -q_2 \end{bmatrix} \quad (B.2)$$

$$\frac{\partial \mathbf{A}}{\partial q_3} = 2 \begin{bmatrix} -q_3 & q_4 & q_1 \\ -q_4 & -q_3 & q_2 \\ q_1 & q_2 & q_3 \end{bmatrix} \quad (B.3)$$

$$\frac{\partial \mathbf{A}}{\partial q_4} = 2 \begin{bmatrix} q_4 & q_3 & -q_2 \\ -q_3 & q_4 & q_1 \\ q_2 & -q_1 & q_4 \end{bmatrix} \quad (B.4)$$

APPENDIX C : COMPONENTS OF THE F MATRIX – 6 STATE

The sub-matrices $\mathbf{F}_{ij}$ are given by

$$\mathbf{F}_{11} = -\begin{bmatrix} 0 & -\omega_z & \omega_y \\ \omega_z & 0 & -\omega_x \\ -\omega_y & \omega_x & 0 \end{bmatrix} \quad (C.1)$$

$$F_{21} = 6\omega_0^2 \begin{bmatrix} \dfrac{(I_y - I_z)}{I_x}(a_{23}^2 - a_{33}^2) & \dfrac{(I_z - I_y)}{I_x}a_{23}a_{13} & \dfrac{(I_y - I_z)}{I_x}a_{33}a_{13} \\ -2\dfrac{m_z B_z + m_y B_y}{I_x} & 2\dfrac{m_y B_x}{I_x} & 2\dfrac{m_z B_x}{I_x} \\ \dfrac{(I_z - I_x)}{I_y}a_{23}a_{13} & \dfrac{(I_x - I_z)}{I_y}(a_{13}^2 - a_{33}^2) & \dfrac{(I_x - I_z)}{I_y}a_{23}a_{33} \\ 2\dfrac{m_x B_y}{I_y} & -2\dfrac{m_z B_z + m_x B_x}{I_y} & 2\dfrac{m_z B_y}{I_y} \\ \dfrac{(I_y - I_x)}{I_z}a_{33}a_{13} & \dfrac{(I_x - I_y)}{I_z}a_{23}a_{33} & \dfrac{(I_x - I_y)}{I_z}(a_{13}^2 - a_{23}^2) \\ 2\dfrac{m_x B_z}{I_z} & 2\dfrac{m_y B_z}{I_z} & -2\dfrac{m_y B_y + m_x B_x}{I_z} \end{bmatrix} \quad (C.2)$$

$$F_{22} = - \begin{bmatrix} 0 & \dfrac{(I_y - I_z)\omega_z - h_z}{I_x} & \dfrac{(I_y - I_z)\omega_y + h_y}{I_x} \\ \dfrac{(I_z - I_x)\omega_z + h_z}{I_y} & 0 & \dfrac{(I_z - I_x)\omega_x + h_x}{I_y} \\ \dfrac{(I_x - I_y)\omega_y - h_y}{I_z} & \dfrac{(I_x - I_y)\omega_x + h_x}{I_z} & 0 \end{bmatrix} \quad (C.3)$$

REFERENCES

[1] Si Mohammed, AM. An Attitude Determination and Control System of the Alsat-1 First Algerian Microsatellite. *IEEE Recent Advances in Space Technologies RAST'03*, Istanbul, Turkey, 2003, 162-167.

[2] Kalman, RE. A New Approach to Linear Filtering and prediction Problems. Transactions of AMSE. *Journal of Basic Engineering. March,* 1960, Vol 82, 35-45.

[3] Kalman, RE; Bucy, RS. New Results in Linear Filtering and Prediction Theory. Journal of *Basic Engineering.* March, 1961, Vol 83, 95-108.

[4] Lefferts, EJ; Markley, FL; Shuster, MD. Kalman filtering for Spacecraft Attitude Estimation. Journal of Guidance, *Control and Dynamics.* Sept-Oct, 1982, Vol 5, N°. 5, 417-429.

[5] Bierman, GJ; Thorton, CL. Numerical Comparison of Kalman Filter Algorithms : Orbit Determination Case Study. *Astronautica.*, 1977, Vol. 13, N°. 1, 23-25.

[6] Psiaki, ML; Huang, L; Fox, SM. Ground Tests of magnetometer based Autonomous MAGNAV for Low Earth Orbiting Spacecraft. *Journal of Guidance, Control, and Dynamics. Jan.-Feb.*, 1993, Vol. 16, N°.1, 206-214.

[7] Gelb, A. *Applied Optimal Estimation*. 1989, Cambridge - Massachusetts – London.

[8] Brown, RG; Hwang, PYC. *Introduction to Random Signals and Applied Kalman Filtering*. 1997, New York – Chichester – Brisbane – Toronto – Singapore.

[9] Wertz, JR. *Space Mission Analysis and Design*; Space Technology Library, Kluwer Academic Publishers, Dordrecht – Boston – London, 1991.

[10] Si Mohammed, AM; Benyettou, M; Sweeting, MN; Cooksley, JR. Full Attitude Determination Specification - Small Libration Version - of the Alsat-1 First Algerian Microsatellite in orbit. *IEEE Recent Advances in Space Technologies*, RAST '05. Istanbul, Turkey, 2005, 150-154.

[11] Si Mohammed, AM; Benyettou, M; Boudjemai, A; Hashida, Y; Sweeting, MN. Simulation of Full Microsatellite Attitude Kalman Filtering in Orbit Results. Journal of Simulation *Practice and Theory – Elsevier*. March, 2008, Vol. 16, N°3, 257-277.

[12] Ickes, BP. A New method for Performing Digital Control System Attitude Computations Using Quaternions. *AIAA Journal*, Jan, 1970, Vol. 8, N°. 1, 13-17.

[13] Frieland, B. Analysis Strapdown Navigation using Quaternions. IEEE Transactions on *Aerospace and Electronic System.*, Sept 1978, Vol. 14, N°. 5, 764-768.

[14] Shuster, MD; Oh, SD. Three Axis Attitude Determination From Vector Observations. *Journal of Guidance, Control and Dynamics. Jan-Fev*, 1981, Vol 4, N°. 1, 70-77.

[15] Gai, E; Daly, K; Lemos, L. Star Sensor Based Satellite Attitude/Attitude Rate Estimator. Journal of Guidance, *Control and Dynamics.*, Sept-Oct 1985, Vol 8, N°. 5, 560-565.

[16] Bar Itzhach, IY; Reiner, J. Attitude Determination From Verctor Observations : Quaternion Estimation. IEEE *Transactions on Aerospace and Electronic System.*, Jan 1985, Vol. 21, N°. 1, 128-135.

[17] Shibata, M. Error Analysis Strapdown Inertial Navigation Using Quaternions. *Journal of Guidance, Control and Dynamics*. May-June 1986, Vol 9, N°. 3, 379-381.

[18] Psiaki, ML; Martel, F; Pal, PK. Three Axis Attitude Determination Via Kalman Filtering of Magnetometer Data. *Journal of Guidance, Control and Dynamics.*, May-June 1990, Vol 13, N°. 3, 506-514.

[19] Markley, FL. New Quaternion Attitude Estimation Method. *Journal of Guidance, Control and Dynamics*. March-April, 1994, Vol 17, N°. 2, 407-409.

[20] Psiaki, ML. Attitude Determination Filtering via Extended Quaternion Estimation. *Journal of Guidance, Control and Dynamics.*, 2000, Vol 23, N°. 2, 206-214.

[21] Pittelkau, ME. Kalman Filtering for Spacecraft System Alignment Calibration. *Journal of Guidance, Control, and Dynamics.*, Nov-Decr 2001, Vol 24, No. 6, 1187-1195.

[22] Si Mohammed, AM; Benyettou, M; Sweeting, MN; Cooksley, JR. Initial Attitude Acquisition Result of the Alsat-1 First Algerian Microsatellite in Orbit. 2005 IEEE International Conference on *Networking, Sensing and Control*, March 19-22, 2005, Tucson, Arizona, USA, 566-571.

[23] Si Mohammed, AM; Benyettou, A; Boudjemai, Y; Hashida. Seven State Kalman Filtering for LEO Microsatellite Attitude Determination. Recent Advances in Signal Processing, Robotics and Automation. Proceedings of the 9th International Conference

on *Signal Processing, Robotics and Automation ISPRA'10*. Cambridge University, Feb. 2010, 151-157.

[24] Si Mohammed, AM; Benyettou, M; Sweeting, MN; Cooksley, JR. Imaging Mode Results of the Alsat-1 First Algerian Microsatellite in Orbit. *IEEE Recent Advances in Space Technologies*, RAST'05. 2005, Istanbul, Turkey, 483-486.

[25] Si Mohammed, AM; Benyettou, M; Bentoutou, Y; Boudjemai, A; Hashida, Y; Sweeting, M.N. Three-Axis Active Control System for *Gravity Gradient Stabilised Microsatellite. Acta Astronautica*. April 2009, Vol. 64, N°7, 796 – 809.

In: Kalman Filtering
Editor: Joaquín M. Gomez

ISBN: 978-1-61761-462-0
© 2011 Nova Science Publishers, Inc.

Chapter 7

DESIGN OF EXTENDED RECURSIVE WIENER FIXED-POINT SMOOTHER AND FILTER IN CONTINUOUS-TIME STOCHASTIC SYSTEMS

Seiichi Nakamori[*]
Department of Technology, Faculty of Education,
Kagoshima University, Kohrimoto, Kagoshima, Japan

ABSTRACT

This paper, at first, designs the extended recursive Wiener fixed-point smoother and filter in continuous-time wide-sense stationary stochastic systems. It is assumed that the signal is observed with the nonlinear mechanism of the signal and with the additional white observation noise. The estimators use the information of the system matrix F for the state vector $x(t)$, the observation vector C for the state vector, the variance $K(t,t) = K(0)$ of the state vector, the nonlinear observation function and the variance R of the white observation noise. F, C and $K(0)$ are usually obtained from the auto-covariance function of the signal.

Secondly, by using the covariance information of the signal and the observation noises, the extended fixed-point smoother and filter for white plus colored observation noise are proposed.

Numerical simulation examples are shown to demonstrate the validity of the proposed estimation algorithms.

Keywords: Continuous-time stochastic systems; Extended recursive Wiener estimators; Covariance information; Fixed-point smoother; Non-linear modulation

1. INTRODUCTION

The estimation problem using the covariance information has been investigated extensively in the area of the signal detection and estimation problems [1]-[5] for communication systems. The estimation technique using the covariance information might be classified into two kinds from the expressions for the auto-covariance function of the signal to be estimated. One kind is expressed in the form of the semi-degenerate kernel [6]-[8]. The semi-degenerate kernel expresses the auto-covariance function of the signal as a finite sum of products of deterministic functions. In Section III of [6], by starting with the Wiener-Hopf integral equation having the semi-degenerate kernel for the process with the rational spectral densities, the linear recursive least-squares (RLS) filtering problem is introduced. In [7], in linear discrete-time stochastic systems, by modelling the signal process in terms of the autoregressive (AR) model and by use of the auto-covariance function of the signal in the form of the semi-degenerate kernel, the RLS filter is derived from the Wiener-Hopf equation. In [8], by using the semi-degenerate kernel expression of the covariance information of the signal and observation noise, least-squares second-order polynomial estimators are designed in systems including uncertain observations. Furthermore, in [9]-[11], for the observation with the nonlinear mechanism of the signal, the extended recursive estimators using the covariance information are designed in stochastic systems. It might be advantageous that the estimation technique using the auto-covariance function of the signal in the form of the semi-degenerate kernel can be applied when the state-space model is not realizable from the covariance information, e.g. for the triangular signal, the rectangular signal [12], etc.

The extended Kalman filter [13], [14],[15],[16] is useful in the wide area of engineering such as signal demodulation problems etc. In [14], the robust extended Kalman filter is proposed for the discrete-time nonlinear systems with norm-bounded parameter uncertainties in Krein space. It is a characteristic that both the Kalman filter and the extended Kalman filter use the full information of the state-space model.

In nonlinear filtering problem, besides the extended Kalman filter, exact nonlinear recursive filter and the batch or non-recursive filters show significant improvement on accuracy of state vector estimate relative to the extended Kalman filter (EKF) for some applications [17]. Also, the unscented Kalman filters [18], [19] often provide a significant improvement relative to EKF, but sometimes it does not [17]. Some classes of dynamics by the exact nonlinear filters are summarized in Table 2 of [20]. In [21], based on Daum's work [20], a nonlinear filter, based on the nonlinear dynamic system and the linear observation equation, is presented.

By the way, the recursive Wiener estimators [3], [6], [22],[23], [24] use the information of the system matrix F, the observation vector C for the state vector, the variance $K_x(0)$ of the state vector and the variance R of the white observation noise. In [24], the RLS Wiener fixed-point smoother and filter are designed in the case of $H(t) = 1$, which indicates that the observation equation (1) does not include the linear modulation. F, C and $K_x(0)$ are obtained from the auto-covariance function of the signal [24]. This paper, at first, in the estimation of the continuous-time wide-sense stationary stochastic signal, newly designs the extended recursive Wiener fixed-point smoother and filter, which use the information of F, C, $K_x(0)$ and R. In this paper, it is assumed that the signal is observed with the additional white noise through the nonlinear mechanism of the linear stochastic signal. In [Theorem 1], the RLS Wiener fixed-point smoothing and filtering algorithms are presented by applying the estimation technique in [22] to the case of the observation equation (1) with the linear modulation. In [Theorem 2], by the first-order approximation of the nonlinear observation function around the filtering estimate of the signal, the extended recursive Wiener fixed-point smoother and filter are proposed. The estimators in [Theorem 2] are derived by extending the RLS Wiener estimators in [Theorem 1] with the linear modulation to the observation equation (17) with the nonlinear modulation of the signal like the derivation of the extended Kalman filter from the RLS Kalman filter. Also, in [Theorem 3], by referring to [13], the recursive Wiener fixed-point smoother and filter are proposed in terms of the second-order approximation of the nonlinear observation mechanism.

Secondly, in [Theorem 4], for the linear modulation of the signal with additional white plus colored observation noise, new RLS fixed-point smoother and filter are designed by using the covariance information of the signal and the observation noises. In [Theorem 5], using the covariance information, for the nonlinear modulation of the signal with additional white plus colored observation noise, the first-order extended fixed-point smoothing and filtering algorithms are proposed.

In a numerical simulation example, the phase demodulation problem [25] is demonstrated to show the validity of the proposed estimation algorithms in this paper.

2. RLS FIXED-POINT SMOOTHING PROBLEM WITH LINEAR MODULATION FOR WHITE OBSERVATION NOISE

Let an m-dimensional time-varying observation equation be given by

$$y(t) = H(t)z(t) + v(t), \quad z(t) = Cx(t), \tag{1}$$

in linear continuous-time stochastic systems. Here, $z(t)$ is an $l \times 1$ signal vector, $H(t)$ is a linear modulation function of $z(t)$ and $x(t)$ is an $n \times 1$ state vector with the wide-sense stationarity. C is an $l \times n$ observation vector transforming $x(t)$ to $z(t)$. $v(t)$ is white observation noise. Also, let the state equation for the state vector $x(t)$ be expressed by

$$\frac{dx(t)}{dt} = Fx(t) + w(t), \quad x(0) = x_0, \tag{2}$$

where F is the system matrix and $w(t)$ is white noise input. It is assumed that the signal and the observation noise are mutually independent and are zero mean. Let the auto-covariance function of $v(t)$ and $w(t)$ be expressed by

$$E[v(t)v^T(s)] = R\delta(t-s), \quad R > 0, \tag{3}$$

$$E[w(t)w^T(s)] = \Psi_w \delta(t-s), \ \Psi_w > 0. \tag{4}$$

Here, $\delta(\cdot)$ denotes the Kronecker δ function.

Let $K_x(t,s) = K_x(t-s)$ represent the auto-covariance function of the state vector $x(t)$ and let $K_x(t,s)$ be expressed in the form of

$$K_x(t,s) = \begin{cases} A(t)B^T(s), & 0 \leq s \leq t \\ B(t)A^T(s), & 0 \leq t \leq s \end{cases} \tag{5}$$

in wide-sense stationary stochastic systems [24]. Here, $A(t) = e^{Ft}$, $B^T(s) = e^{-s}K_x(s,s)$. $K_x(s,s)$ represents the variance of $x(s)$.

Let a fixed-point smoothing estimate $\hat{x}(t|T)$ of $x(t)$ be expressed by

$$\hat{x}(t|T) = \int_0^T h(t,s,T)y(s)ds, \ 0 \leq t \leq T, \tag{6}$$

as a linear transformation of the observed value $y(s)$ during the time interval $0 \leq s \leq T$. In (6), $h(t,s,T)$ is a time-varying impulse response function and t is the fixed point respectively.

Let us consider the estimation problem, which minimizes the mean-square value

$$J = E[\|x(t) - \hat{x}(t|T)\|^2] \tag{7}$$

of the fixed-point smoothing error. From an orthogonal projection lemma [13],

$$x(t) - \int_0^T h(t,s,T)y(s)ds \perp y(s), \ 0 \leq s,t \leq T, \tag{8}$$

the optimal impulse response function satisfies the Wiener-Hopf integral equation

$$E[x(t)y^T(s)] = \int_0^T h(t,\tau,T)E[y(\tau)y^T(s)]d\tau. \tag{9}$$

Here, '$\perp$' denotes the notation of the orthogonality. Substituting (1) and (3) into (9), we obtain

$$h(t,s,T)R = K_x(t,s)C^T H^T(s) - \int_0^T h(t,\tau,T)H(\tau)CK_x(\tau,s)d\tau C^T H^T(s). \tag{10}$$

3. RLS Wiener Estimation Algorithms with Linear Modulation for White Observation Noise

Under the linear least-squares estimation problem of the signal $z(t)$ in Section 2, [Theorem 1] shows the RLS Wiener fixed-point smoothing and filtering algorithms, which use the covariance information of the signal and observation noise.

[Theorem 1]
Let the observation equation, concerned with the linear modulation for the signal $z(t)$, be given by (1). Let the auto-covariance function of the state vector $x(t)$ be expressed by (5) and let the variance of the white observation noise be R in wide-sense stationary stochastic systems. Then, the linear RLS algorithms, using the information of the system matrix F, the observation vector C and the variance $K_x(0)$ of the state vector, for the fixed-point smoothing and filtering estimates consist of (11)-(16).

Fixed-point smoothing estimate of $z(t)$ at the fixed point t: $\hat{z}(t|T) = C\hat{x}(t|T)$
Fixed-point smoothing estimate of $x(t)$ at the fixed point t: $\hat{x}(t|T)$

$$\frac{\partial \hat{x}(t|T)}{\partial T} = h(t,T,T)(y(T) - H(T)\hat{z}(T|T)), \; \hat{z}(T|T) = C\hat{x}(T|T) \quad (11)$$

Smoother gain: $h(t,T,T)$

$$h(t,T,T) = (K_x(t,T)C^T H^T(T) - Q(t,T)C^T H^T(T))R^{-1} \quad (12)$$

$$\frac{\partial Q(t,T)}{\partial T} = h(t,T,T)(H(T)CK_x(T,T) - H(T)CG(T)) + Q(t,T)F^T,$$
$$Q(t,t) = G(t) \quad (13)$$

Filtering estimate of $z(t)$: $\hat{z}(t|t) = C\hat{x}(t|t)$
Filtering estimate of state vector $x(t)$: $\hat{x}(t|t)$

$$\frac{d\hat{x}(t|t)}{dt} = F\hat{x}(t,t) + h(t,t,t)(y(t) - H(t)C\hat{x}(t|t)), \; \hat{x}(0|0) = 0 \quad (14)$$

Auto-variance function of $\hat{x}(t|t)$: $G(t)$

$$\frac{dG(t)}{dt} = FG(t) + G(t)F^T + h(t,t,t)H(t)C(K_x(t,t) - G(t)), \; G(0) = 0 \quad (15)$$

Filter gain: $h(t,t,t)$

$$h(t,t,t) = (K_x(t,t)C^T H^T(T) - G(T)C^T H^T(t))R^{-1} \qquad (16)$$

Proof. In [24], the RLS Wiener fixed-point smoother and filter, using the information of F, C, $K_x(0)$ and R, are designed in the case of $H(t) = 1$ which does not include the linear modulation in the observation equation (1). The RLS Wiener fixed-point smoothing and filtering algorithms in [Theorem 1] are derived by applying the estimation technique in [24] to the case of the observation equation (1) with the linear modulation. The algorithms use the information of F, C, $K_x(0)$ and R. Since $\hat{x}(0|0) = 0$, the initial value of the variance $G(t)$ of $\hat{x}(t|t)$ at $t = 0$ is $G(0) = 0$. (Q.E.D.).

4. Extended Recursive Wiener Estimation Algorithms for White Observation Noise

Let an observation equation with the nonlinear mechanism be given by

$$y(t) = f(z(t),t) + v(t), \quad z(t) = Cx(t), \qquad (17)$$

where the signal $z(t)$ and the observation noise $v(t)$ have the same stochastic properties as those in Section 2.

Like the design of the extended Kalman filter, in the design of the extended recursive Wiener estimators using the covariance information, the modulating function is put as $H(t) = \left. \dfrac{\partial f(z(t),t)}{\partial z(t)} \right|_{z(t)=\hat{z}(t|t)}$ in [Theorem 1] after expanding the nonlinear observation function in a first-order Taylor series about $\hat{z}(t|t)$ [13]. Here, $\hat{z}(t|t) = C\hat{x}(t|t)$ represents the filtering estimate of the signal $z(t)$. Also, $H(T)\hat{z}(T|T)$ and $H(t)C\hat{x}(t|t)$ in [Theorem 1] are replaced with $f(\hat{z}(T|T),T)$ and $f(\hat{z}(t,t),t)$ respectively.

Consequently, the first-order extended recursive Wiener fixed-point smoothing and filtering algorithms in case of the observation equation (17), which has the nonlinear mechanism of the signal $z(t)$, is summarized in [Theorem 2]. It is noted that the proposed extended recursive Wiener estimators are sub-optimal because of the Taylor series approximation of the modulating function.

[Theorem 2]
Let the observation equation, which has nonlinear mechanism of the signal, be given by (17). Let the auto-covariance function of the state vector $x(t)$ be expressed by (5) and let the variance of white observation noise be R in wide-sense stationary stochastic systems. Then, the first-order extended recursive Wiener fixed-point smoothing and filtering algorithms, using the covariance information of the signal and observation noise, consist of (18)-(26).

Fixed-point smoothing estimate of the signal $z(t)$ at the fixed point t: $\hat{z}(t|T)$

$$\hat{z}(t|T) = C\hat{x}(t|T) \tag{18}$$

Fixed-point smoothing estimate of the state vector $x(t)$ at the fixed point t: $\hat{x}(t|T)$

$$\frac{\partial \hat{x}(t|T)}{\partial T} = h(t,T,T)(y(T) - f(\hat{z}(T|T),T)), \; \hat{z}(T|T) = C\hat{x}(T|T) \tag{19}$$

Smoother gain: $h(t,T,T)$

$$h(t,T,T) = (K_x(t,T)C^T H^T(T) - Q(t,T)C^T H^T(T))R^{-1} \tag{20}$$

$$\frac{\partial Q(t,T)}{\partial T} = h(t,T,T)(H(T)CK_x(T,T) - H(T)CG(T)) + Q(t,T)F^T,$$
$$Q(t,t) = G(t) \tag{21}$$

Filtering estimate of the signal $z(t)$:

$$\hat{z}(t|t) = C\hat{x}(t|t) \tag{22}$$

Filtering estimate of the state vector $x(t)$: $\hat{x}(t|t)$

$$\frac{d\hat{x}(t|t)}{dt} = F\hat{x}(t,t) + h(t,t,t)(y(t) - f(\hat{z}(t|t),t)), \; \hat{x}(0|0) = 0 \tag{23}$$

Auto-variance function of $\hat{x}(t|t)$: $G(t)$

$$\frac{dG(t)}{dt} = FG(t) + G(t)F^T + h(t,t,t)H(t)C(K_x(t,t) - G(t)), \; G(0) = 0 \tag{24}$$

Filter gain: $h(t,t,t)$

$$h(t,t,t) = (K_x(t,t)C^T H^T(T) - G(T)C^T H^T(t))R^{-1} \tag{25}$$

Here, the function $H(t)$ is given by

$$H(t) = \left.\frac{\partial f(z(t),t)}{\partial z(t)}\right|_{z(t)=\hat{z}(t|t)}. \tag{26}$$

The extended recursive Wiener estimators, in comparison with the extended Kalman estimators, require less information. The extended recursive Wiener estimators use the information of F, C, $K_x(0)$ and R. The extended Kalman estimators use the information of F, C and the variance of the white noise input, Ψ_w, in (4). Both estimators use the information of nonlinear modulation function. Since $G(t)$ is the auto-variance function of the filtering estimate $\hat{x}(t|t)$, the algorithm for the filtering error variance function $P_{\tilde{x}}(t|t)$ is obtained by substituting $G(t) = K_x(t,t) - P_{\tilde{x}}(t|t)$ into (24) in the extended recursive Wiener estimation algorithms of [Theorem 2].

In [Theorem 3], in terms of the second-order approximation of the nonlinear observation function, the extended recursive Wiener fixed-point smoother and filter are presented.

[Theorem 3]

Let the nonlinear observation equation of the signal be given by (17). Let the same covariance information of the signal and the observation noise be given as described in Section 2 in wide-sense stationary stochastic systems. Then the extended recursive Wiener fixed-point smoothing and filtering algorithms, in terms of the second-order approximation of the nonlinear observation function, consist of (27)-(34).

Fixed-point smoothing estimate of the signal $z(t)$ at the fixed point t: $\hat{z}(t|T)$

$$\hat{z}(t|T) = C\hat{x}(t|T) \tag{27}$$

Fixed-point smoothing estimate of the state vector $x(t)$ at the fixed point t: $\hat{x}(t|T)$

$$\frac{\partial \hat{x}(t|T)}{\partial T} = h(t,T,T)(y(T) - f(\hat{z}(T|T),T) - \frac{1}{2}\frac{\partial^2 h(\hat{z}(T|T),T)}{\partial^2 \hat{z}(T|T)} : C(K_x(0) - G(T))C^T)$$
$$\hat{z}(T|T) = C\hat{x}(T|T) \tag{28}$$

Smoother gain: $h(t,T,T)$

$$h(t,T,T) = (K_x(t,T)C^T H^T(T) - Q(t,T)C^T H^T(T))R^{-1} \tag{29}$$

$$\frac{\partial Q(t,T)}{\partial T} = h(t,T,T)(H(T)CK_x(T,T) - H(T)CG(T)) + Q(t,T)F^T,$$
$$Q(t,t) = G(t) \tag{30}$$

Filtering estimate of the signal $z(t)$:

$$\hat{z}(t|t) = C\hat{x}(t|t) \tag{31}$$

Filtering estimate of the state vector $x(t)$: $\hat{x}(t|t)$

$$\frac{d\hat{x}(t,t)}{dt} = F\hat{x}(t|t) + h(t,t,t)(y(t) - f(\hat{z}(t,t),t) - \frac{1}{2}\frac{\partial^2 h(\hat{z}(t|t),t)}{\partial^2 \hat{z}(t|t)^2} : C(K_x(0) - G(t))C^T),$$
$$\hat{x}(0|0) = 0 \tag{32}$$

Auto-variance function of $\hat{x}(t|t)$: $G(t)$

$$\frac{dG(t)}{dt} = FG(t) + G(t)F^T + h(t,t,t)H(t)C(K_x(t,t) - G(t)) - \Xi(t), \; G(0) = G_0 \tag{33}$$

$$\Xi(t)_{kl} = \left\{ \frac{1}{2}\sum_{i,j=1}^{N}[(K_x(0) - G(t))_{ik}(K_x(0) - G(t))_{lj} + (K_x(0) - G(t))_{kj}(K_x(0) - G(t))_{li}\frac{\partial^2 h(\hat{z}(t|t),t)}{\partial \hat{x}_i(t|t)\partial \hat{x}_j(t|t)} \right\}^T$$

$$\times R^{-1}\left\{ y(t) - h(\hat{z}(t|t),t) - \frac{\partial^2 h(\hat{z}(t|t),t)}{\partial^2 \hat{z}(t|t)} : C(K_x(0) - G(t))C^T \right\}$$

Filter gain: $h(t,t,t)$

$$h(t,t,t) = (K_x(t,t)C^T H^T(T) - G(T)C^T H^T(t))R^{-1} \tag{34}$$

Here, the function $H(t)$ is given by

$$H(t) = \frac{\partial f(z(t),t)}{\partial z(t)}\bigg|_{z(t)=\hat{z}(t|t)}.$$

Proof. The extended filtering algorithm in [Theorem 3] is obtained by substituting the filtering error variance function $P_{\tilde{x}}(t|t)$ expressed by $P_{\tilde{x}}(t|t) = K_x(0) - G(t)$ into the extended Kalman filter based on the second-order approximation [13]. The extended fixed-point smoothing algorithm is obtained by applying the second-order approximation to the nonlinear observation function in [Theorem 2]. (Q.E.D.)

5. RLS ESTIMATORS FOR LINEAR MODULATION AND ADDITIONAL WHITE PLUS COLORED OBSERVATION NOISE

Let an m-dimensional time-varying observation equation be given by

$$y(t) = H(t)z(t) + v(t) + v_c(t), \; z(t) = Cx(t), \tag{35}$$

in linear continuous-time stochastic systems. Here, $z(t)$ is an $l \times 1$ signal vector, $H(t)$ is a linear modulation function of $z(t)$ and $x(t)$ is an $n \times 1$ state vector with the wide-sense stationary property. C is an $l \times n$ observation vector transforming $x(t)$ to $z(t)$. $v(t)$ is white observation noise and $v_c(t)$ is colored observation noise. The state vector $x(t)$ satisfies (2). It is assumed that $x(\cdot)$ and $v(\cdot)$ are mutually independent, and $x(t)$, $v(t)$ and $v_c(t)$ are zero mean. Let the auto-covariance function of $v(t)$ be given by (3).

Let $K_Z(t,s) = K_Z(t-s)$ represent the auto-covariance function of

$$Z(t) = H(t)z(t) + v_c(t) \tag{36}$$

and let $K_Z(t,s)$ be expressed in the form of

$$K_Z(t,s) = \begin{cases} \Lambda(t)B^T(s), & 0 \leq s \leq t, \\ B(t)\Lambda^T(s), & 0 \leq t \leq s, \end{cases} \tag{37}$$

in wide-sense stationary stochastic systems.

Similarly to [Theorem 1], from [26], the recursive algorithm for the linear RLS estimate $\hat{Z}(t|t)$ of $Z(t)$ is obtained as (38)-(41).

$$\hat{Z}(t|t) = \Lambda(t)e(t) \tag{38}$$

$$\frac{de(t)}{dt} = J(t,t)(y(t) - \hat{Z}(t|t)), \quad \hat{Z}(t|t) = 0 \tag{39}$$

$$J(t,t) = (B^T(t) - r(t)\Lambda^T(t))R^{-1} \tag{40}$$

$$\frac{dr(t)}{dt} = J(t,t)(B(t) - \Lambda(t)r(t)), \quad r(0) = 0 \tag{41}$$

For the innovations process $\upsilon(s) = y(s) - \hat{Z}(s|s)$, it is seen that

$$E[x(t)\upsilon^T(s)] = E[x(t)(y(s) - \hat{Z}(s|s))^T], \quad 0 \leq s \leq t. \tag{42}$$

Let $\hat{Z}(t|t)$ be given by $\hat{Z}(t|t) = \int_0^t h_Z(t,s)y(s)ds$. Then

$$E[x(t)\upsilon^T(s)] = E[x(t)y^T(s)] - \int_0^s E[x(t)y^T(s')]h_Z^T(s,s')ds'. \qquad (43)$$

Now, let the cross-covariance function of $x(t)$ with $y(s)$ be expressed by

$$K_{xy}(t,s) = \begin{cases} \alpha(t)\beta^T(s), & 0 \le s \le t, \\ \varepsilon(t)\zeta^T(s), & 0 \le t \le s. \end{cases} \qquad (44)$$

Since $h_Z(t,s) = \Lambda(t)J(t,s)$ [26],

$$E[x(t)\upsilon^T(s)] = \alpha(t)(\beta^T(s) - \int_0^s \beta^T(s')J^T(s,s')ds'\Lambda^T(s)). \qquad (45)$$

By introducing

$$q(t) = \int_0^t J(t,s')\beta(s')ds', \qquad (46)$$

(45) is written as

$$E[x(t)\upsilon^T(s)] = \alpha(t)(\beta^T(s) - q^T(s)\Lambda^T(s)). \qquad (47)$$

From [5], the impulse response function based on the innovations theory is given by

$$g(t,s) = E[x(t)\upsilon^T(s)]R^{-1}$$
$$= \begin{cases} g_1(t,s), & 0 \le s \le t, \\ g_2(t,s), & 0 \le t \le s. \end{cases}$$

Differentiating (46) with respect to t and using $\dfrac{\partial J(t,s)}{\partial t} = -J(t,t)\Lambda(t)J(t,s)$, $0 \le s \le t$, [26], we obtain

$$\frac{dq(t)}{dt} = J(t,t)(\beta(t) - \Lambda(t)q(t)), \; q(0) = 0. \qquad (48)$$

The impulse response function $g_2(t,s)$, $0 \le s \le t$, is developed, from (45), as

$$g_2(t,s) = E[x(t)\upsilon^T(s)]R^{-1}$$

$$= E[x(t)(y(s) - \hat{Z}(s\mid s))^T]R^{-1}$$

$$= (\varepsilon(t)\zeta^T(s) - \int E[x(t)y^T(s')]h_Z^T(s,s')ds' - \int E[x(t)y^T(s')]h_Z^T(s,s')ds')R^{-1}$$

$$= (\varepsilon(t)\zeta^T(s) - \alpha(t)D^T(s,t)\Lambda^T(s) - \varepsilon(t)E^T(s,t)\Lambda^T(s))R^{-1}. \tag{49}$$

here

$$D(s,t) = \int_0^s J(s,s')\beta(s')ds', \tag{50}$$

$$E(s,t) = \int J(s,s')\zeta(s')ds'. \tag{51}$$

From [5], the fixed-point smoothing estimate is given by

$$\hat{x}(t\mid T) = \int_0^t g_1(t,s)\upsilon(s)ds + \int_t^T g_2(t,s)\upsilon(s)ds. \tag{52}$$

Differentiating (52) with respect to T and using (49), we have the equation for the fixed-point smoothing estimate $\hat{x}(t\mid T)$ of $x(t)$ as

$$\frac{\partial \hat{x}(t\mid T)}{\partial T} = \varepsilon(t)(\zeta^T(T) - E^T(T,t)\Lambda^T(t))R^{-1}(y(T) - \hat{Z}(T\mid T))$$
$$- \alpha(t)D^T(T,t)\Lambda^T(t)R^{-1}(y(T) - \hat{Z}(T\mid T)). \tag{53}$$

From (47), the filtering estimate $\hat{x}(t\mid t)$ of $x(t)$ is obtained as

$$\hat{x}(t\mid t) = \int_0^t g_1(t,s)\upsilon(s)ds$$

$$= \alpha(t)\int_0^t (\beta^T(s) - q^T(s)\Lambda^T(s))R^{-1}\upsilon(s)ds$$

$$= \alpha(t)Q(t). \tag{54}$$

Here,

$$Q(t) = \int_0^t (\beta^T(s) - q^T(s)\Lambda^T(s))R^{-1}\upsilon(s)ds. \tag{55}$$

Differentiating (55) with respect to t, we obtain

$$\frac{dQ(t)}{dt} = (\beta^T(t) - q^T(t)\Lambda^T(t))R^{-1}(y(t) - \hat{Z}(t|t)), \quad Q(0) = 0. \tag{56}$$

From (51) and $\dfrac{\partial J(T,s)}{\partial T} = -J(T,T)\Lambda(T)J(T,s)$, we obtain

$$\frac{\partial E(T,t)}{\partial T} = J(T,T)\zeta(T') - \int^T \frac{\partial J(T,s')}{\partial T}\zeta(s')ds'$$

$$= J(T,T)\zeta(T') - J(T,T)\Lambda(T)\int^T J(T,s')\zeta(s')ds'$$

$$= J(T,T)(\zeta(T) - \Lambda(T)E(T,t)), \quad E(t,t) = 0. \tag{57}$$

Also, from (50), we obtain

$$\frac{\partial D(T,t)}{\partial T} = \int_0^t \frac{\partial J(T,s')}{\partial T}\beta(s')ds'$$

$$= -J(T,T)\Lambda(T)D(T,t). \tag{58}$$

Likewise, from (50), we obtain

$$\frac{dD(t,t)}{dt} = J(t,t)\beta(t) + \int_0^t \frac{\partial J(t,s')}{\partial t}\beta(s')ds'$$

$$= J(t,t)\beta(t) - J(t,t)\Lambda(t)\int_0^t J(t,s')\beta(s')ds'$$

$$= J(t,t)(\beta(t) - \Lambda(t)D(t,t)). \tag{59}$$

From (54) and (55) with $\dfrac{d\alpha(t)}{dt} = F\alpha(t)$, the filtering estimate $\hat{x}(t|t)$ is calculated by

$$\frac{d\hat{x}(t|t)}{dt} = F\hat{x}(t|t) + (\alpha(t)\beta^T(t) - \alpha(t)q^T(t)\Lambda^T(t))R^{-1}\upsilon(t)$$

$$= F\hat{x}(t|t) + (K_{xy}(t,t) - S(t))R^{-1}(y(t) - \hat{Z}(t|t)), \quad \hat{x}(0|0) = 0, \tag{60}$$

here

$$S(t) = \alpha(t)q^T(t)\Lambda^T(t). \tag{61}$$

Let us summarize the above results in [Theorem 4].
[Theorem 4]
Let the observation equation, concerned with the linear modulation of the signal be given by (35) for white plus colored observation noise. Let the auto-covariance function of $Z(t) = H(t)z(t) + v_c(t)$ be given by (37) and let the cross-covariance function $K_{xy}(t,s)$ of the state vector $x(t)$ with the observed value $y(s)$ be given by (44). Then the RLS fixed-point smoothing and filtering algorithms of the signal $z(t)$ consist of (62)-(74).

Fixed-point smoothing estimate of the signal $z(t)$ at the fixed point t: $\hat{z}(t|T)$

$$\hat{z}(t|T) = C\hat{x}(t|T) \tag{62}$$

Fixed-point smoothing estimate of the state vector $x(t)$ at the fixed point t: $\hat{x}(t|T)$

$$\frac{\partial \hat{x}(t|T)}{\partial T} = \varepsilon(t)(\zeta^T(T) - E^T(T,t)\Lambda^T(t))R^{-1}(y(T) - \hat{Z}(T|T)) \\ - \alpha(t)D^T(T,t)\Lambda^T(t)R^{-1}(y(T) - \hat{Z}(T|T)) \tag{63}$$

$$\frac{\partial E(T,t)}{\partial T} = J(T,T)(\zeta(T) - \Lambda(T)E(T,t)), \quad E(t,t) = 0 \tag{64}$$

$$\frac{\partial D(T,t)}{\partial T} = -J(T,T)\Lambda(T)D(T,t) \tag{65}$$

$$\frac{dD(t,t)}{dt} = J(t,t)(\beta(t) - \Lambda(t)D(t,t)), \quad D(0,0) = 0 \tag{66}$$

Filtering estimate of the signal $z(t)$:

$$\hat{z}(t|t) = C\hat{x}(t|t) \tag{67}$$

Filtering estimate of the state vector $x(t)$: $\hat{x}(t|t)$

$$\frac{d\hat{x}(t|t)}{dt} = F\hat{x}(t|t) + (K_{xy}(t,t) - S(t))R^{-1}(y(t) - \hat{Z}(t|t)), \quad \hat{x}(0|0) = 0 \tag{68}$$

Filtering estimate of $Z(t) = H(t)z(t) + v_c(t)$: $\hat{Z}(t|t)$

$$\hat{Z}(t|t) = \Lambda(t)e(t) \tag{69}$$

$$\frac{de(t)}{dt} = J(t,t)(y(t) - \hat{Z}(t|t)), \ e(0) = 0 \tag{70}$$

$$S(t) = \alpha(t)q^T(t)\Lambda^T(t) \tag{71}$$

$$\frac{dq(t)}{dt} = J(t,t)(\beta(t) - \Lambda(t)q(t)), \ q(0) = 0 \tag{72}$$

$$J(t,t) = (\mathbf{B}^T(t) - r(t)\Lambda^T(t))R^{-1} \tag{73}$$

$$\frac{dr(t)}{dt} = J(t,t)(\mathrm{B}(t) - \Lambda(t)r(t)), \ r(0) = 0 \tag{74}$$

Now, let us consider deriving the estimation equations for the colored observation noise $v_c(t)$.

In terms of the impulse response function $g_c(t,s) = E[v_c(t)\upsilon^T(s)]R^{-1}$, $0 \le s \le t$, the filtering estimate $\hat{v}_c(t|t)$ of $v_c(t)$ is given by

$$\hat{v}_c(t|t) = \int_0^t g_c(t,s)\upsilon(s)ds$$

$$= \int_0^t E[v_c(t)\upsilon^T(s)]R^{-1}\upsilon(s)ds. \tag{75}$$

Let $K_{cy}(t,s)$ represent the cross-covariance function of the colored observation noise $v_c(t)$ with the observed value $y(s)$ as $K_{cy}(t,s) = \alpha_c(t)\beta_c^T(s)$, $0 \le s \le t$. From $\hat{Z}(t|t) = \int_0^t h_Z(t,s)y(s)ds$ and $h_Z(t,s) = \Lambda(t)J(t,s)$,

$$E[v_c(t)\upsilon^T(s)] = E[v_c(t)y^T(s)] - \int_0^s E[v_c(t)y^T(s')]h_Z^T(s,s')ds'$$

$$= \alpha_c(t)(\beta_c^T(s) - \int_0^s \beta_c^T(s')J^T(s,s')ds'\Lambda^T(s))$$

$$= \alpha_c(t)(\beta_c^T(s) - q_c^T(s)\Lambda^T(s)). \tag{76}$$

Here,

$$q_c(t) = \int_0^t J(t,s')\beta_c(s)ds. \tag{77}$$

Differentiating (77) with respect to t, from $\dfrac{\partial J(t,s)}{\partial t} = -J(t,t)\Lambda(t)J(t,s)$, $0 \le s \le t$, we obtain

$$\frac{dq_c(t)}{dt} = J(t,t)(\beta_c(t) - \Lambda(t)q_c(t)), \quad q_c(0) = 0. \tag{78}$$

Substituting (76) into (75), we obtain

$$\hat{v}_c(t|t) = \alpha_c(t)\int_0^t (\beta_c^T(s) - q_c^T(s)\Lambda^T(s))R^{-1}\upsilon(s)ds.$$

By introducing

$$e_c(t) = \int_0^t (\beta_c^T(s) - q_c^T(s)\Lambda^T(s))R^{-1}\upsilon(s)ds, \tag{79}$$

the filtering estimate $\hat{v}_c(t|t)$ is given by

$$\hat{v}_c(t|t) = \alpha_c(t)e_c(t). \tag{80}$$

Differentiating (79) with respect to t, we obtain

$$\frac{de_c(t)}{dt} = (\beta_c^T(t) - q_c^T(t)A^T(t))R^{-1}(y(t) - \hat{Z}(t|t)), \quad e_c(0) = 0. \tag{81}$$

6. EXTENDED ESTIMATION ALGORITHMS FOR WHITE PLUS COLORED OBSERVATION NOISE

Let an observation equation with the nonlinear mechanism be given by

$$y(t) = f(z(t),t) + v_c(t) + v(t), \quad z(t) = Cx(t), \tag{82}$$

for white plus colored observation noise.

Let the state vector $x(\cdot)$, the colored observation noise $v_c(\cdot)$ and the white observation noise $v(\cdot)$ be uncorrelated mutually. Namely, let the functions related with the covariance information, used in [Theorem 4], now be expressed, from (5) as $\alpha(t) = A(t)$, $\beta^T(s) = B^T(s)C^T H^T(s)$, $\varepsilon(t) = B(t)$, $\zeta^T(s) = A^T(s)C^T H^T(s)$. Let the auto-covariance function of $v_c(t)$ be given by $K_c(t,s) = E[v_c(t)v_c^T(s)] = \theta(t)\pi^T(s)$, $0 \le s \le t$. Then, from $K_{cy}(t,s) = \alpha_c(t)\beta_c^T(s)$, $0 \le s \le t$, and the un-correlation property of $v_c(\cdot)$ with $x(\cdot)$ and $v(\cdot)$, it is found that $\alpha_c(t) = \theta(t)$, $\beta_c^T(s) = \pi^T(s)C^T H^T(s)$. Also, $K_Z(t,s)$ is expressed as

$$K_Z(t,s) = H(t)CK_x(t,s)C^T H^T(s) + K_c(t,s) = H(t)CA(t)B^T(s)C^T H^T(s) + \theta(t)\pi^T(s)$$
$$= \Lambda(t)\mathbf{B}^T(s), \quad 0 \le s \le t. \text{ Here, } \Lambda(t) = [H(t)CA(t) \quad \theta(t)], \; \mathbf{B}^T(s) = [H(s)CB(s) \quad \pi(s)]$$

Based on the above statistical assumptions and the linear estimation algorithms in section 5, like the design of the extended recursive Wiener estimators using the covariance information in [Theorem 2], the first-order extended fixed-point smoothing and filtering algorithms for the white plus colored observation noise are summarized in [Theorem 5].

[Theorem 5]
Let the observation equation with the nonlinear mechanism be given by (90) for white plus colored observation noise. Let the state vector $x(\cdot)$ and the colored observation noise $v_c(\cdot)$ and the white observation noise $v(\cdot)$ be uncorrelated mutually. Let the auto-covariance function of the state vector $x(t)$ be given by (5). Let the auto-covariance function of $v_c(t)$ be given by $K_c(t,s) = E[v_c(t)v_c^T(s)] = \theta(t)\pi^T(s)$, $0 \le s \le t$. Then the first-order extended fixed-point smoothing and filtering algorithms of the signal $z(t)$ consist of (83)-(96).

Fixed-point smoothing estimate of the signal $z(t)$ at the fixed point t: $\hat{z}(t|T)$

$$\hat{z}(t|T) = C\hat{x}(t|T) \tag{83}$$

Fixed-point smoothing estimate of the state vector $x(t)$ at the fixed point t: $\hat{x}(t|T)$

$$\frac{\partial \hat{x}(t\mid T)}{\partial T} = \varepsilon(t)(\zeta^T(T) - E^T(T,t)\Lambda^T(t))R^{-1}(y(T) - f(\hat{z}(T\mid T),T) - \hat{v}_c(T\mid T))$$
$$-\alpha(t)D^T(T,t)\Lambda^T(t)R^{-1}(y(T) - f(\hat{z}(T\mid T),T) - \hat{v}_c(T\mid T)) \quad (84)$$

$$\frac{\partial E(T,t)}{\partial T} = J(T,T)(\zeta(T) - \Lambda(T)E(T,t)), \quad E(t,t) = 0 \quad (85)$$

$$\frac{\partial D(T,t)}{\partial T} = -J(T,T)\Lambda(T)D(T,t) \quad (86)$$

$$\frac{dD(t,t)}{dt} = J(t,t)(B^T(t)C^T H^T(t) - \Lambda(t)D(t,t)), \quad D(0,0) = 0 \quad (87)$$

Filtering estimate of the signal $z(t)$:

$$\hat{z}(t\mid t) = C\hat{x}(t\mid t) \quad (88)$$

Filtering estimate of the state vector $x(t)$: $\hat{x}(t\mid t)$

$$\frac{d\hat{x}(t\mid t)}{dt} = F\hat{x}(t\mid t) + (A(t)B^T(t)C^T H^T(t) - S(t))R^{-1}(y(t) - f(\hat{z}(t\mid t),t) - \hat{v}_c(t\mid t)),$$
$$\hat{x}(0\mid 0) = 0 \quad (89)$$

Filtering estimate of the colored noise $v_c(t)$: $\hat{v}_c(t\mid t)$

$$\hat{v}_c(t\mid t) = \theta(t)e_c(t) \quad (90)$$

$$S(t) = A(t)q^T(t)\Lambda^T(t). \quad (91)$$

$$\frac{dq(t)}{dt} = J(t,t)(H(t)C\pi(t) - \Lambda(t)q(t)), \quad q(0) = 0 \quad (92)$$

$$J(t,t) = (\mathrm{B}^T(t) - r(t)\Lambda^T(t))R^{-1} \quad (93)$$

$$\frac{dr(t)}{dt} = J(t,t)(\mathrm{B}(t) - \Lambda(t)r(t)), \quad r(0) = 0 \quad (94)$$

$$\frac{de_c(t)}{dt} = (\pi^T(t)C^T H^T(t) - q_c^T(t)\Lambda^T(t))R^{-1}(y(t) - f(\hat{z}(t|t),t) - \hat{v}_c(t|t)),$$
$$e_c(0) = 0 \qquad (95)$$

$$\frac{dq_c(t)}{dt} = J(t,t)(H(t)C\pi(t) - \Lambda(t)q_c(t)), \quad q_c(0) = 0 \qquad (96)$$

Here, $\Lambda(t) = [H(t)CA(t) \quad \theta(t)]$, $\mathbf{B}^T(s) = [H(s)CB(s) \quad \pi(s)]$.
The function $H(t)$, as in the first-order extended estimators in [Theorem 2], is given by

$$H(t) = \left.\frac{\partial f(z(t),t)}{\partial z(t)}\right|_{z(t)=\hat{z}(t|t)}$$

after expanding the nonlinear observation function in a first-order Taylor series about $\hat{z}(t|t)$.

7. NUMERICAL SIMULATION EXAMPLES

7-1. Nonlinear Observation Mechanism for White Observation Noise

Let a scalar observation equation with the nonlinear mechanism be given by

$$y(t) = f(z(t),t) + v(t), \quad z(t) = Cx(t),$$

$$f(z(t),t) = \cos(2\pi f_c t + m_A z(t)), \quad f_c = 10,000(Hz), \quad m_A = 1.2. \qquad (97)$$

The nonlinear function in (97) appears on the phase modulation in analogue communication systems [25]. Here, f_c and m_A represent the carrier frequency and the phase sensitivity respectively. The observation function is given by

$$H(t) = \left.\frac{\partial f(z(t),t)}{\partial z(t)}\right|_{z(t)=\hat{z}(t|t)} = -m_A \sin(2\pi f_c t + m_A \hat{z}(t|t)). \qquad (98)$$

Let $v(t)$ be white Gaussian observation noise having the mean zero and the variance R, which is expressed by $N(0, R)$.

Let the signal $z(t)$ be expressed in terms of the state vector $x(t)$, which consists of the state variables $x_1(t) = z(t)$, $x_2(t) = \dfrac{dz(t)}{dt}$, as

$$z(t) = Cx(t), \ x(t) = [x_1(t) \ x_2(t)]^T, \ z(t) = x_1(t), \ C = [1 \ 0]. \tag{99}$$

Let the linear state differential equations for the state vector $x(t)$ be given by

$$\frac{dx(t)}{dt} = Fx(t) + w(t), \ F = \begin{bmatrix} 0 & 1 \\ -\omega_n^2 & -2\zeta\omega_n \end{bmatrix}, \ w(t) = \begin{bmatrix} 1 \\ \omega_n^2 \end{bmatrix} u(t),$$

$$E[u(t)u(s)] = \delta(t-s). \tag{100}$$

The transfer function $N(s)$ from $u(t)$ to $z(t)$ is given by $N(s) = \dfrac{\omega_n^2}{s^2 + 2\zeta\omega_n s + \omega_n^2}$, where ζ is the damping ratio and ω_n is the resonant frequency. Here, $\zeta = 0.7$, $\omega_n = 16$.

The auto-covariance function $K(t,s)$ of the state vector $x(t)$ is given by

$$K(t,s) = e^{t-s} K(s,s), \ K(s,s) = \begin{bmatrix} \dfrac{1 + \omega_n^2 + 4\zeta^2 + 4\zeta\omega_n}{4\zeta\omega_n} & -0.5 \\ -0.5 & \dfrac{\omega_n + \omega_n^3}{4\zeta} \end{bmatrix}. \tag{101}$$

The auto-covariance function $K_z(t,s)$ of the signal $z(t)$ is given by $K_z(t,s) = CK(t,s)C^T$.

By substituting F, C and $K(s,s) = K(0)$ into the extended recursive Wiener estimation algorithms of [Theorem 2], the fixed-point smoothing estimate $\hat{z}(t|T)$ at the fixed point t and the filtering estimate $\hat{z}(t|t)$ of the signal are calculated recursively.

Figure 1 illustrates the signal $z(t)$ and the fixed-point smoothing estimate $\hat{z}(t|t+0.005)$ vs. t, $0 \le t \le 1.3$, for the observation noise $N(0, 0.3^2)$.

Figure 2 depicts the mean-square values (MSVs) of the filtering error $z(t) - \hat{z}(t|t)$ and the fixed-point smoothing error $z(t) - \hat{z}(t|t+L)$, $T = t + L$, by the first-order extended recursive Wiener estimators in [Theorem 2] and the first-order extended Kalman estimators respectively vs. L, $\Delta \le L \le 10\Delta$, for the observation noises $N(0, 0.3^2)$, $N(0, 0.5^2)$, $N(0, 0.7^2)$ and $N(0,1)$.

Figure 3 illustrates the MSVs of the filtering error $z(t) - \hat{z}(t|t)$ and the fixed-point smoothing error $z(t) - \hat{z}(t|t+L)$ by the second-order extended recursive Wiener estimators in [Theorem 3] and the second-order extended Kalman estimators respectively vs. L, $\Delta \le L \le 10\Delta$, for the observation noises $N(0, 0.3^2)$, $N(0, 0.5^2)$, $N(0, 0.7^2)$ and $N(0,1)$.

In Figure 3, the initial value of the filtering error variance is $\begin{bmatrix} 5 & 2 \\ 2 & 1 \end{bmatrix}$ for the first-order and second-order extended Kalman filters.

The MSV of the filtering error $z(t) - \hat{z}(t\,|\,t)$ corresponds to the case of $L = 0$ in Figure 2 and Figure 3. As the value of the variance of the observation noise becomes large, there is a tendency that the MSVs of the fixed-point smoothing and filtering errors become large. As the value of L becomes large, the MSV of the smoothing errors becomes gradually small except the graphs (b), (c) and (d) in Figure 3. The MSV of the fixed-point smoothing errors is less than that of the filtering errors in these figures except graphs (b), (c) and (d) in Figure 3. From Figure 2 and Figure 3, it is seen that the estimation accuracy is feasible in the order of the first-order extended recursive Wiener estimators, the first-order extended Kalman estimators, the second-order extended Kalman estimators and the second-order extended recursive Wiener estimators.

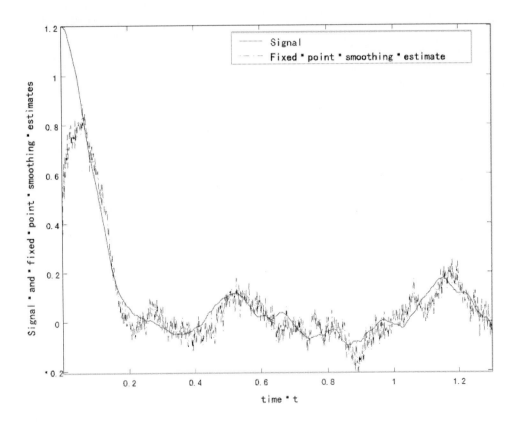

Figure 1. Signal $z(t)$ and the fixed-point smoothing estimate $\hat{z}(t\,|\,t+0.005)$ vs. t, $0 \le t \le 1.3$, for the observation noise $N(0, 0.3^2)$.

By the way, the MSVs of the filtering errors by the unscented filter [18] are 0.2015966131, 1.3534898746 and 21.9927418234 for the observation noises $N(0, 0.1^2)$,

$N(0,0.3^2)$, and $N(0,0.5^2)$ respectively. The unscented filter, in this simulation example, is inferior in estimation accuracy to the extended recursive Wiener filter of [Theorem 2] and [Theorem 3].

Here, the MSVs of the filtering and fixed-point smoothing errors are calculated by $\dfrac{\sum_{i=1}^{1300}(z(i)-\hat{z}(i,i))^2}{1300}$ and $\dfrac{\sum_{i=1}^{1300}\sum_{j=1}^{L}(z(i)-\hat{z}(i,i+j))^2}{1300\cdot L}$ respectively.

As a numerical integration method of the differential equations, the fourth-order Runge-Kutta method with the step-size $\Delta = 0.001$ is used.

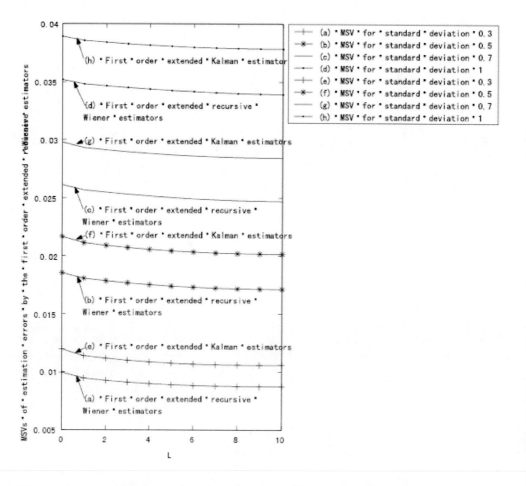

Figure 2. Mean-square values of the filtering error $z(t)-\hat{z}(t\mid t)$ and the fixed-point smoothing error $z(t)-\hat{z}(t\mid t+L)$ by the first-order extended recursive Wiener estimators in [Theorem 2] and the first-order extended Kalman estimators respectively vs. L, $\Delta \leq L \leq 10\Delta$, for the observation noises $N(0,0.3^2)$, $N(0,0.5^2)$, $N(0,0.7^2)$ and $N(0,1)$.

Design of Extended Recursive Wiener Fixed-Point Smoother... 219

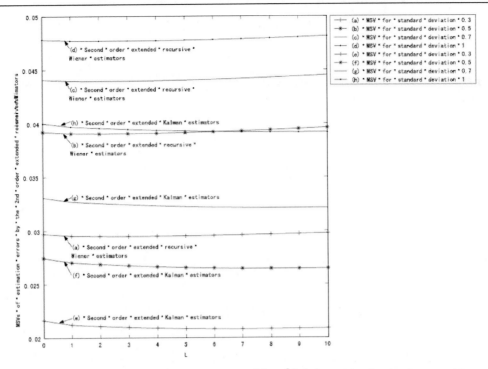

Figure 3. Mean-square values of the filtering error $z(t) - \hat{z}(t \mid t)$ and the fixed-point smoothing error $z(t) - \hat{z}(t \mid t + L)$ by the second-order extended recursive Wiener estimators in [Theorem 3] and the second-order extended Kalman estimators respectively vs. L, $\Delta \leq L \leq 10\Delta$, for the observation noises $N(0, 0.3^2)$, $N(0, 0.5^2)$, $N(0, 0.7^2)$ and $N(0,1)$.

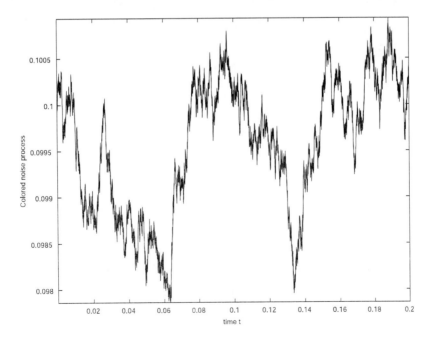

Figure 4. Colored observation noise process $v_c(t)$ vs. t, $0 \leq t \leq 0.2$.

7.2. Nonlinear Observation Mechanism for White Plus Colored Observation Noise

Let a scalar observation equation with the nonlinear mechanism for white plus colored observation noise be given by

$$y(t) = f(z(t),t) + v_c(t) + v(t), \ z(t) = Cx(t),$$
$$f(z(t),t) = \cos(2\pi f_c t + m_A z(t)), \ f_c = 10{,}000(Hz), \ m_A = 1.2. \quad (102)$$

Let $v(t)$ be white Gaussian observation noise $N(0,R)$.
Let the signal $z(t)$ be expressed in terms of the state variable $x(t)$ as

$$z(t) = Cx(t), \ C = 1. \quad (103)$$

Let the state differential equation for the state variable $x(t)$ be given by

$$\frac{dx(t)}{dt} = Fx(t) + u(t), \ F = -k,$$
$$E[u(t)u(s)] = 2kP\delta(t-s), \ E[x^2(0)] = P, \ k = 10, \ P = 10. \quad (104)$$

The auto-covariance function of the signal $z(t)$ is given by

$$K_z(t,s) = K_x(t,s) = Pe^{-k(t-s)} \ [27] \text{ for } 0 \leq s \leq t. \text{ Thus, } A(t) = Pe^{-kt}, \ B(t) = e^{kt}.$$

Also, let the colored noise process be a Wiener process, which satisfies

$$\frac{dv_c(t)}{dt} = w(t), \ E[w(t)w(s)] = \sigma^2\delta(t-s), \ E[v_c^2(0)] = 0, \ \sigma^2 = 5. \quad (105)$$

The autocovariance function of the colored noise process is given by

$$K_c(t,s) = \theta(t)\pi(s) = \sigma^2 s, \ 0 \leq s \leq t \ [27]. \text{ Hence, } \theta(t) = \sigma^2, \ \pi(t) = t.$$

By substituting C, $A(t)$, $B(t)$, $\theta(t)$, $\pi(t)$, $H(t)$ and F into the first-order extended estimation algorithms of [Theorem 5], the fixed-point smoothing estimate $\hat{z}(t|T)$ at the fixed point t and the filtering estimate $\hat{z}(t|t)$ of the signal are calculated recursively.

Figure 4 illustrates the colored observation noise process $v_c(t)$ vs. t, $0 \leq t \leq 0.2$. Figure 5 illustrates the signal $z(t)$ and the fixed-point smoothing estimate

$\hat{z}(t|t+0.00005)$ vs. t, $0 \leq t \leq 0.2$, for the observation noise $N(0,0.3^2)$. From Figure 5, it is seen that the fixed-point smoothing estimate approaches the signal as time passes.

Figure 6 depicts the MSVs of the fixed-point smoothing error $z(t) - \hat{z}(t|t+L)$ by the first-order extended estimators in [Theorem 5] vs. L, $\Delta \leq L \leq 10\Delta$, for the observation noises $N(0,0.3^2)$, $N(0,0.4^2)$ and $N(0,0.5^2)$. Here, the MSVs of the filtering and fixed-point smoothing errors are calculated by $\dfrac{\sum\limits_{i=1}^{20000}(z(i)-\hat{z}(i,i))^2}{20000}$ and $\dfrac{\sum\limits_{i=1}^{20000}\sum\limits_{j=1}^{L}(z(i)-\hat{z}(i,i+j))^2}{20000 \cdot L}$ respectively.

From Figure 6, as the Lag L becomes large, the rate of the decrease of the MSV is very small for each white observation noise.

Here, as a numerical integration method of the differential equations, the fourth-order Runge-Kutta method with the step-size $\Delta = 0.00001$ is used.

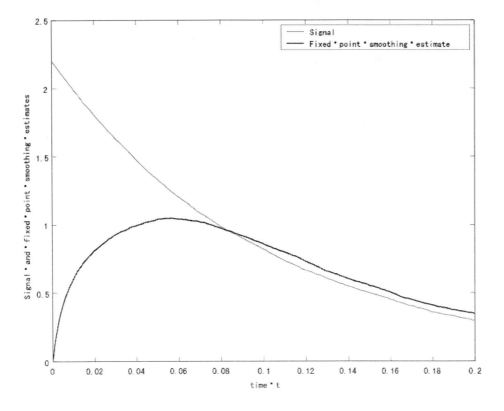

Figure 5. Signal $z(t)$ and the fixed-point smoothing estimate $\hat{z}(t|t+0.00005)$ vs. t, $0 \leq t \leq 0.2$, for the observation noise $N(0,0.3^2)$.

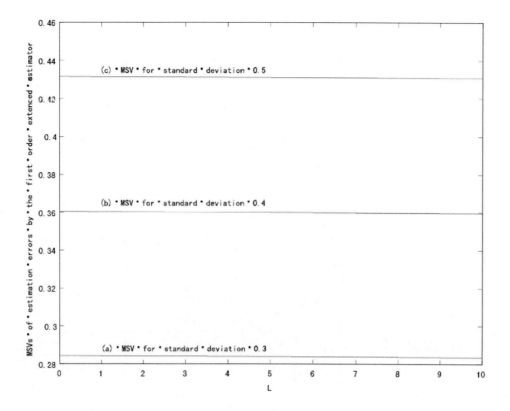

Figure 6. MSVs of the filtering error $z(t) - \hat{z}(t|t)$ and the fixed-point smoothing error $z(t) - \hat{z}(t|t+L)$ by the first-order extended estimators in [Theorem 5] vs. L, $\Delta \leq L \leq 10\Delta$, for the observation noises $N(0,0.3^2)$, $N(0,0.4^2)$ and $N(0,0.5^2)$.

CONCLUSIONS

The numerical simulation example has shown the validity of the extended recursive Wiener fixed-point smoothing and filtering algorithms in [Theorem 2] and [Theorem 3] for white observation noise. In the numerical simulation example, it is indicated that the estimation accuracy of the first-order extended recursive Wiener estimators is preferable to the first-order extended Kalman estimators, the second-order extended Kalman estimators and the second-order extended recursive Wiener estimators. Also, the simulation result has shown that the first-order extended fixed-point smoother and filter in [Theorem 5] are feasible for white plus colored observation noise.

Finally, from the nature of the extended estimators, the recursive extended estimators proposed in this paper might be applicable to the estimation problem with the nonlinear observation mechanism, which changes with time smoothly, in continuous-time stochastic systems.

REFERENCES

[1] Van Trees, H. L. (1968). *Detection, Estimation, and Modulation Theory, Part 1*, New York: Wiley.

[2] Nakamori, S. (1991). Design of a fixed-point smoother based on an innovations theory for white gaussian plus coloured observation noise, *Int. J. Systems Science, Vol. 22*, 2573-2584.

[3] Kailath, T. (1976). *Lectures on Linear Lest-Squares Estimation*, New York: Springer-Verlag.

[4] Nakamori, S. (1997). Estimation technique using covariance information with uncertain observations in linear discrete-time systems, *Signal Processing, Vol.58*, 309-317.

[5] Nakamori, S. (1991). Design of a fixed-point smoother based on an innovations theory for white gaussian plus coloured observation noise, *Int. J. Systems Science, Vol. 22*, 2573-2584.

[6] Kailath, T. (1974). A view of three decades of linear filtering theory, *IEEE Trans. Information Theory, Vol. IT-20*, 146-181.

[7] Nakamori, S. (1995). Estimation technique using covariance information in linear discrete-time systems, *Signal Processing, Vol.43*, 169-179.

[8] Nakamori, S., Caballero-'Aguilla, R., Hermoso-Carazo, A. & Linares-P'erez, A. (1995). Second-order polynomial estimators from non-independent uncertain observations using covariance information, IEICE Trans. Fundamentals of Electronics, *Communications and Computer Sciences, Vol. E86*-A, 1240-1248.

[9] Nakamori, S. (1999). Design of estimators using covariance information in discrete-time stochastic systems with nonlinear observation mechanism, IEICE Trans. Fundamentals of Electronics, *Communications and Computer Sciences, Vol.E82*-A, 1292-1304.

[10] Nakamori, S. (1998). Design of filter using covariance information in continuous-time stochastic systems with nonlinear observation mechanism, IEICE Trans. Fundamentals of Electronics, *Communications and Computer Sciences, Vol. E81*-A, 904-912.

[11] Nakamori, S. (1998). Design of predictor using covariance information in continuous-time stochastic systems with nonlinear observation mechanism, *Signal Processing, Vol.68*, 183-193.

[12] Nakamori, S. (1981). Linear Continuous Sequential Estimators using Covariance Information and its Applications to Estimation Problems of Air Pollution Levels, *Dissertation for the Degree of Doctor of Engineering*, Kyoto University.

[13] Sage, A. P. & Melsa, J. L. (1971). *Estimation Theory with Applications to Communications and Control*, New York: McGraw-Hill.

[14] Hoon, L. T., Sang, R. W., Hee, J. S., Sung, Y. T. & Bae, P. J. (2004). Robust extended Kalman filtering via Krein space estimation, IEICE Transactions on Fundamentals of Electronics, *Communications and Computer Sciences, Vol.E87*-A, 243-250.

[15] Yeh, H. D. & Huang, Y. C. (2005). Parameter estimation for leaky aquifers using the extended Kalman filter, and considering model and data measurement uncertainties, *Journal of Hydrology, Vol.302*, 28-45.

[16] Wang, Y. & Papageorgiou, M. (2005). Real-time freeway traffic state estimation based on extended Kalman filter: a general approach, *Transportation Research Part B*: *Methodological, 39*, 141-167.

[17] Daum, F. (2005). Nonlinear filters: beyond the Kalman filter, *IEEE A&E Systems Magazine, Vol.20*, 57-69.

[18] Julier, S., Uhlmann, J. & Durrant-Whyte, H. F. (2000). A new method for the nonlinear transformation of means and covariances in filters and estimators, IEEE Trans. *Automatic Control, Vol.45*, 477-482..

[19] Ito, K. & Xiong, K. (2000). Gaussian filters for nonlinear filtering problems, IEEE Trans. *Automatic Control, Vol.45*, 910-927.

[20] Daum, F. E. (1988). New exact nonlinear filters. In @Spall, J. C. editor, Bayesian *Analysis of Time Series and Dynamic Models.* New York: Marcel Dekker, ch 8 (199-226).

[21] Schmit, G. C. (1993). Designing nonlinear filters based on Daum's theory, Journal of Guidance, *Control, and Dynamics, Vol.16*, 371-376.

[22] Nakamori, S. (1995). Recursive estimation technique of signal from output measurement data in linear discrete-time systems, IEICE Trans. Fundamentals of Electronics, *Communications and Computer Sciences, Vol.E82*-A, 600-607.

[23] Nakamori, S. (1997). Reduced-order estimation technique using covariance data of observed value in linear discrete-time systems, *Digital Signal Processing, Vol.7*, 55-62.

[24] Nakamori, S. (1996). Design of recursive Wiener smoother given covariance information, IEICE Trans. Fundamentals of Electronics, *Communications and Computer Sciences, Vol.E79*-A, 864-872.

[25] Blahut, R. E. (1990). *Digital Transmission of Information*, MA: Addison-Wesley Publishing Company.

[26] Nakamori, S. & Sugisaka, M. (1977) Initial-value system for linear smoothing problems by covariance information, *Automatica, Vol.13*, 623-627.

[27] Baggeroer, B. (1970). State Variables and Communication Theory. *Research Monograph No.61,* New York: MIT Press.

In: Kalman Filtering
Editor: Joaqun M. Gomez

ISBN: 978-1-61761-462-0
© 2011 Nova Science Publishers, Inc.

Chapter 8

KALMAN FILTERING APPROACH TO BLIND SEPARATION OF INDEPENDENT SOURCE COMPONENTS

Andreas Galka[1]*and Tohru Ozaki*[2]
[1]Department of Neuropediatrics, University of Kiel,
Schwanenweg 20, 24105 Kiel, Germany
[2]Tohoku University, 28 Kawauchi,
Aoba-ku, Sendai 980-8576, Japan

Abstract

The problem of extracting a set of independent components from given multivariate time series data under the assumption of linear instantaneous mixing can be addressed within a Kalman filtering framework. For this purpose, we introduce a new class of state space models, the Independent Components State Space Model (IC-SSM). The resulting algorithm has several attractive features: It takes temporal correlations within the data into account; it allows for the presence of observation noise; it can deal with both gaussian and non-gaussian source distributions; it provides a representation for the main frequencies present in the data; and it succeeds in distinguishing between dependencies which are introduced by the mixing step, and dependencies which represent coincidental dependencies resulting from finite time series length. Analysis by fitting IC-SSM is compared to five well-known algorithms for Independent Component Analysis (ICA). Through simulations we show that the ICA algorithms, in most cases, produce estimates of the source components for which the residual mutual information is too small, as compared to the correct value; this problem does not arise for IC-SSM.

1. Introduction

The task of analyzing multivariate data sets may be approached by explicitly specifying the probability distribution of all variables involved, including observed and unobserved variables; unobserved variables are also known as *latent variables* (Bartholomew, 1987; West *et*

---

[*]E-mail address: andreas.galka@googlemail.com

al., 1999). Perhaps the simplest possible latent variable model is given by a separable distribution, corresponding to the latent variables being mutually independent (Mackay, 2003). The problem of extracting independent source components from multivariate time series represents an example of *Blind Signal Separation* (BSS). Recent experience has shown that modeling multivariate data sets by sets of underlying independent components represents a useful approach to data analysis in numerous fields, such as geophysics, neuroscience, telecommunications and audio signal processing (Hyvärinen *et al.*, 2001).

Independent Component Analysis (ICA) has been introduced in the 1990s (Jutten & Hérault, 1991; Comon, 1994) as a new approach to analysing multivariate data sets, but latent variable models have a much longer history, including classical statistical methods such as Principal Component Analysis (PCA) and Factor Analysis (FA) (Basilevsky, 1994). When applied to time series data, these methods ignore the temporal ordering of the data; however, a number of authors has proposed generalizations of PCA, FA and ICA based on inclusion of temporal information. The probably earliest work to be mentioned is Molenaar's *Dynamic Factor Analysis* (Molenaar, 1985); later algorithms such as *Dynamic Component Analysis* (Attias & Schreiner, 1998), *Independent Factor Analysis* (Attias, 1999) and *Temporal Factor Analysis* (Xu, 2000; Cheung & Xu, 2003a; Cheung & Xu, 2003b; Cheung, 2006) were proposed. The resulting algorithms belong to a class of algorithms which bear close similarity to classical linear *State Space modeling* (SSM) (Aoki, 1987; Durbin & Koopman, 2001). It is well known that also Autoregressive (AR) models and Autoregressive Moving-Average (ARMA) models for time series prediction (Box & Jenkins, 1976) can be rewritten as state space models.

In this chapter we discuss a new class of state space models for extracting independent source components from time series data, thereby reinterpreting and generalizing the earlier work mentioned above. We will put particular emphasis on the aspect of *whitening* the data through prediction: the data are mapped to prediction errors, also known as *innovations* (Kailath, 1968), which have, for optimum prediction, a white power spectrum. The innovation approach to time series modeling provides a crucial link between parametric predictive modeling and maximum-likelihood estimation (Galka *et al.*, 2006).

It is typically assumed by many PCA, FA and ICA algorithms that the data were generated by linear instantaneous mixing of the underlying source components; neither the source components nor the mixing matrix are known, but it is assumed that the source components were uncorrelated (FA) or even independent (ICA). Independent source components are uncorrelated by definition.

Independent source components can be correctly reconstructed, except for residual indeterminacies with respect to scaling, polarity and ordering, provided they fulfill certain additional assumptions. For ICA, the most commonly employed assumption is given by the assumption of non-gaussian (i.e., non-normal) distributions of the source components; this assumption is employed in the original ICA algorithm (Comon, 1994). At most one component with gaussian distribution is permitted; PCA and FA, on the contrary, assume gaussian distributions for all source components.

As an alternative (or addition) to the assumption of non-gaussianity, it may be assumed that the source components have mutually different temporal second-order auto-correlations (Cardoso, 2001). A third alternative is given by the assumption that the variances of the source components are smoothly changing with time, corresponding to nonstationary behavior (Cardoso, 2001).

When reviewing these three alternatives, non-gaussian distributions, mutually different auto-correlations and nonstationary variances, we note that any ICA algorithm based solely on the assumption of non-gaussianity will ignore the temporal ordering of the data; we therefore call such algorithms *static* ICA algorithms. However, it can be easily seen that, potentially, the vast majority of information contained in a multivariate time series is given by correlations between the data vectors at different time points, therefore ignoring the temporal ordering corresponds to discarding the majority of available information; only if the true source components would consist of pure white noise, there would be no meaningful temporal correlations to be lost. This consideration leads us to prefer algorithms based on temporal second-order, and/or higher-order, auto-correlations, such as AMUSE (Tong et al., 1991; Molgedey & Schuster, 1994), SOBI (Belouchrani et al., 1997) and TDSEP (Ziehe & Müller, 1998).

The presence of temporal correlations in multivariate time series, both auto-correlations and cross-correlations, renders it possible to predict future data. Predictive (or, equivalently, causal) modeling represents the core concept of state space modeling. AR and ARMA models are well-known as classic linear models for time series prediction, both for the univariate and the multivariate case. Second-order correlations may be described by the autocovariance function, i.e., by a set of covariance matrices (for time lags $\tau = 0, 1, 2, \ldots$); the parameters of AR models represent an alternative parametrization of these correlations. The close relationship between autocovariance matrices and AR models is well known (Box & Jenkins, 1976).

With respect to the description of temporal correlations, ARMA models and state space models represent alternatives to AR models and to the set of covariance matrices; they aim at providing the same or even better predictive performance by using a considerably smaller set of model parameters. The task of keeping the number of model parameters as low as possible is known as the principle of *parsimony*.

The structure of this chapter is as follows. In section 2 we will discuss ICA and state space modeling within the framework of linear observation models, then the state space approach to blind separation of independent source components will be derived, and the practical fitting of these models will be discussed. Some more technical aspects will be defered to an Appendix. In section 3 we will present two simulation studies, including detailed comparisons between state space modeling and five well-known ICA algorithms. Further discussion and conclusions are provided in section 4; there we will also discuss the actual meaning of extracting independent source components from real-world data.

2. Models for Separating Independent Source Components

2.1. Linear Instantaneous Observation Models

Let the multivariate time series be denoted by $\mathbf{y}(t) = \bigl(y_1(t), \ldots, y_N(t)\bigr)^\dagger$, $t = 1, \ldots, T$, where N denotes the dimension of the data vectors and T the length of the time series, i.e., the number of time points at which the data were sampled; the superscript $\dagger$ denotes matrix transposition.

A widely employed observation model for multivariate data is given by a linear instantaneous *observation equation*

$$\mathbf{y}(t) = \mathsf{C}\mathbf{x}(t) + \boldsymbol{\epsilon}(t) , \qquad (1)$$

where $\mathbf{x}(t) = \bigl(x_1(t), \ldots, x_M(t)\bigr)^\dagger$, $t = 1, \ldots, T$, denotes an input signal, given by a time series of not directly observed M-dimensional vectors, C denotes a constant $(N \times M)$-dimensional *observation matrix* or *mixing matrix*, and $\boldsymbol{\epsilon}(t) = \bigl(\epsilon_1(t), \ldots, \epsilon_N(t)\bigr)^\dagger$, $t = 1, \ldots, T$, denotes an observation noise component, given by a time series of not directly observed N-dimensional vectors.

We shall assume that $\mathbf{y}(t)$, $\mathbf{x}(t)$ and $\boldsymbol{\epsilon}(t)$ have zero mean with respect to time t. Let S_x and S_ϵ denote the covariance matrices of $\mathbf{x}(t)$ and $\boldsymbol{\epsilon}(t)$, respectively.

Both Independent Component Analysis (ICA) (Hyvärinen *et al.*, 2001) and Factor Analysis (FA) (Bartholomew, 1987) regard the components of $\mathbf{x}(t)$ as the source components (or *independent components*, or *common factors*) to be estimated from given data $\mathbf{y}(t)$, while the elements of C, S_x and S_ϵ represent the model parameters.

2.2. Independent Component Analysis (ICA)

Starting from the linear instantaneous observation model of Eq. (1), a typical set of model assumptions of ICA algorithms is given by (Hyvärinen *et al.*, 2001):

- $M = N$ (mixing matrix is square)

- the probability distributions $p(x_i)$ of the independent components x_i are *non-gaussian* (except for at most one component)

- $\mathsf{S}_x = \mathsf{I}_M$ (uncorrelated components, standardized to zero-mean and unit-variance)

- $p(x_1 \ldots, x_M) = \prod_{i=1}^{M} p(x_i)$ (independence of components)

- $\mathsf{S}_\epsilon = 0$ (observation noise is negligible or absent)

Numerous ICA algorithms have been proposed so far; in this chapter we select five algorithms, namely *extended InfoMax* (Bell & Sejnowski, 1995; Lee *et al.*, 1999), *FastICA* (Hyvärinen, 1999), *JADE* (Cardoso, 1999), *MILCA* (Stögbauer *et al.*, 2004) and *TDSEP* (Ziehe & Müller, 1998). The first four of these algorithms are static algorithms, ignoring the temporal ordering of the data; only TDSEP is a non-static algorithm, since it is based on joint diagonalization of instantaneous and delayed covariance matrices. TDSEP thereby represents an example of an algorithm based on the assumption that the temporal autocorrelations of the source components are mutually different; other examples are AMUSE (Tong *et al.*, 1991; Molgedey & Schuster, 1994) and its generalization, SOBI (Belouchrani *et al.*, 1997). We remark that a non-static version of MILCA, employing time-delay embedding vectors, has been proposed (Stögbauer *et al.*, 2004). The acronym "MILCA" stands for *Mutual Information Least-dependent Component Analysis*, reminding us of the fact that in practice we cannot expect the separated components to be perfectly independent, corresponding to a mutual information of zero, but only to be least dependent.

2.3. State Space Modeling (SSM)

As mentioned above, state space models are based on time series prediction, therefore they take the temporal ordering of the data into account, including the *direction of time*; i.e., state space modeling is a non-static algorithm. In linear state space modeling, Eq. (1) serves as observation equation, as in ICA; but now the components of the vector $\mathbf{x}(t)$ are no longer regarded as input signals, but rather as latent variables or *states*, following an explicit first-order multivariate autoregressive dynamics:

$$\mathbf{x}(t) = \mathbf{A}\mathbf{x}(t-1) + \boldsymbol{\eta}(t) \; ; \tag{2}$$

consequently, $\mathbf{x}(t)$ is called the *state vector* of the model; the dimension of $\mathbf{x}(t)$ is denoted by M. $\boldsymbol{\eta}(t) = (\eta_1(t), \ldots, \eta_M(t))^\dagger$, $t = 1, \ldots, T$, denotes an input signal, given by a time series of unobserved M-dimensional vectors, representing *dynamical noise* (also called *driving noise*), with covariance matrix $\mathbf{S}_\eta$, in contrast to the *observation noise* term $\boldsymbol{\epsilon}(t)$. The $(M \times M)$-dimensional parameter matrix A represents the *state transition matrix* of the dynamics. The elements of the four matrices A, C, $\mathbf{S}_\eta$ and $\mathbf{S}_\epsilon$ form the set of main model parameters of the linear state space model, to be summarized by the parameter vector ϑ. Eq. (2) represents the *state dynamics equation* of the state space model.

For the purpose of "independent component analysis", we propose to identify the components $x_i(t)$ of the state vector, or a subset of these components, with the independent source components to be estimated, while we regard $\boldsymbol{\eta}(t)$ as a gaussian white noise input, driving the dynamics. In this point, our approach differs fundamentally from the work of Zhang & Cichocki (2000) and Waheed & Salem (2005); these authors identify the components of the unobserved input signal $\boldsymbol{\eta}(t)$ with the independent source components, while a driving noise input signal may be separately provided to the state dynamics equation, Eq. (2). Therefore, their approach can be interpreted as employing a linear state space model as a model for a complicated observation process, corresponding to the case known as *convolutive mixing* in the BSS field, while the temporal correlation structure of the independent sources themselves is not explicitly modeled. In contrast, we choose to retain the *instantaneous mixing* observation equation, Eq. (1), while the state dynamics equation, Eq. (2) serves the purpose of explicitly modeling the temporal correlation structure of the independent source components.

Predictive modeling of time series depends on the correlation structure present in the data. In a linear state space model the covariance matrix of the dynamical noise $\mathbf{S}_\eta$ describes instantaneous correlations between the components $x_i(t)$ of the state vector $\mathbf{x}(t)$, while the state transition matrix A describes *delayed* correlations between the components $x_i(t)$; diagonal elements of A describe the auto-correlations of the $x_i(t)$, while off-diagonal elements describe cross-correlations. The correlation structure of the data $\mathbf{y}(t)$ then follows from multiplication with the observation matrix C.

2.4. Independent Components State Space Model (IC-SSM)

Eqs. (1) and (2) describe a general linear state space model; by choosing a specific structure for the parameter matrices A, C, $\mathbf{S}_\eta$ and $\mathbf{S}_\epsilon$ we may design models for specific purposes. If the purpose is separation of independent (or "least-dependent") components, we should choose a block-diagonal structure for A and S_η:

$$A = \begin{pmatrix} A_1 & 0 & \cdots & 0 \\ 0 & A_2 & \cdots & 0 \\ \vdots & \vdots & \ddots & \vdots \\ 0 & 0 & \cdots & A_{N_c} \end{pmatrix}, \quad S_\eta = \begin{pmatrix} S_1 & 0 & \cdots & 0 \\ 0 & S_2 & \cdots & 0 \\ \vdots & \vdots & \ddots & \vdots \\ 0 & 0 & \cdots & S_{N_c} \end{pmatrix} \quad (3)$$

where N_c denotes the number of blocks and thereby also the number of independent components; note that the actual state space dimension M will in general be larger than N_c. A_j and S_j denote $(n_j \times n_j)$-dimensional square matrices, where the n_j, $j = 1, \ldots, N_c$, are a set of "local" state space dimensions, with $\sum_j n_j = M$; we remark that in this chapter only the cases $n_j = 1$ and $n_j = 2$ will be employed. The block-diagonal structure generates a corresponding partition of the state vector

$$\mathbf{x}(t) = \left(\mathbf{x}_1^\dagger(t), \mathbf{x}_2^\dagger(t), \ldots, \mathbf{x}_{N_c}^\dagger(t)\right)^\dagger \quad (4)$$

and of the observation matrix

$$C = (C_1, C_2, \ldots, C_{N_c}) \;. \quad (5)$$

By choosing the block-diagonal structure for the "global" state space model, we describe the data as a superposition of N_c non-interacting "local" state space models, defined by their parameter matrices A_j, C_j and S_j. The local state space models are unable to interact, since both A and S_η were chosen as block-diagonal, such that all parameters connecting two different local state space models are zero; this is a convenient way to implement the independence assumption of ICA indirectly, on the level of the structure of the predictive model, instead directly on the level of the estimated independent source components. Finally, the covariance matrix of the observation noise S_ϵ is chosen as a diagonal matrix with non-zero diagonal elements which also form part of the set of model parameters. The block-diagonal structure proposed here represents a generalization over the *Temporal Factor Analysis* algorithm (Xu, 2000; Cheung, 2006) where the matrices corresponding to A and S_η were defined as diagonal.

For the local state space models we choose a structure corresponding to ARMA$(n_j, n_j - 1)$ models, where n_j is an integer model order; this model order is chosen equal to the local state space dimension introduced above. For a univariate variable $x_j(t)$, an ARMA$(n_j, n_j - 1)$ model would be given by

$$x_j(t) = \sum_{\tau=1}^{n_j} a_j(\tau) x_j(t - \tau) + \sum_{\tau=0}^{n_j-1} b_j(\tau) \eta_j(t - \tau) \;, \quad (6)$$

where $\eta_j(t)$ denotes a driving noise term, assumed to be zero-mean unit-variance gaussian noise. $a_j(\tau)$, $\tau = 1, \ldots, n_j$ and $b_j(\tau)$, $\tau = 0, \ldots, n_j - 1$ denote the sets of AR parameters and MA parameters, respectively; as a scaling convention for the driving noise term, it is advisable to set $b_j(0) = 1$ (see Appendix for details). Among various possible state space representations of ARMA$(n_j, n_j - 1)$ models, we choose the *observer canonical form* model (Franklin et al., 1998):

$$A_j = \begin{pmatrix} a_j(1) & 1 & 0 & \cdots & 0 \\ a_j(2) & 0 & 1 & \cdots & 0 \\ \vdots & \vdots & \vdots & \ddots & \vdots \\ a_j(n_j-1) & 0 & 0 & \cdots & 1 \\ a_j(n_j) & 0 & 0 & \cdots & 0 \end{pmatrix}, \qquad (7)$$

$$S_j = \begin{pmatrix} b_j^2(0) & b_j(0)b_j(1) & \cdots & b_j(0)b_j(n_j-1) \\ b_j(1)b_j(0) & b_j^2(1) & \cdots & b_j(1)b_j(n_j-1) \\ \vdots & \vdots & \ddots & \vdots \\ b_j(n_j-2)b_j(0) & b_j(n_j-2)b_j(1) & \cdots & b_j(n_j-2)b_j(n_j-1) \\ b_j(n_j-1)b_j(0) & b_j(n_j-1)b_j(1) & \cdots & b_j^2(n_j-1) \end{pmatrix}. \qquad (8)$$

Within the observer canonical form state space model, the observation matrix corresponding to Eq. (6) would be given by the $(1 \times n_j)$-dimensional matrix $C_j = (1\ 0\ \cdots\ 0)$; if we intend to employ this local state space model for predicting multivariate time series data within a global state space model, we need to replace this observation matrix by

$$C_j = \begin{pmatrix} c_{1j} & 0 & \cdots & 0 \\ c_{2j} & 0 & \cdots & 0 \\ \vdots & \vdots & \ddots & \vdots \\ c_{Nj} & 0 & \cdots & 0 \end{pmatrix}, \qquad (9)$$

such that for each local state space model j (i.e., for each independent source component) and each data channel i a corresponding scale factor c_{ij} is provided. According to the observer canonical form state space model, the remaining state components of each local state space model remain unobserved, therefore the corresponding elements of the local observation matrix in Eq. (9) are zero. The resulting global state space model will be called the *Independent Components State Space Model* (IC-SSM).

We repeat that the state space model of Eqs. (3 – 9) represents a special case of the general model given by Eqs. (1) and (2); the general model contains sets of non-interacting (or possibly interacting) ARMA models as special cases.

2.5. Comparison between ICA and IC-SSM

We will now provide a brief comparison of the assumptions on which the independent components state space model, as outlined in the previous section, is based, with the assumptions underlying typical ICA algorithms, as listed above in section 2.2.

- Unlike many ICA algorithms, the state space model does not assume the dimension M of the latent variable, or rather the number of independent components N_c, to be equal to the dimension N of the data. In principle, there are no constraints on the relative values of M and N. Whether for given model and given data the latent variables, i.e., the components of the state vector, can be uniquely and correctly reconstructed, does not depend on the relative values of M and N, but it depends on whether the state space model is *observable* (Kailath, 1980). Also for $M > N$, or $N_c > N$, a model may be observable. We will return to this point below.

- Since the state space model is linear, the probability distributions $p(x_i)$ of the independent components x_i are implicitly assumed to be gaussian, in contrast to the non-gaussianity assumption of ICA. But practical experience has shown that also non-gaussian source components can be reconstructed by linear state space models, as will be demonstrated below.

- ICA assumes the covariance matrix of the independent source components to be an identity matrix; linear state space models do not regard the elements of the covariance matrix of the latent variables, S_x, as constant model parameters which could become subject to specific assumptions. However, a similar role is played by the covariance matrix of the driving noise, S_η, which is assumed to be block-diagonal, with a value of one at position $(1, 1)$ of each block. Since these positions represent those state components which are observed, this assumption can be regarded as the assumption of S_η being an identity matrix.

- In state space models there is no direct correspondence to the independence assumption of ICA; however, independence of source components is implemented indirectly through the structure of the matrices A and S_η, as presented above.

- Unlike many ICA algorithms, state space models allow for the presence of observation noise, which is assumed to be uncorrelated between data channels, i.e., S_ϵ is assumed to be a diagonal matrix, as in Factor Analysis. We remark that there exist generalizations of ICA which allow for observation noise, such as *Independent Factor Analysis* (Attias, 1999) and *Probabilistic ICA* (Beckmann & Smith, 2004).

2.6. Estimation of States and of Model Parameters

Fitting a state space model to given data represents a twofold estimation problem: both the unknown states, i.e., the latent variables, and the model parameters need to be estimated. The estimation of states for known model parameters can be done by a forward pass through the time series by a Kalman filter (Kalman, 1960; Sorenson, 1970; Grewal & Andrews, 2001); already Molenaar (1985) has suggested to employ Kalman filtering for *Dynamical Factor Analysis*, and later also Cheung & Xu (2003) have resorted to Kalman filtering.

In addition to the forward pass of the Kalman filter, we propose to perform a backward pass by a smoother, thereby obtaining further improvement of the state estimates. In this chapter we use the Rauch-Tung-Striebel (RTS) smoother (Rauch *et al.*, 1965).

The estimation of the model parameters poses a considerably more difficult and time-consuming task than the estimation of states. Following the well established procedures of state space modeling (Mehra, 1974; Åström, 1980; Otter, 1986; Durbin & Koopman, 2001), we choose a maximum-likelihood approach. In order to compute the likelihood of the data with respect to the parameters of the proposed state space model, as given by Eqs. (1) and (2), we may conveniently employ the *innovation approach*, according to which for *causal* models the likelihood of a time series of prediction errors, or *innovations*, is equal to the likelihood of the corresponding original data (Kailath, 1968; Galka *et al.*, 2006). The predictions are provided by the state space model; if the model is able to remove most correlations from the data, both instantaneous and delayed, the innovations will approximately be distributed as white gaussian noise, therefore we say that the data have been whitened. For the case of a Markov process with continuous dynamics, this claim has been proven rigorously (see theorem 41 of Protter, 1990).

Denoting the innovations by $\boldsymbol{\nu}(t)$ and the complete set of data vectors $\mathbf{y}(1), \ldots, \mathbf{y}(N)$ by $\mathbf{Y}$, the logarithmic likelihood of the data follows as

$$\log L(\mathbf{Y}|\vartheta) = -\frac{1}{2} \sum_{t=1}^{T} \Big(\log |\mathsf{S}_\nu(t)| + \boldsymbol{\nu}^\dagger(t) \mathsf{S}_\nu^{-1}(t) \boldsymbol{\nu}(t) + N \log(2\pi) \Big) , \qquad (10)$$

where $\mathsf{S}_\nu(t)$ denotes the covariance matrix of the innovations, provided by the Kalman filter. The logarithmic likelihood needs to be maximized by suitable numerical optimization procedures or learning rules. In this chapter we employ two standard algorithms, the Broyden-Fletcher-Goldfarb-Shanno (BFGS) quasi-Newton algorithm (Dyrholm, 2007) and the Nelder-Mead simplex algorithm (Baldick, 2006); in most cases the quasi-Newton algorithm is the faster choice, but in situations where it fails due to numerical problems, the simplex algorithm may provide further improvement.

The resulting model provides a decomposition of the data into source components, as estimated by the Kalman filter and, possibly, the RTS smoother. Note that only the forward pass of the Kalman filter is required for computing the likelihood, while the backward pass of the RTS smoother has no effect on the likelihood; during the forward pass the predictions are performed, whereby the original time series is mapped to the innovations time series. Only after final estimates of parameters have been obtained, do we apply the RTS smoother, since it further improves state estimates.

The success of state space modeling by numerical maximization of the logarithmic likelihood depends on choosing suitable initial estimates for the model parameters; the initial estimates should already represent the correlation structure of the data, i.e., its spectral properties. A simple, but efficient approach to finding suitable initial estimates is described in the Appendix.

Furthermore, we have not yet commented on the issue of how the block dimensions n_j and the number of components N_c should be chosen; also this discussion is postponed to the Appendix.

3. Simulations

3.1. Measures of Performance

We demonstrate the separation of independent source components by state space modeling through three simulation studies and compare the results with results obtained by several well-known ICA algorithms. The performance of the algorithms will be quantified by two measures.

First, we estimate the mutual information (MI) of the set of reconstructed source components and compare it with the mutual information of the set of true source components; MI is a standard measure of dependency within a set of time series. Even for the set of true source components, MI will be non-zero; this is an effect resulting from the presence of coincidental dependencies which are due to finite time series length. We estimate MI by the k-nearest-neighbour estimator proposed by Kraskov *et al.* (2004), where $k = 6$; this estimator is convenient especially for estimating MI within sets of more than two time series.

Second, we directly compare reconstructed and true source components by forming the cross-correlation matrix between these two sets of time series. All time series are standardized to zero mean and unit variance for this purpose. For perfect reconstruction, we would expect the cross-correlation matrix to be a permutation matrix with elements $+1$ or -1 at the non-zero positions, corresponding to the unavoidable residual indeterminacies with respect to polarity and ordering. For the actual cross-correlation matrix, the closest permutation matrix with $+1/-1$ elements is determined, and the difference between these two matrices is quantified by the Frobenius norm of the difference matrix. For perfect reconstruction, we would expect this Frobenius norm to be zero.

3.2. AR(1) and AR(2) Processes

We begin by creating independent source components from linear autoregressive processes of first or second order, denoted by AR(1) or AR(2); such processes have been employed by various authors as source processes for simulations (Hyvärinen, 2005).

Design of simulation

First, source time series are generated by six AR(1) processes, given by

$$\xi(t) = a\xi(t-1) + e(t) \, , \tag{11}$$

where for the six processes the parameter a assumes the values $0.70, 0.80, 0.90, 0.75, 0.85, 0.95$, and $e(t)$ is a driving noise term. The first three processes are driven by Laplacian noise, the remaining three by gaussian noise; driving noises are mutually independent, ensuring independent sources.

The length of the simulated time series is chosen as $N = 4096$. The six source time series are individually standardized to unit variance, then they are instantaneously mixed by a random (6×6)-dimensional matrix, drawn from gaussian noise of zero mean and unit variance. 2% of gaussian observation noise is added to each mixture. A set of 100 simulated time series is created in this way, each of which created from different realizations of the driving and observation noise terms and of the mixing matrix.

For the independent source components created from autoregressive processes of second order, AR(2), the process equation is

$$\xi(t) = a_1\xi(t-1) + a_2\xi(t-2) + e(t) \, . \tag{12}$$

Parameters a_1, a_2 are chosen according to the spectral properties of the resulting dynamics, i.e., according to frequency φ and damping coefficient r, as will be explained now. The six AR(2) processes are created with frequencies $\varphi = 0.05, 0.08, 0.12, 0.17, 0.23, 0.3$ (where the Nyquist limit is at $\varphi = 0.5$) and damping coefficients, i.e., moduli of complex eigenvalues, $r = 0.95, 0.75, 0.80, 0.70, 0.85, 0.90$ (where the limit of stable dynamics is at $r = 1.0$), respectively. Parameters then follow from

$$a_1 = 2r\cos(2\pi\varphi) \quad , \quad a_2 = -r^2 \, . \tag{13}$$

Details of driving and observation noises, and of mixing matrices, are the same as for AR(1) processes. Again a set of 100 time series of length $N = 4096$ is created.

Data analysis

We apply extended InfoMax, FastICA, JADE, MILCA, TDSEP and IC-SSM to each simulated data set. For InfoMax, default options of Revision 1.27 of the EEGLab toolbox (http://www.sccn.ucsd.edu/eeglab) are used; note that for a simulation without subgaussian sources the extended version of the InfoMax algorithm is actually not necessary. For FastICA, the "deflation" approach with cubic nonlinearity is chosen, and stabilization is switched on; FastICA code was downloaded from http://www.cis.hut.fi/projects/ica/fastica For MILCA, the rectangular estimator of MI with $k = 6$ nearest neighbours is chosen; MILCA code was downloaded from http://www.klab.caltech.edu/~kraskov/MILCA. For TDSEP, joint diagonalization of covariance matrices for lags 0 and 1 is chosen; including higher lags does not improve results, as should be expected for AR(1) source processes. In the case of AR(2) source processes, covariance matrices for lags 0, 1 and 2 are diagonalized. TDSEP code was downloaded from http://ida.first.fraunhofer.de/~ziehe/download.html.

While for the ICA algorithms the number of independent source components is equal to the data dimension $N = 6$ by definition, for IC-SSM the number of components N_c could be chosen independently of N; but in order to facilitate comparison of results, we choose again $N_c = 6$. The structure of the state space models is chosen as $n_j = 1$, $j = 1, \ldots, 6$, for the six AR(1) source processes, and as $n_j = 2$, $j = 1, \ldots, 6$, for the six AR(2) source processes, corresponding to state space dimensions of $M = 6$ and $M = 12$. Initial parameter estimates are obtained from MAR models of order $p_{MAR} = 7$, as described in the Appendix. The observation noise covariance matrix S_ϵ is initialized with $\exp(-10)I_N$, i.e., with very small values; here, I_N denotes the $(N \times N)$-dimensional identity matrix. In order to reduce the number of model parameters, moving-average terms are not employed. Reconstructed source components are provided by state estimates; state estimates obtained solely by application of the Kalman filter – for optimized models – and state estimates obtained by additional application of the RTS smoother, in addition to the Kalman filter, are separately evaluated. Note that linear state space models implicitly assume all probability distributions to be gaussian, although three of the true sources were driven by Laplacian, i.e., supergaussian, noise.

The resulting sets of reconstructed source components are evaluated by MI and Frobenius norm, as explained above. More precisely, MI is estimated first for each set of true source components; due to finite length, in most cases a non-zero value is obtained. Then for each set of reconstructed source components the corresponding estimate of residual MI is subtracted from this "true" value of MI. For the set of 100 simulated mixed time series, the resulting estimates of MI difference and of Frobenius norm are evaluated by forming histograms; results are shown in Figs. 1 and 2 for AR(1) and AR(2) processes, respectively. In the figures mean values of the distributions of results are also shown by dotted lines and by their numerical values.

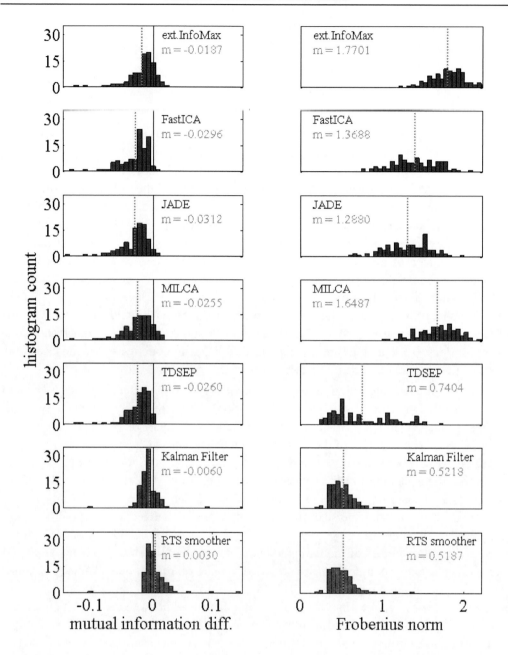

Figure 1. Difference of mutual information between true source components and reconstructed source components (left column of subfigures) and correlation between true source components and reconstructed source components quantified by Frobenius norm (right column of subfigures, see text for details) for true source components created by six AR(1) processes. Histograms refer to a set of 100 simulated mixtures. *Ext.InfoMax*, *FastICA*, *JADE*, and *MILCA* are static ICA algorithms, *TDSEP* is a non-static ICA algorithm, and *Kalman Filter* and *RTS smoother* both refer to the Independent Components State Space modeling (IC-SSM) algorithm.

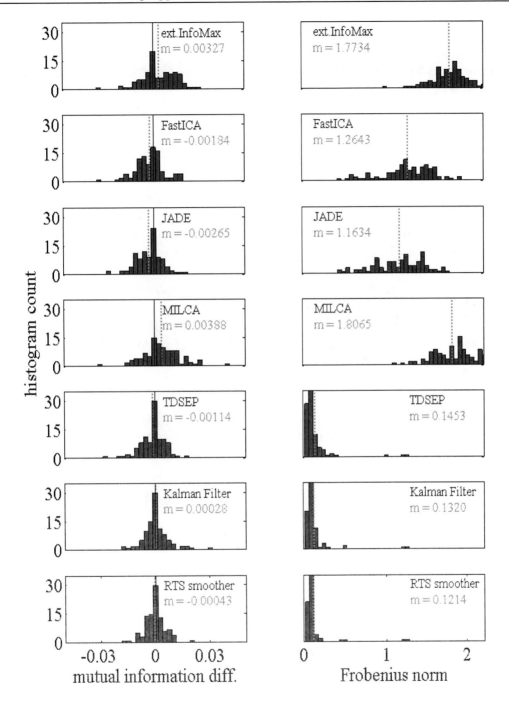

Figure 2. Same as Fig. 1, but with six AR(2) processes instead of six AR(1) processes. Note that the horizontal scale for the left column of subfigures is different from Figs. 1 and 5.

Results

For AR(1) processes, the averaged estimates of MI for the true source components and for the mixed time series are 0.1498 and 2.2142, respectively. Keeping these values in mind, it can be seen from Fig. 1 that ext.InfoMax, FastICA, JADE, MILCA and TDSEP were successful in reducing the residual MI of the estimated source components to very small values; in most cases these values were even smaller than the "true" value for MI, therefore the difference is negative (left side of Fig. 1). On the other hand, IC-SSM provides source components with approximately the correct amount of residual MI; this can be seen for both the Kalman filter and the RTS smoother.

When directly comparing reconstructed and true source components by the Frobenius norm (right side of Fig. 1), it can be seen that TDSEP and IC-SSM achieve considerably lower values than ext.InfoMax, FastICA, JADE and MILCA. We remark that for the design of this simulation a Frobenius norm of about 0.7 or smaller represents an essentially successful separation of all six source components, while for larger values at least two or more source components remain mixed. When assessed with respect to this threshold, about half of the TDSEP results and most of the IC-SSM results represent successful reconstructions.

The almost complete failure of ext.InfoMax, FastICA, JADE, MILCA for this simulation is to be expected, since there are three gaussian source components; these four algorithms are static algorithms and therefore unable to separate more than one gaussian source component.

AR(1) processes represent "weak" dynamics, since they impose only a limited amount of correlation structure onto the driving noise; the smaller the parameter a, the weaker the correlation structure. Considerably more correlation structure can be imposed by AR(2) processes.

For AR(2) processes, the averaged estimates of MI for the true source components and for the mixed time series are 0.0069 and 2.1013, respectively. From Fig. 2, it can be seen that in this case, for all algorithms, differences of MI between reconstructed and true source components are much smaller than in the case of AR(1) processes, possibly corresponding to the much smaller value of MI of the true source components. In this case there is no clear direction of the bias, with ext.InfoMax and MILCA producing too large residual MI, and FastICA and JADE producing too small residual MI. Still, on average IC-SSM comes closest to the true values, for both Kalman filter and RTS smoother. We remark that for very small values of MI the estimator may sometimes become numerically unreliable; occasionally negative values for MI are produced, which have to be set to zero.

The results for the Frobenius norm (right side of Fig. 2) are similar to the case of AR(1) processes. The static algorithms fail again, while TDSEP and IC-SSM demonstrate even better performance than in the AR(1) case, the reason being that AR(2) processes impose stronger correlation structure onto the driving noise, compared to AR(1) processes. In this case, IC-SSM performs only slightly better than TDSEP.

3.3. Deterministic Nonlinear Processes

Generating simulated time series from AR(1) and AR(2) processes represents a favourable situation for analysis by IC-SSM, since this model class is itself based on ARMA models.

We will now choose a different class of processes which deviates considerably from the assumptions underlying linear state space models. For this simulation, true sources will have strongly nonlinear dynamics and overlapping broad-band power spectra, and the crucial question will be, whether linear state space models are still able to separate these sources.

Design of simulation

We choose five nonlinear deterministic processes evolving on fractal attractors:

- Lorenz system
- Thomas cyclically symmetric system
- Hadley circulation system
- Hénon-Heiles system
- Mackey-Glass differential delay system at a delay of 55 time units.

Equations and references for these systems can be found in Appendix A of Sprott (2003); all model parameters are fixed to the "usual parameters", as listed by Sprott (2003). The Mackey-Glass system is numerically integrated by the method used by Ding *et al.* (1993); the other four systems are integrated by the Local Linearization method (Ozaki, 1992). Time discretization is 0.01 time units for Lorenz, Hadley and Hénon-Heiles systems, and 0.1 time units for Thomas and Mackey-Glass systems. After integration, the simulated time series are subsampled by factors of 4 (Lorenz), 12 (Thomas), 20 (Hadley and Mackey-Glass) and 40 (Hénon-Heiles).

In addition to these five nonlinear deterministic processes, a linear stochastic AR(8) process driven by white gaussian noise is employed as another source component process; the model parameters of this process are chosen by fitting an AR(8) model to a human electroencephalogram (EEG) time series displaying pronounced alpha activity, sampled at 256 Hz.

The length of the simulated time series is chosen as $N = 8192$. From the sets of state variables of the Lorenz, Thomas, Hadley and Hénon-Heiles systems, the first variable is chosen (x in the notation used by Sprott (2003)), while the remaining variables are discarded; Mackey-Glass and AR(8) produce univariate time series by definition. The six source time series are standardized individually to unit variance, then they are instantaneously mixed by a random (6×6)-dimensional matrix, drawn from gaussian noise of zero mean and unit variance. 2% of gaussian observation noise is added to each mixture. A set of 100 simulated time series is created in this way, each of which created from different realizations of the driving and observation noise terms and of the mixing matrix; note that only the AR(8) system requires driving noise.

In Fig. 3 a sub-interval from the simulated source components is shown; the corresponding mixtures, for the same sub-interval, are shown in Fig. 4.

The five nonlinear deterministic systems produce time series with broad distribution of power over frequencies; unlike linear AR processes, their dynamics cannot be characterized by a set of constant eigenfrequencies. Furthermore, these time series display sub-gaussian distributions of amplitudes. The AR(8) system represents the only gaussian source; typical static ICA algorithms allow for one gaussian source component.

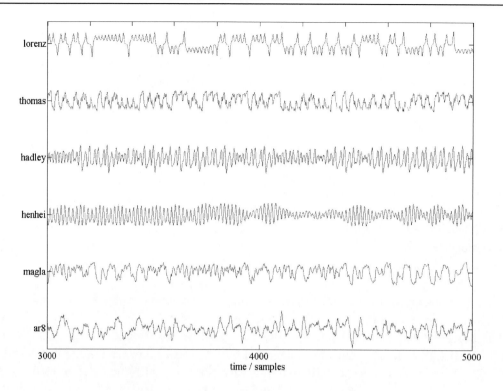

Figure 3. Subset of 2000 samples from true source components created by five deterministic nonlinear processes plus one AR(8) process. The processes are (from top to bottom): Lorenz system, Thomas cyclically symmetric system, Hadley circulation system, Hénon-Heiles system, Mackey-Glass differential delay system and linear stochastic AR(8) system driven by white gaussian noise.

Data analysis

We apply extended InfoMax, FastICA, JADE, MILCA, TDSEP and IC-SSM to each simulated data set. Options are the same as in the case of AR(1) and AR(2) processes, except for TDSEP; for this algorithm we now choose joint diagonalization of covariance matrices for lags $0, 1, 2, \ldots, 50$, since we observe that for this particular data this order of lag is required for optimum performance.

For IC-SSM, the structure of the state space models is again chosen as $n_j = 2$, $j = 1, \ldots, 6$; initial parameters are again obtained from a MAR model of order $p_{MAR} = 7$, and S_ϵ is again initialised with $\exp(-10)I_N$. For this simulation we allow moving-average terms, since we find that on average ARMA(2,1) models provide better performance than pure AR(2) models. The price to pay for this improved performance is increased time for numerical maximization of log-likelihood, due to the increase of the number of model parameters by six additional parameters.

The resulting sets of reconstructed source components are evaluated in the same way as in the case of AR(1) and AR(2) processes; results are shown in Fig. 5.

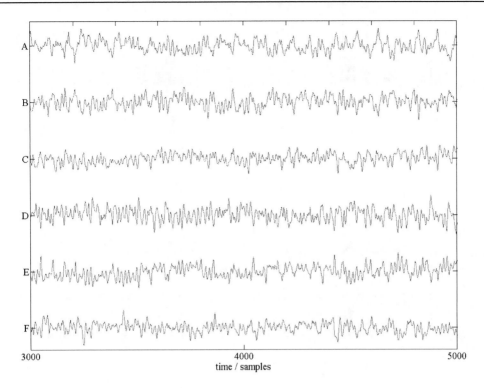

Figure 4. Subset of 2000 samples from a linear instantaneous mixing of the source components shown in Fig. 3.

Results

For this simulation, the averaged estimates of MI for the true source components and for the mixed time series are 0.6474 and 2.7148, respectively. Keeping these values in mind, it can be seen from Fig. 5 that ext.InfoMax, FastICA, JADE, MILCA and TDSEP were again successful in reducing the residual MI of the reconstructed source components to very small values; again in most cases these values were smaller than the "true" value for MI. Since positive values of the MI difference almost never occur, distributions of results become very asymmetric. On the other hand, IC-SSM provides source components with somewhat too large residual MI; this result is more pronounced for the RTS smoother than for the pure Kalman filter.

Direct comparison of reconstructed and true source components by the Frobenius norm (right side of Fig. 5) reveals that in most cases ext.InfoMax, FastICA, JADE, MILCA and TDSEP successfully reconstruct the true source components. There are only a few data sets for which larger values of Frobenius norm are obtained; closer inspection reveals that the data sets with larger Frobenius norm correspond to the tail of larger negative values of MI difference (left side of Fig. 5).

In this simulation IC-SSM performs on average worse than the five ICA algorithms, although for the majority of the simulated data sets the separation of all six source components was still successful, as can be seen from the values of Frobenius norm. The distribution of

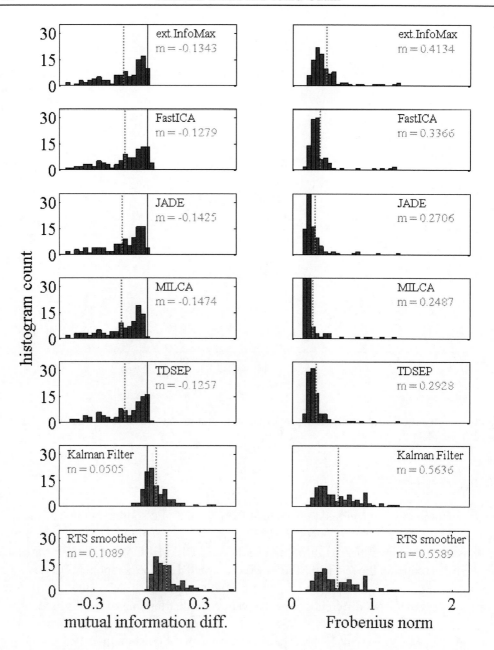

Figure 5. Same as Fig. 1, but with five deterministic nonlinear processes plus one AR(8) process, instead of six AR(1) processes. Note that the horizontal scale for the left column of subfigures is different from Figs. 1 and 2.

Frobenius norm results shows a peak around a value of 0.4, comprising about half of the data sets, while the other half forms a tail at larger values; part of this tail corresponds to partly failed separation of source components. Closer inspection reveals that in this case larger values of Frobenius norm correspond to larger positive values of MI difference.

At first sight, it may seem hopeless to find a useful predictor for nonlinear deterministic time series on the basis of low-order linear ARMA models; the results presented in this section show that the Kalman filter is sufficiently flexible with respect to poor predictive models, such that for the majority of simulated data sets separation of source components becomes possible nevertheless. Also the AR(8) process is sufficiently well represented by one ARMA(2,1) process to allow for separation.

3.4. Discussion of Simulation Results

The results of the simulations presented in this chapter show that state space modeling is capable of separating independent source components from time series, if a suitable model structure is employed, namely the IC-SSM structure. In the case of mixtures of linear stochastic processes of first or second order with gaussian and super-gaussian distributions, we found that static ICA algorithms were unable to accomplish separation; this is the expected result. Furthermore, the separation performance of IC-SSM was better than of TDSEP, which represents a non-static ICA algorithm.

In the case of mixtures of nonlinear deterministic processes with sub-gaussian distributions, plus one linear stochastic process of higher order with gaussian distribution, separation was still successfully performed by IC-SSM for the majority of the data sets, although the performance of TDSEP and of the static ICA algorithms was superior; note that this simulation can be regarded as a "classic" situation for the application of static ICA algorithms. This result confirms that, in principle, linear state space models are capable of unmixing even sources with strongly nonlinear dynamics and overlapping broad-band power spectra.

As a main result of this chapter, we emphasize the behavior of residual MI. Both static ICA algorithms and TDSEP show a clear tendency to provide "over-independent" reconstructed source components, as shown on the left side of Figs. 1 and 5; however, in the case of AR(2) processes residual MI is very small, such that this tendency cannot be observed for some of the algorithms. For all simulations, estimation of MI for the sets of true source components usually produces non-zero positive values, resulting from the presence of coincidental dependencies within finite-length time series; we find that static ICA algorithms and also TDSEP fail to distinguish between dependencies introduced by mixing of source components and coincidental dependencies within the true source components.

Therefore, we would expect that algorithms based on estimating source components by directly minimizing the residual MI within the set of estimated components, such as static ICA algorithms and TDSEP, should display a tendency towards over-independent reconstructed source components. MILCA is an example of an algorithm which directly minimizes an estimate of MI; other algorithms maximize non-gaussianity, but it has been shown that these two tasks are closely related (Hyvärinen, 2001). It has to be expected that the tendency towards over-independency will lead to distortions of the reconstructed source components; this can be concluded from the result that negative values of MI difference are correlated with larger Frobenius norm.

On the other hand, state space modeling of time series is based on predictive performance within the maximum-likelihood framework; residual MI plays no role in the model fitting algorithm. Independence of source components is indirectly imposed through the model structure which does not contain terms connecting different source components. The simulation results presented in this chapter indicate that this indirect approach of imposing independence does not suffer from the tendency of producing over-independent source components; this represents an attractive feature of state space modeling.

In this chapter, we have not presented results for the case of two source components having identical autocorrelation functions; this case would correspond to two AR(1) processes having the same parameter a, or two AR(2) processes having the same parameters a_1 and a_2. If these two processes were driven by noises with different distribution, such as one gaussian noise and one Laplacian noise, separation would still be possible, also for IC-SSM; but this would be a difficult situation, since IC-SSM represents an intrinsically linear approach, based on the implicit assumption of gaussianity, therefore model fitting would become more difficult. In such case, small differences of the optimal predictors for different finite-length time series from the same system would need to be exploited by the state space model, requiring very careful maximization of likelihood.

4. Conclusion

In this chapter we have explored the application of Kalman filtering to the blind signal separation problem, i.e., the problem of reconstructing independent source components from linear mixtures. Linear mixtures of unknown components (or *factors*) form the observation model of classical latent variable models, such as Factor Analysis and linear state space models. While Factor Analysis does not take the temporal ordering of time series data into account, i.e., it is a *static* algorithm, state space modeling is based on exploiting temporal correlations for the purpose of optimal prediction. Dynamical generalizations of Factor Analysis have been proposed by several authors (Molenaar, 1985; Attias & Schreiner, 1998; Attias, 1999; Xu, 2000; Cheung & Xu, 2003a; Cheung & Xu, 2003b; Cheung, 2006); some of these generalizations have approached the shape of linear state space models. So far, the blind signal separation problem has mostly been approached by Independent Component Analysis (ICA); like Factor Analysis, most ICA algorithms are static algorithms. In this chapter, we have presented a dynamic generalization of ICA.

We do not claim that the presented algorithm based on Kalman filtering and state space modeling could easily outperform and replace the available ICA algorithms; rather we have aimed at providing a comparison of some particular strong and weak points of these different algorithms, and to attract attention to the potential that state space methods have to offer for further work on this field.

We have introduced a particular state space model, denoted as the Independent Component State Space Model (IC-SSM), which was designed for the purpose of separation of independent source components from multivariate time series. Through simulations, we have demonstrated two attractive features of IC-SSM: the ability to deal both with gaussian and non-gaussian amplitude distributions, and the absence of over-independency of reconstructed source components. While IC-SSM shares the former feature with non-static ICA algorithms, like AMUSE, SOBI and TDSEP, the latter feature seems to be a unique characteristic of state space modeling, resulting from the approach of indirectly imposing the independence constraint through the model structure, as opposed to directly imposing it by minimization of residual MI.

Further strong points of IC-SSM are given by its capacity to deal with observation noise and by the representation of the main frequencies of the data directly as model parameters. Static ICA algorithms ignore the temporal information in the data, and even AMUSE, SOBI and TDSEP access this information only indirectly. As a consequence, components with strong periodicity, such as stochastic or deterministic oscillations, can be extracted better by IC-SSM than by ICA algorithms. This is an inherent advantage of a linear predictive model.

Since IC-SSM is based on a parametric predictor for the time series data, the natural direction towards improving its performance is given by improving the predictor. The AR(1) and ARMA(2,1) predictors which we have employed, represent but the most basic choices, corresponding to real eigenvalues and pairs of complex conjugated eigenvalues of the state transition matrix, respectively. The simulation with nonlinear deterministic source processes has shown the need for further improvement, since for a part of the simulated mixtures IC-SSM failed to reconstruct all sources correctly.

A natural generalization would be given by higher-order ARMA models; however, correct identification of such models in the absence of prior information poses considerable difficulties. Two related problems are the estimation of correct model orders, and the estimation of the correct number of independent source components.

As an example for the first problem, in the simulations with AR(1) or AR(2) source processes we have tacitly assumed that we know that all local state space models should be chosen as first-order or second-order models. In this particular case, this model design decision could be supported by observing corresponding peaks in the power spectrum of the mixtures, or absence of such peaks. However, if the true structure would be a heterogeneous set of AR(1) and AR(2) processes, the choice of the model design would become more difficult, and there would possibly be a need for lengthy model comparison procedures. Different models would need to be compared with respect to their prediction performance, e.g. by comparing their likelihood values; model complexity would need to be taken into account as well, which could be done by correction terms, leading to criteria such as AIC and BIC (see additional discussion in the Appendix).

Also the second problem, the problem of estimation of the correct number of independent source components, could be addressed by model comparison procedures. In the simulations presented in this chapter we have assumed that the number of independent source component was equal to the number of available mixtures, since this is a standard assumption also of typical ICA algorithms. If the true number of source components was larger, $N_c > N$, we would be facing the *overcomplete basis problem*, i.e., the problem of a non-invertible mixing process; if it was smaller, $N_c < N$, there would be the risk of creating spurious source components.

In principle, state space modeling can deal with both cases, since it does not require the observation matrix to be square or to obey some other constraint relating to its dimensions. If the correct model is known, and if observation matrix and state transition matrix fulfill the condition of observability (Kailath, 1980), the Kalman filter can reconstruct the state vector, after an initial transient. The question of whether the model itself can be estimated correctly from given data alone is considerably more difficult and has been subject of extensive theoretical work (Tong et al., 1991). Practical experience with simulations has shown that it depends very much on the properties of the underlying source processes whether the model can be correctly estimated. Without showing details, we mention that for the case of fewer mixtures than sources, for AR(1) source processes (i.e., "weak" dynamics), so far we have been unable to estimate source components correctly, while for AR(2) source processes this has been at least partly possible, since these processes represent "stronger" dynamics. As an example from practice, in the analysis of time series from economy it is not unusual to extract several oscillatory processes from even a univariate time series, $N = 1$, by state space methods (Kato, 1996).

We believe that questions regarding the "true" number of independent source components, or the "true" model order of a predictive model, are of limited relevance for most practical time series analysis problems, the reason being that for many natural systems no true model with simple structure exists; this is the main difference between simulated and real-world time series. As an example, the "true" dynamical model of human brain is certainly inaccessible for any practical purpose, and all models which are fitted to data measured from brain dynamics have to be regarded as massively simplified and approximative. The same remark applies to the "true" number of independent source components represented by such data.

In some cases one may even doubt whether it is reasonable to expect the presence of even just *two* mutually independent source components within multivariate time series obtained from spatially extended and highly interconnected systems like human brain (unless one is exclusively interested in removing artifacts of exogenous origin). With respect to this doubt, practical experience has shown that the assumption of the presence of a set of independent components is a useful "zeroth-order" approximation, in order to obtain a first impression of the properties of the data. Here it becomes obvious that ICA, and also state space modeling based on the independence assumption, is a generalization of exploratory techniques like PCA and projection pursuit. Unlike with simulations, independent source components may be estimated for real-world data with the expectation that they might be useful, but not with the expectation that they must necessarily represent underlying "truth" in a deeper sense.

For the data analyst who intends to extract independent source components from given time series data by IC-SSM, the decision of model orders and number of components then becomes mainly a question of how large a model can be handled. Typically, real-world data will always be willing to reveal additional components, if such are asked for, since there probably is no low-dimensional "true" model underlying the data. Any additional component, and any increase of the model order of an existing component, will increase the state space dimension M and also increase the number of model parameters. Increased state space dimension will slow down the Kalman filter and RTS smoother steps; increased number of model parameters will cause the maximum-likelihood model fitting step to consume more computation time.

Larger number of model parameters also increases the risk that the maximum-likelihood optimization produces solutions of insufficient quality; as an example, consider the IC-SSM results shown in Fig. 5. The distribution of Frobenius norm results for Kalman filter and RTS smoother shows a peak around a value of 0.4, comprising about half of the simulated data sets, while the other half forms a tail at larger values. We believe that this tail is due to poor performance of optimization, and that these data sets could also be moved into the peak at lower values of Frobenius norm by an improved optimization scheme.

We also remark that the strongly nonlinear/non-gaussian and deterministic properties of the dynamical systems employed in Section 3.3 represent an extreme behavior which is unlikely to be found in real-world time series. It is not unexpected that on average the ICA algorithms perform better for this simulation than IC-SSM.

Without doubt, low computational time consumption is the strongest advantage of ICA algorithms like InfoMax, FastICA, JADE and TDSEP; MILCA consumes somewhat more time, but still less than full numerical optimization of likelihood for state space models, which can easily consume hours of CPU time for a single data set of moderate size. We have not yet optimized our implementation with respect to computational efficiency; future work should focus on developing improved optimization algorithms for this specific application.

Other topics for future work will be the overcomplete basis problem, nonlinear predictors and components with nonstationary behaviour. Improvements with respect to the overcomplete basis problem will again require improved optimization algorithms, since in this situation the task to be addressed by the optimization procedure is even harder.

Acknowledgments

The work of A. Galka was supported by the German Research Foundation (Deutsche Forschungsgemeinschaft, DFG) through Sonderforschungsbereich SFB 855 "Biomagnetic Sensing" and Sonderforschungsbereich SFB 654 "Plasticity and Sleep", and by the Japanese Society for the Promotion of Science (JSPS) through fellowship ID No. P 03059. The work of T. Ozaki was supported by JSPS through grants KIBAN B No. 173000922301 and WAKATE B No. 197002710002.

Appendix: Finding suitable initial estimates for model parameters

The success of state space modeling by numerical maximization of the logarithmic likelihood depends on choosing suitable initial estimates for the model parameters; for this reason we will describe a simple, but efficient approach to finding suitable initial estimates.

Our approach is based on the modal decomposition of a multivariate autoregressive (MAR) model. Let the N-dimensional data vectors be denoted by $\mathbf{y}(t)$, $t = 1, \ldots, T$, then the MAR model is given by

$$\mathbf{y}(t) = \sum_{\tau=1}^{p_{\text{MAR}}} \Phi(\tau) \mathbf{y}(t-\tau) + \mathbf{v}(t) , \qquad (14)$$

where $\Phi(\tau)$, $\tau = 1, \ldots, p_{\text{MAR}}$ denotes a set of $(N \times N)$-dimensional parameter matrices, p_{MAR} denotes the model order, and $\mathbf{v}(t)$ denotes a gaussian driving noise term. MAR models can be fitted to time series data efficiently by an algorithm developed by Levinson (1947) and later by Whittle (1963). Also an estimate for the driving noise covariance matrix S_v^{MAR} is provided; this matrix is not to be confused with the driving noise covariance matrix S_η of the state space model, Eq. (2).

The MAR model, with model order p_{MAR} can be transformed into the following first-order *controller canonical form* state space model (Akaike, 1974):

$$\mathcal{X}(t) = \boldsymbol{\Phi}\mathcal{X}(t-1) + \mathcal{H}(t) , \tag{15}$$

where we have defined

$$\mathcal{X}(t) = \begin{pmatrix} \mathbf{y}(t) \\ \mathbf{y}(t-1) \\ \vdots \\ \mathbf{y}(t-p_{\text{MAR}}+1) \end{pmatrix} , \quad \mathcal{H}(t) = \begin{pmatrix} \boldsymbol{v}(t) \\ \mathbf{0}_N \\ \vdots \\ \mathbf{0}_N \end{pmatrix} \tag{16}$$

and

$$\boldsymbol{\Phi} = \begin{pmatrix} \Phi(1) & \Phi(2) & \cdots & \Phi(p_{\text{MAR}}-1) & \Phi(p_{\text{MAR}}) \\ \mathsf{I}_N & \mathsf{0}_N & \cdots & \mathsf{0}_N & \mathsf{0}_N \\ \vdots & \vdots & \ddots & \vdots & \vdots \\ \mathsf{0}_N & \mathsf{0}_N & \cdots & \mathsf{I}_N & \mathsf{0}_N \end{pmatrix} \tag{17}$$

Here I_N, $\mathsf{0}_N$ and $\mathbf{0}_N$ denote the $(N \times N)$-dimensional identity matrix, the $(N \times N)$-dimensional matrix of zeros and the $(N \times 1)$-dimensional vector of zeros, respectively.

The relationship between the state vector $\mathcal{X}(t)$ and the original data vector is given by

$$\mathbf{y}(t) = \begin{pmatrix} \mathsf{I}_N & \mathsf{0}_N \ldots \mathsf{0}_N \end{pmatrix} \mathcal{X}(t) =: \mathcal{C}\mathcal{X}(t) , \tag{18}$$

where there are $(p_{\text{MAR}} - 1)$ $(N \times N)$-dimensional matrices of zeros $\mathsf{0}_N$ in $\mathcal{C}$; this equation can formally be regarded as the observation equation of the state space model. An observation noise term may be added. The $(Np_{\text{MAR}} \times Np_{\text{MAR}})$-dimensional covariance matrix of the augmented driving noise vector $\mathcal{H}(t)$ is given by

$$\mathsf{S}_{\mathcal{H}} = \begin{pmatrix} \mathsf{S}_v^{\text{MAR}} & \mathsf{0}_N & \cdots & \mathsf{0}_N \\ \mathsf{0}_N & \mathsf{0}_N & \cdots & \mathsf{0}_N \\ \vdots & \vdots & \ddots & \vdots \\ \mathsf{0}_N & \mathsf{0}_N & \cdots & \mathsf{0}_N \end{pmatrix} . \tag{19}$$

In order to approach an independent components state space model, the $(Np_{\text{MAR}} \times Np_{\text{MAR}})$-dimensional state transition matrix $\boldsymbol{\Phi}$ is diagonalized, yielding a set of d_r real eigenvalues λ_i and another set of d_c pairs of complex conjugated eigenvalues (ψ_j, ψ_j^*), where $Np_{\text{MAR}} = d_r + 2d_c$; this diagonalization corresponds to a particular linear transformation of state space. We then reorder the state space dimensions such that in the new state vector all dimensions corresponding to real eigenvalues are accommodated first, followed by the pairs of dimensions corresponding to pairs of complex conjugated eigenvalues. Furthermore real eigenvalue dimensions shall be ordered according to increasing size of the eigenvalues, and complex eigenvalue dimensions shall be ordered according to increasing phase of the complex eigenvalues.

Since we intend to avoid complex numbers in the transition matrix, a further linear transformation is applied to each pair of complex conjugated eigenvalues (ψ_j, ψ_j^*); within the diagonalized state transition matrix, each (2×2)-dimensional block

$$\begin{pmatrix} \psi_j & 0 \\ 0 & \psi_j^* \end{pmatrix} \tag{20}$$

is transformed into the corresponding "left companion" form

$$\begin{pmatrix} a_j(1) & 1 \\ a_j(2) & 0 \end{pmatrix} \tag{21}$$

This transformation step may be performed according to Eq. (13); then $2\pi\varphi$ and r denote phase and modulus of ψ_j, respectively. If this transformation is applied to all complex eigenvalues, the resulting state transition matrix $\tilde{\Phi}$ will be block-diagonal with d_r (1×1)-dimensional blocks, representing the real eigenvalues, and d_c (2×2)-dimensional blocks in *observer canonical form*, representing the pairs of complex conjugated eigenvalues. Since the eigenvalues of the MAR model represent the main spectral features of the data, they provide suitable initial values for the corresponding model parameters.

Let the linear transformation comprising the diagonalization of the initial state transition matrix Φ and the subsequent steps of reordering of eigenvalues, of transformation to left companion form and of rescaling (see below) be denoted by D, then the transformations for the state vector, the state transition matrix, the driving noise covariance matrix and the observation matrix are given by:

$$\tilde{\mathcal{X}}(t) = D\,\mathcal{X}(t) \tag{22}$$

$$\tilde{\Phi} = D\,\Phi\,D^{-1} \tag{23}$$

$$\tilde{S}_\mathcal{H} = D\,S_\mathcal{H}\,D^\dagger \tag{24}$$

$$\tilde{\mathcal{C}} = \mathcal{C}\,D^{-1} \,; \tag{25}$$

The transformed driving noise covariance matrix $\tilde{S}_\mathcal{H}$ will not yet have the block-diagonal shape which is required for an IC-SSM model; in general, it is impossible to find a transformation that would diagonalize Φ and $S_\mathcal{H}$ simultaneously. We therefore set zero all elements of $S_\mathcal{H}$ that do not belong to the blocks on the diagonal. This step represents an act of major violence against the model, which will massively deteriorate its performance in data prediction and thereby reduce its likelihood; subsequent numerical maximization of likelihood is needed in order to recover good performance. Furthermore, $\tilde{S}_\mathcal{H}$ will not yet follow the scaling convention for the driving noise term, according to which we should have $b_j(0) = 1$ in Eq. (6); this convention corresponds to setting the element (1,1) of each block on the diagonal of $\tilde{S}_\mathcal{H}$ to a value of one (compare with Eq. (8)). We therefore need to rescale each state space dimension accordingly; the rescaling is incorporated into the transformation matrix D. This step completes the generation of an initial IC-SSM model from a MAR model: $\tilde{\Phi}$, $\tilde{S}_\mathcal{H}$ and $\tilde{\mathcal{C}}$ serve as initial estimates for A, S_η and C, respectively.

Choice of Model Order and Block Dimensions

For the model order p_{MAR} of the initial MAR model a sufficiently large value should be chosen (e.g. $p_{MAR} = 20$ or larger), such that the spectral properties of the data can be sampled by a sufficiently large number of eigenvalues; note that not all eigenvalues will have to be retained for the initial IC-SSM model. As an example, eigenvalues with small modulus r will correspond to very broad, flat peaks in the spectrum, which may be of minor relevance for modeling the data. The decision of which eigenvalues to retain could be based on some kind of quantitative criterion, or on subjective assessment of relevant components (i.e., by visual inspection); since this step only serves for providing initial parameter estimates, certain subjective elements can be tolerated. For the simulations presented in this chapter, smaller MAR model orders of $p_{\text{MAR}} = 7$ were already sufficient for reliably identifying the underlying true source components.

The number of retained eigenvalues determines the number of components N_c within the model; it depends on the number of independent components that are to be extracted from the data. In principle, this number could be obtained by minimizing appropriately defined measures, such as the Akaike Information Criterion (AIC) (Akaike 1974), the Bayesian Information Criterion (BIC) (Schwarz 1978) or other similar criteria; but such procedure would be very costly in terms of computational time expense since many competing IC-SSM models would have to be fitted to the data. In practise, a compromise needs to be found between the quality of the description of the data and the required time for model fitting.

Block dimensions n_j, $j = 1, \ldots, N_c$ follow from the algorithm presented so far as $n_j = 1$ for components derived from real eigenvalues, and as $n_j = 2$ for components derived from pairs of complex conjugated eigenvalues. In many cases it will not be necessary to apply any further changes, such that the choice of block dimensions is completely determined by the initialisation process. An ARMA(2,1) model describes a stochastic oscillation with one eigenfrequency, and the simulations presented in this chapter show that a wide class of components can be extracted from mixtures by fitting a single eigenfrequency. Blocks of dimension $n_j > 2$ could be created by merging blocks with dimensions $n_j = 1$ and/or $n_j = 2$; however, in this chapter we do not present cases where such transformation would be useful or necessary.

References

Akaike, H. (1974). Markovian representation of stochastic processes and its application to the analysis of autoregressive moving average processes. *Ann. Inst. Stat. Math,* 26, 363–387.

Aoki, M. (1987). *State Space Modeling of Time Series.* Berlin: Springer.

Attias, H., & Schreiner, C.E. (1998). Blind source separation and deconvolution: The dynamic component analysis algorithm. *Neural Computation, 10,* 1373–1424.

Attias, H. (1999). Independent Factor Analysis. *Neural Computation, 11,* 803–851.

Baldick, R., editor (2006). *Applied Optimization: Formulation and Algorithms for Engineering Systems.* Cambridge: Cambridge University Press.

Bartholomew, D., editor (1987). *Latent variable models and factor analysis.* London: Griffin.

Basilevsky, A.T. (1994). *Statistical Factor Analysis and Related Methods: Theory and Applications.* New York: Wiley-Interscience.

Beckmann, C.F., & Smith, S.M. (2004). Probabilistic independent component analysis for functional magnetic resonance imaging. *IEEE Trans. Medical Imaging, 23,* 137–152.

Bell, A.J., & Sejnowski, T.J. (1995). An information-maximization approach to blind separation and blind deconvolution. *Neural Computation, 7,* 1129–11591.

Belouchrani, A., Abed-Meraim, K., Cardoso, J.-F., & Moulines, E. (1997). A blind source separation technique using second order statistics. *IEEE Trans. Signal Processing, 45*, 434–444.

Box, G.E.P., & Jenkins, G.M. (1976). *Time series analysis, forecasting and control.* San Francisco: Holden-Day, 2nd edition.

Cardoso, J.-F. (1999). High-order contrasts for independent component analysis. *Neural Computation, 11*, 157–192.

Cardoso, J.-F. (2001). The three easy routes to independent component analysis: contrasts and geometry. In: Lee, T.-W., editor, *Proc. Int. Workshop on Independent Component Analysis and Blind Signal Separation (ICA2001), San Diego*, pages 1–6.

Cheung, Y.-M., & Xu, L. (2003a). Dual multivariate auto-regressive modeling in state space for temporal signal separation. *IEEE Trans. Syst. Man Cyb., 33*, 386–398.

Cheung, Y.-M., & Xu, L. (2003b). Further studies on temporal factor analysis: comparison and Kalman-filter-based algorithm. *Neurocomputing, 50*, 87–103.

Cheung, Y.-M. (2006). A maximum likelihood approach to temporal factor analysis in state-space model. *Signal Processing, 86*, 2966–2980.

Comon, P. (1994). Independent component analysis, a new concept? *Signal Processing, 36*, 287–314.

Ding, M., Grebogi, C., Ott, E., Sauer, T., & Yorke, J.A. (1993). Plateau onset of correlation dimension: when does it occur? *Phys. Rev. Lett., 70*, 3872–3875.

Durbin, J., & Koopman, S.J. (2001). *Time Series Analysis by State Space Methods.* Oxford: Oxford University Press.

Dyrholm, M., Makeig, S., & Hansen, L.K. (2007). Model selection for convolutive ICA with an application to spatiotemporal analysis of EEG. *Neural Computation, 19*, 934–955.

Franklin, G.F., Powell, J.D., & Workman, M.L. (1998). *Digital Control of Dynamic Systems.* Englewood Cliffs: Prentice-Hall, 3rd edition.

Galka, A., Ozaki, T., Bosch-Bayard, J., & Yamashita, O. (2006). Whitening as a tool for estimating mutual information in spatiotemporal data sets. *J. Stat. Phys., 124*, 1275–1315.

Grewal, M.S., & Andrews, A.P. (2001). *Kalman Filtering: Theory and Practice Using MATLAB.* New York: Wiley-Interscience.

Hyvärinen, A. (1999). Fast and robust fixed-point algorithms for independent component analysis. *IEEE Trans. Neural Networks, 10*, 626–634.

Hyvärinen, A., Karhunen, J., & Oja, E. (2001). *Independent Component Analysis.* New York: Wiley.

Hyvärinen, A. (2005). A unifying model for blind separation of independent sources. *Signal Processing, 85*, 1419–1427.

Jutten, C., & Hérault, J. (1991). Blind separation of sources, part I: An adaptive algorithm based on neuromimetic architecture. *Signal Processing, 24*, 1–10.

Kailath, Th. (1968). An innovations approach to least-squares estimation – Part I: Linear filtering in additive white noise. *IEEE Trans. Autom. Control, 13*, 646–655.

Kailath, Th. (1980). *Linear systems.* Information and system sciences series. Englewood Cliffs: Prentice-Hall.

Kalman, R.E. (1960). A new approach to linear filtering and prediction problems. *J. Basic Engin., 82*, 35–45.

Kato, H., Naniwa, S.,& Ishiguro, M. (1996). A bayesian multivariate nonstationary time series model for estimating mutual relationships among variables. *J. Econometrics, 75*, 147–161.

Kraskov, A., Stögbauer, H., & Grassberger, P. (2004). Estimating mutual information. *Phys. Rev. E, 69*, 066138.

Lee, T.-W., Girolami, M., & Sejnowski, T.J. (1999). Independent component analysis using an extended infomax algorithm for mixed sub-gaussian and super-gaussian sources. *Neural Computation, 11*, 417–441.

Levinson, N. (1947). The Wiener RMS error criterion in filter design and prediction. *J. Math. Phys., 25*, 261–278.

Mackay, D.J.C. (2003). *Information theory, inference, and learning algorithms*. Cambridge: Cambridge University Press.

Mehra, R.K. (1974). Identification in control and econometrics: similarities and differences. *Annals of Economic and Social Measurement, 3*, 21–47.

Molenaar, P.C. (1985). A dynamic factor model for the analysis of multivariate time series. *Psychometrika, 50*, 181–202.

Molgedey, L., & Schuster, H.G. (1994). Separation of a mixture of independent signals using time delayed correlations. *Phys. Rev. Lett., 72*, 3634–3637.

Otter, P.W. (1986). Dynamic structural systems under indirect observation: identifiability and estimation aspects from a system theoretic perspective. *Psychometrika, 51*, 415–428.

Ozaki, T. (1992). A bridge between nonlinear time series models and nonlinear stochastic dynamical systems: a local linearization approach. *Statistica Sinica, 2*, 113–135.

Protter, P. (1990). *Stochastic Integration and Differential Equations*. Berlin: Springer.

Rauch, H.E., Tung, G., & Striebel, C.T. (1965). Maximum likelihood estimates of linear dynamic systems. *American Inst. Aeronautics Astronautics (AIAA) Journal, 3*, 1445–1450.

Schwarz, G. (1978). Estimating the dimension of a model. *Ann. Stat., 6*, 461–464.

Sorenson, H.W. (1970). Least-squares estimation: from Gauss to Kalman. *IEEE Spectrum, 7*, 63–68.

Sprott, J.C. (2003). *Chaos and Time-Series Analysis*. Oxford: Oxford University Press.

Stögbauer, H., Kraskov, A., Astakhov, S.A., & Grassberger, P. (2004). Least-dependent-component analysis based on mutual information. *Phys. Rev. E, 70*, 066123.

Tong, L., Liu, R., Soon, V.C., & Huang, Y. (1991). Indeterminacy and identifiability of blind separation. *IEEE Trans. Circuits and Systems, 38*, 499–509.

Waheed, K., & Salem, F.M. (2005). Linear state space feedforward and feedback structures for blind source recovery in dynamic environments. *Neural Processing Letters, 22*, 325–344.

West, M., Prado, R., & Krystal, A.D. (1999). Evaluation and comparison of EEG traces: Latent structure in nonstationary time series. *J. Amer. Stat. Assoc., 94*, 1083–1095.

Whittle, P. (1963). On the fitting of multivariate autoregressions, and the approximate canonical factorization of a spectral density matrix. *Biometrika, 50*, 129–134.

Xu, L. (2000). Temporal BYY learning for state space approach, hidden Markov model and blind source separation. *IEEE Trans. Signal Process., 48*, 2132–2144.

Zhang, L., & Cichocki, A. (2000). Blind deconvolution of dynamical systems: A state space approach. *Journal of Signal Processing, 4*, 111–130.

Ziehe, A., & Müller, K.-R. (1998). TDSEP – an efficient algorithm for blind separation using time structure. In: Niklasson, L., Bodén, M., & Ziemke, T., editors, *Proc. 8th Int. Conf. Artificial Neural Networks (ICANN'98)*, pages 675–680. Springer, Berlin, Heidelberg, New York.

Åström, K.J. (1980). Maximum likelihood and prediction error methods. *Automatica, 16*, 551–574.

In: Kalman Filtering
Editor: Joaquín M. Gomez

ISBN: 978-1-61761-462-0
© 2011 Nova Science Publishers, Inc.

Chapter 9

USING A RESTRICTED KALMAN FILTERING APPROACH FOR THE ESTIMATION OF A DYNAMIC EXCHANGE-RATE PASS-THROUGH

Rafael Martins de Souza[1], *LuizFelipe Pires Maciel*[2] *and Adrian Pizzinga*[3,*]

[1]National School of Statistical Sciences (ENCE) -
Brazilian Institute of Geography and Statistics (IBGE).
Sponsored by CNPq and CAPES

[2]Graduate School of Economics (EPGE) and Brazilian Institute of Economics (IBRE) - Getulio Vargas Foundation (FGV). Sponsored by CAPES and IBRE.

[3]Financial and Actuarial Risk Management Institute of Pontifical Catholic University of Rio de Janeiro (IAPUC)

Abstract

In this paper we propose linear state space models to estimate the time-varying pass-through of Brazilian price indexes against the US Dollar/Real exchange rate from 1999 to 2007. The methodological framework encompasses the restricted Kalman filtering under a reduced modeling approach, under which it becomes possible to check whether some economic hypotheses are supported by the data. The paper has three main targets. The first is to decide whether models of null (or of full) pass-through are acceptable to the price indexes investigated here. The second is to carry out likelihood ratio tests for the significance of some economic exogenous variables, which are termed determinants in this paper and are theoretically associated with the pass-through. The third is to analyze the behavior of the Kalman filter estimates of the pass-through from the best models.

Keywords: determinants, pass-through, restricted Kalman filtering, state space models.

1. Introduction

In an open economy, domestic prices can be affected by external shocks, whether from currency relative prices adjustment or from movements of international supply and demand.

---
[*](Corresponding author). adrianhpster@gmail.com. Sponsored by FAPERJ (PDR scholarship).

The exchange rate is a quite volatile economic variables in macroeconomic policy. How much the exchange rate affects the economy? One of the faster channels is into prices. This channel is called *(exchange rate) pass-through*. There are few studies for this effect in Brazil in which the response of the prices to a change in exchange rate is suitably tackled.

The importance of past-through estimation has increased since the adoption of inflation targeting regime (cf. Fraga, Goldfajn & Minella, 2003) and the recognition that it is crucial for inflation forecasting (cf. Goldfajn & Werlang, 2000). In addition to these motivations above, we have some evidence of a time-varying pass-through, even though there are few studies considering this assumption. Indeed, as Parsley (1995) points out, the stability of exchange rate pass-through is not well tested in common econometric specifications of pass-through equations.

In this paper we use a linear state space modeling approach to estimate exchange rate pass-through into some Brazilian price indexes. This framework allows us to consider the time-varying structure of the exchange rate pass-through in the estimation process, besides turning possible the investigation about the impact from some *determinants* of the pass-through, i.e., we can test whether some exogenous variables can be used to explain movements of the pass-through along time. We shall make use of the *reduced restricted Kalman filtering* (cf. Pizzinga, 2010), a technique that allows us to check whether models of null and of full exchange rate pass-through are adequate to describe the behavior of the price indexes series used in this work.

The paper is structured as follows. Section 2 reviews the literature on the subject and discusses candidates for the determinants of the past-through. In Section 3, we present the econometric model setting and inference, which are entirely based on linear state space models and the Kalman filter. Section 4 presents the data and shows our empirical results. Section 5 concludes the paper.

2. Exchange Rate Pass-Through

2.1. A Glimpse at the Literature

The pass-through degree is the elasticity exchange rate – domestic prices. In other words, it is the percentage of change in domestic prices generated of 1% change in exchange rate into. Since 1980, the concern on exchange rate pass-through has been intensified. The literature is focused on the behavior of the impact of exchange rate over prices and on its determinants. Although, the real impulse was the study about the *Purchase Power Parity Puzzle* (PPP). The PPP assumption says that all variation in exchange rate is passed-through into prices. For more than twenty years, the major conclusion taken from several empirical studies is that the PPP is not valid in the short run: the exchange rate pass-through into prices is less than one. However, there is some evidence in favor of the validity of PPP in the long run.

The pass-through into prices is one of the main drivers of optimal monetary policies. According to Betts & Devereux (2000), the trade-off between output volatility and inflation volatility depends on how sensitive prices are to exchange rate variations. They begin their argument explaining that the nature of trade-off between different exchange rate regimes is quite distinct in industrial countries as compared with emerging ones. Using a Dynamic

General Equilibrium Model (DGE) model, they argue that the critical distinction is the exchange rate pass-through into prices. With very high exchange rate pass-through, policies that stabilize output require high exchange rate volatility, which implies high inflation volatility. But with limited or delayed pass-through, this trade-off is less pronounced, and a flexible exchange rate policy that stabilizes output can be done so without high inflation volatility.

Another study that emphasizes the importance of exchange rate pass-through in an inflation targeting regime is due to Fraga, Goldfajn & Minella (2003). They have shown that the problem of having a high exchange rate pass-through degree is that it implies a greater difficulty for their attainment of inflation targets. The bigger is the exchange rate pass-through, the domestic economy becomes more sensitive to external shocks, which implies that the impact of exogenous shocks into domestic prices is amplified.

Another importance of exchange rate pass-through is related to inflation forecasting. According to Goldfajn & Werlang (2000), the pass-through into prices is directly associated with inflation forecasting error. Under a smaller pass-through, the domestic economy is more stable and less affected by external facts. So, a smaller pass-through means that the distance between inflation expectations and inflation target is shorter. Consequently, a small pass-through is associated with a major transparency of inflation path and a minor volatility in price variations in this economy, rising social welfare and monetary policy efficiency.

A main issue in pass-through estimation is the difficulties due to using aggregated data and the known problem of aggregation bias. This fact suggests a disaggregation process for prices, which aims at trying to capture the exchange rate pass-through for each good or for each market. Campa & Goldberg (1995) presented results where the past-through estimates are better-behaved across industries than across countries with aggregate data. These authors also say that the major source of pass-through variations are competition issues at each sector. Yang (1996) and Olivei (2002) have shown that pass-through varies significantly across industries. Menon (1996) support this point of view shedding light over aggregation bias and, therefore, indicating that disaggregated data provide more accurate estimates and capture more efficiently the exchange rate impact over commodities prices. Other studies that followed this tendency of disaggregated estimation of pass-through are the papers by Campa & Goldberg (2002) and by Pollard & Coughlin (2005).

Some results about pass-through estimation can be seen in Goldberg & Knetter (1997), where the exchange rate pass-through to US inflation was near 50% after 6 months. Campa & Goldberg (2002) estimated the pass-through for 25 countries belonging to OECD. They found a pass-through of 26%, in short term, and 41% in the long term for US. The average pass-through estimated for OECD countries in short and long run was 61% and 77%, respectively.

Sekine (2006) estimated exchange rate pass-through for six developed countries (United States, Japan, Germany, United Kingdom, France and Italy) by taking in account their time-varying natures. The author incorporated that nature allowing permanent shifts in pass-through parameters. He found that pass-through has declined over time in all major industrial countries and, in most cases, the pass-through did not show the parameter shift envisaged by split sample estimations.

Calvo & Reinhart (2000) have shown that emerging countries display a four times higher pass-through degree than developed countries. Additionally, these authors calculated that the variance of inflation compared with the variation of exchange rate is 43% for emerging countries and 13% for developed ones.

2.2. Pass-Through Determinants

According to Menon (1996), Goldfajn & Werlang (2000), Taylor (2000) and Campa & Goldberg (2002), the main drivers of price sensibility to exchange rate changes can be inferred. In face of literature with macroeconomic approach, the pass-through depends on: *inflation persistence*, *openness degree of the economy*, the *output gap*, and *real exchange rate disalignments*. From the standpoint of disaggregated analysis, the exchange rate pass-through is also associated with *the competition degree of each industry* and with *firm's market power* (with the elasticity price-demand).

2.2.1. Inflation Environment

According to Goldfajn & Werlang (2000), the variable inflation environment means the frequency under which agents remark their prices based on past inflation. On countries with an inflation environment, the agents have more facility to pass-through costs changes and rise prices. By this way, the bigger the inflationary environment and the inflation persistence are, it becomes easier to agents pass-through exchange rate increases into prices. This thought is corroborated by Taylor(2000), who suggests a relation between inflation and exchange rate pass-through using the inflation persistence as a channel of transmission.

2.2.2. Output Gap

The output gap is defined by the deviation of product in relation to his long term value. The evidence of past studies is a positive relation between the pass-through and the output gap. As larger the length between GNP and his potential turn, bigger the demand pressure over prices becomes. This fact generates an inflation environment, rising increased probabilities that firms pass-through changes in costs into prices. Therefore, in an environment where the output gap increases, the exchange rate pass-through effect to inflation is intensified.

2.2.3. Real Exchange Rate Disalignment

According to Goldfajn & Valdés(1999), an overvaluation of the real exchange rate signifies a mean factor for the future inflation composition. If real exchange rate is below his long term value, agents make-up the expectation of future devaluations, adjusting relative prices. Although, if exchange rate variation do not be adjusted by relative prices, it will imply a rise in internal inflation as compared with to external. The agents will assume this expectation of future depreciation, amplifying the effects over prices. Then the exchange rate pass-through will be negative associated with the difference of real exchange rate and its log run value.

2.2.4. Openness Degree

The openness degree of a economy means the presence of tradable goods in na economy, indicating the intensity which prices can respond to changes in exchange rate. This degree can be defined as the sumo of imports and exports as a proportion of GNP. In an more open economy, we expect that the presence of goods more sensitive to exchange rate will be bigger, which implies larger exchange rate pass-through to inflation.

3. Econometric model setting and estimation

3.1. Linear State Space Models under Restrictions

We define a *linear Gaussian state space model* by the following *measurement* equation, *state* equation and initial state vector:

$$Y_t = Z_t\beta_t + d_t + \varepsilon_t \quad , \quad \varepsilon_t \sim NID(0, H_t)$$
$$\beta_{t+1} = T_t\beta_t + c_t + \eta_t \quad , \quad \eta_t \sim NID(0, Q_t) \tag{1}$$
$$\beta_1 \sim N(b_1, P_1).$$

The former equation linearly relates the observed time series Y_t to the unobserved state β_t and the latter gives the state evolution through a Markovian structure. The random errors ε_t and η_t are independent (between each other and of β_1), and the system matrices Z_t, d_t, H_t, T_t, c_t and Q_t are deterministic. Notice that d_t and c_t are generally reserved to the inclusion of exogenous explanatory variables.

For a given time series of size n and any t,j, set $\mathcal{F}_j \equiv \sigma(Y_1, \ldots, Y_j)$, $\hat{\beta}_{t|j} \equiv E(\beta_t|\mathcal{F}_j)$ and $\hat{P}_{t|j} \equiv Var(\beta_t|\mathcal{F}_j)$. The *Kalman filtering* consists of recursive equations for these first- and second-order conditional moments. The formulae and their respective derivations corresponding to *predicting* ($j = t-1$), *filtering* ($j = t$) and *smoothing* ($j = n$), and details on the estimation of unknown parameters in the system matrices by (*quasi*) maximum likelihood can be found in Harvey (1989) and Durbin & Koopman (2001).

Now, suppose the following: for each t, $A_t\beta_t = q_t$, where A_t is a known $k \times m$ fixed matrix and $q_t = (q_{t1}, \ldots, q_{tk})'$ is a $k \times 1$ observable random vector (possibly degenerated). Also, suppose that q_t is $\mathcal{F}_t$-measurable. A restricted estimation of this type can be achieved under the *reduced restricted Kalman filtering*, presented in Pizzinga (2010) and resumed in the following algorithm:

Let t be an arbitrary time period.

1. Re-write the linear restrictions as

$$A_{t,1}\beta_{t,1} + A_{t,2}\beta_{t,2} = [A_{t,1} \ A_{t,2}] (\beta'_{t,1}, \beta'_{t,2})' = q_t, \tag{2}$$

where $A_{t,1}$ is a $k \times k$ full rank matrix.

2. Solve (2) for $\beta_{t,1}$:

$$\beta_{t,1} = A_{t,1}^{-1}q_t - A_{t,1}^{-1}A_{t,2}\beta_{t,2}. \tag{3}$$

3. Take (3) and replace it in the measurement equation of model (1):

$$Y_t = Z_{t,1}\beta_{t,1} + Z_{t,2}\beta_{t,2} + \varepsilon_t$$
$$= Z_{t,1}\left(A_{t,1}^{-1}q_t - A_{t,1}^{-1}A_{t,2}\beta_{t,2}\right) + Z_{t,2}\beta_{t,2} + \varepsilon_t$$

$$\begin{aligned} &= Z_{t,1}A_{t,1}^{-1}q_t - Z_{t,1}A_{t,1}^{-1}A_{t,2}\beta_{t,2} + Z_{t,2}\beta_{t,2} + \varepsilon_t \\ \Rightarrow Y_t^* &\equiv Y_t - Z_{t,1}A_{t,1}^{-1}q_t = \left(Z_{t,2} - Z_{t,1}A_{t,1}^{-1}A_{t,2}\right)\beta_{t,2} + \varepsilon_t \\ &\equiv Z_{t,1}^*\beta_{t,2} + \varepsilon_t. \end{aligned}$$

4. Postulate a transition equation for the unrestricted state vector $\beta_{t,2}$ and finally get the following *reduced* linear state space model:

$$\begin{aligned} Y_t^* &= Z_{t,2}^*\beta_{t,2} + \varepsilon_t \ , \ \varepsilon_t \sim (0, H_t) \\ \beta_{t+1,2} &= T_{t,2}\beta_{t,2} + c_{t,2} + R_{t,2}\eta_{t,2} \ , \ \eta_{t,2} \sim (0, Q_{t,2}) \\ \beta_{1,2} &\sim (a_{1,2}, P_{1,2}). \end{aligned} \quad (4)$$

5. Apply the usual Kalman filter to the model in (4) and obtain $\hat{\gamma}_{t,2|j}$, for all $j \geq t$.

6. Reconstitute the estimates $\hat{\beta}_{t,2|j}$:

$$\hat{\beta}_{t,1|j} = A_{t,1}^{-1}q_t - A_{t,1}^{-1}A_{t,2}\hat{\beta}_{t,2|j}. \quad (5)$$

As Pizzinga (2010) claims, an interesting feature of this approach is that there is no need to worry about the state equation specifications until the reduced form is achieved in the 4th step of the described algorithm. This avoids any risk of obtaining a measurement equation that is theoretically inconsistent with the original state equation. Another good property that should be mentioned, and that will be evoked later in this paper, is that the reduced restricted Kalman filtering enables us to investigate the plausibility of the assumed linear restrictions by using information criteria (*e.g. AIC* and *BIC*).

3.2. The Model Setting and Inference

We now present our state space model for the exchange rate pass-through for a given index price as follows:

$$\Delta log\ p_t = \sum_{k=1}^{m} \beta_{kt}\Delta log\ e_{t-k} + \psi_0 + \psi_1 log\ ap_{jt} + \epsilon_t, \ \epsilon_t \sim NID(0, \sigma^2), \quad (6)$$

$$\beta_{t+1} = \beta_t + \gamma_1 IPA_{t-1}\mathbf{1}_{q\times 1} + \gamma_2 ip_{t-1}\mathbf{1}_{q\times 1} + \gamma_3 re_t\mathbf{1}_{q\times 1} + \gamma_4 o_t\mathbf{1}_{q\times 1} + \xi_t, \ \xi_t \sim NID(0, Q). \quad (7)$$

The former equation linearly relates the observed monthly log-variation of price to the log-variation of exchange rate until time $t - m$ and to an exogenous variable, the American price index, ap_t. The coefficients of $\Delta log\ e_{t-k}$ in equation (6) are the state coordinates and their dynamics are given in equation (7), which also sets the impact from the following determinants: IPA_t represents the inflationary environment; ip_t is the industrial production index; re_t is the log of exchange rate disalignment; and o_t is the log of the openness of the economy. The matrix $Q_{m \times m}$ is set diagonal, even though the components from the past-through (*i.e.* the state coordinates) do maintain degrees of dependency due to the presence of common determinants in the state equation.

The reducing method from the preceding subsection has been evoked in order to make the restrictions of *full* past-through ($\sum_{i=1}^{m} \beta_{it} = 1$) and of *null* past-through ($\sum_{i=1}^{m} \beta_{it} = 0$) attainable and testable. The completeness of the exchange rate passing-through (the first restriction) means that all the variation of the exchange rate is passed to the domestic prices. This is key for economic theory standpoint, since it means that the PPP hypothesis is acceptable. On the other hand, should the acceptance of a null exchange rate pass-through model be accepted as the most adequate scenario, it follows that the exchange rate movements do not have any effect on the domestic prices, and so, the monetary authority needs not be concerned with exchange rate movements to make monetary policy with such price indexes.

3.3. Model Selection and Inference

One of the purposes of this work is to identify the most adequate number of lags of the exchange rate, which is denoted by m. The hypotheses of completeness (or absence) of exchange rate passing-through would also be checked. For such, we use the following steps:

1. Diagnostic tests with the (standardized) residuals.

2. Information criteria, such as AIC and BIC.

3. Predictive power by comparing $PseudoR^2$ and MSE measures.

Finally, the significance of the parameters $\psi_0, \psi_1, \gamma_1, \gamma_2, \gamma_3$ and γ_4 will be tested under a likelihood ratio (LR) testing approach. Since both the reduced and the complete model maintain the standards for good properties of maximum likelihood estimation (cf. Pagan, 1980), it follows that, asymptotically, $LR \equiv 2\left[logL_{Max,Comp} - logL_{Max,Red}\right] \sim \chi_1^2$, in which $logL_{Max,Red}$ represents the maximum of the log-likelihood for a model with a particular explanatory variable dropped from the specification.

4. Results

4.1. Data and Computational Aspects

he data that we consider in this paper contain monthly observations from August of 1999 to January of 2007 of the Brazilian Wholesale Price Index (IPA), the IPC administered prices index, the American price index, the exchange rate between the Brazilian Real and the American Dollar, the Brazilian industrial production index and a measure of openness, which is the sum of imports and exports as a proportion of GDP. The decision of using data since August of 1999 is justified by the inflation target system adopted by the Banco Central (the institution in Brazil corresponding to the American Federal Reserve) in June of 1999. The data have been obtained from IPEA Data (www.ipeadata.gov.br).

The initialization of the restricted Kalman filtering has been carried out by the *exact initial Kalman filter* as presented in Durbin & Koopman (2001), ch.5. The unknown parameters, in turn, were estimated by maximum likelihood, in which we adopted the *exact log-likelihood function*; see Durbin & Koopman (2001), ch.7. The state space implementations have been carried out using the Ox 3.0 language (cf. Doornick, 2001) together with the Ssfpack 3.0 library for linear state space modeling (cf. Koopman *et al.*, 2002). For all the implementations, our computer was an 1.60Hz Intel Pentium M, with 2 Gb-RAM. Each of the estimations has taken less than 2 seconds, something that highlights the computational efficiency of the adopted state space framework.

4.2. Overall IPA

The most adequate model for the IPA series is the model with 7 lags on the exchange rate. Although only the 4 first states have confidence intervals that do not contain zero, this decision has been based on the lack of serial correlation for the residuals. Figure 1 shows the evolution of the coefficients along time. The $PseudoR^2 = 0.64$ suggests that the model provides a reasonable adjustment for the IPA. The long run pass-through given in Figure 2 has some variation when we compare the beginning of the sample with the end with a edge at the 2002, the year of elections preceding the Lula's administration in Brazil, a period of great volatility in the exchange rate.

We estimated the restricted models to verify whether the hypothesis of null and full exchange rate pass-through have some support from the data. The information criteria shown in Table 1 do not provide any evidence that these extremes allow a better fit. The LR significance tests are given in table 2. The p-values reveal no evidence that the proposed determinants help to explain the behavior of the pass-through.

Table 1. IPA-DI information criteria of the unrestricted and the restricted models.

Criterium	unrestricted	$\sum_{i=1}^{7} \beta_{i,t} = 0$	$\sum_{i=1}^{7} \beta_{i,t} = 1$
AIC	3.000	3.746	4.326
BIC	3.583	4.274	4.854

Table 2. IPA-DI estimated parameters and corresponding p-values in parenthesis.

ψ_1	γ_1	γ_2	γ_3	γ_4
0.001	0.000	0.001	0.001	0.000
(0.988)	(0.991)	(0.956)	(0.505)	(0.905)

4.3. First Level IPA Disaggregation

In order to evaluate the disaggregation effects on the exchange rate pass-through, we consider the estimation for some groups of products. The first level of disaggregation splits the

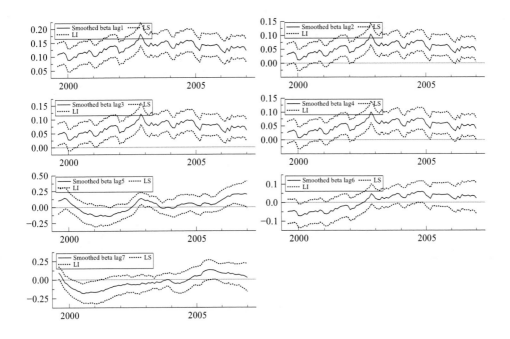

Figure 1. IPA-DI smoothed betas.

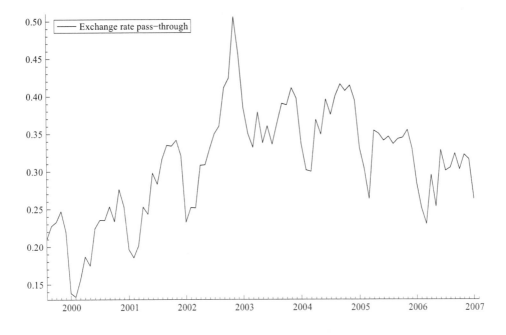

Figure 2. IPA-DI long run exchange rate pass-through.

overall IPA-DI into two main groups: consumption and production goods.

4.3.1. IPA-DI Consumption

The more adequate model for the IPA-DI consumption series has only a lag of the exchange rate, since it has the lower information criteria values and its residuals show no serial correlation. The $PseudoR^2 = 0.615434$ provides evidence in favor of goodness-of-fit. Since we decided to have only one lag for the exchange rate, the short and long run exchange rate pass-through are the same. Its variation over time is depicted in Figure 3. During the year of 2002, the exchange rate pass-through presented higher values compared to the rest of the sample period, probably due to the same explanations already given. Also, there is some indication of seasonal patterns, since the pass-through seems to be close to zero in the very begging of each year.

As shown in Table 3, the LR significance tests reveal that three proposed determinants are supported by the data. We also observe that inertial parameter ψ_1 is statistically significant for the measurement equation.

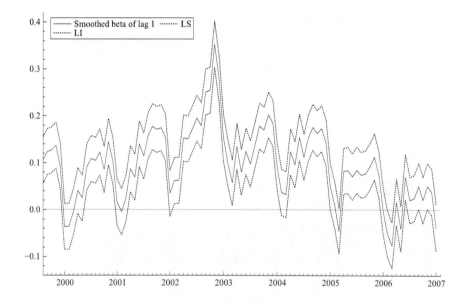

Figure 3. IPA-DI consumption smoothed betas.

Table 3. IPA-DI consumption estimated parameters and corresponding p-values in parenthesis.

	ψ_1	γ_1	γ_2	γ_3	γ_4
	0.561	-0.0004	0.012	0.005	-0.002
	(0.000)	(0.000)	(0.061)	(0.006)	(0.347)

4.3.2. IPA-DI production

Once again, the more adequate model for the IPA-DI production series was the model with only one lag of exchange rate pass-through. Also, the $PseudoR^2 = 0.729$ provides us with some confidence that the model fits the data in a proper way. The pass-through variation over time can be seen in Figure 4. This remarks some aspects similar to those found in the previous analysis, except for the lack of evidence on seasonality.

The LR significance tests shown in Table 4 provide us with two statistically significant determinants. We still have the inertial parameter ψ_1 statistically significant at the measurement equation.

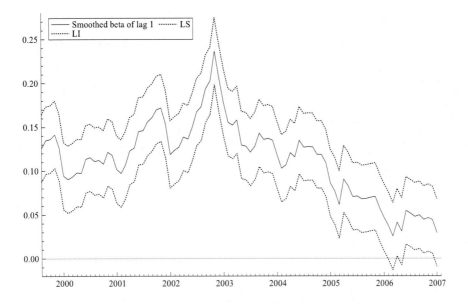

Figure 4. IPA-DI production smoothed betas.

Table 4. IPA-DI production estimated parameters and corresponding p-values in parenthesis.

ψ_1	γ_1	γ_2	γ_3	γ_4
0.590	-0.001	0.002	0.002	-0.001
(0.000)	(0.699)	(0.000)	(0.000)	(0.823)

4.4. IPC Administered Prices Index

The model adjusted with 2 lags of the exchange rate shows that the IPC administered prices index seems to be not responding to the exchange rate movements. As we can see in Figure 5, the states corresponding to all lags are varying around zero within the whole sample period. The long-run pass-through presented in figure 6 is also oscillating around zero. This shall be taken as the first symptom of absence of passing-through, and this is reinforced by the application of the restricted Kalman filtering, since the model which has the null pass-through restriction has the best information criteria; see Table 5.

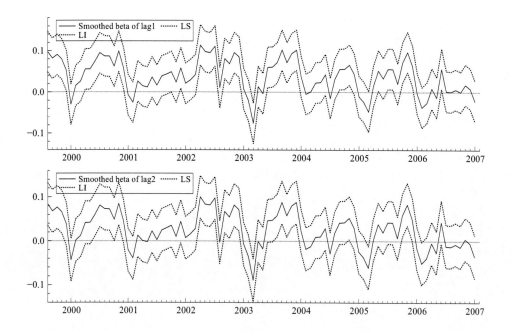

Figure 5. IPC administered prices smoothed betas.

Table 5. IPC administered prices information criteria of the unrestricted and the restricted models.

Criterium	unrestricted	$\sum_{i=1}^{2}\beta_{i,t}=0$	$\sum_{i=1}^{2}\beta_{i,t}=1$
AIC	2.838	2.801	5.291
BIC	3.144	3.051	5.541

5. Conclusions

This paper has focused on analyzing the Brazilian exchange rate pass-through, its plausible time-varying nature and the statistical significance of potential determinants of its dynamics. Another contribution of the paper was the the implementation of a restricted Kalman filtering technique to investigate whether the PPP and the null long-run exchange rate pass-through are reasonable scenarios. We applied these methods to some Brazilian price indexes series that, to our knowledge, had not been properly explored. Even though the results cannot be considered definitive, we have gathered some directions. The proposed econometric

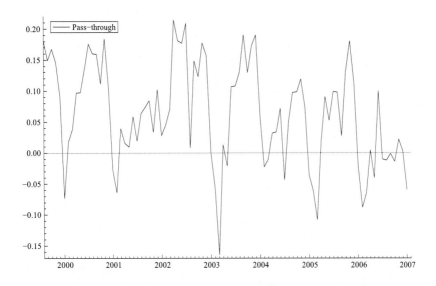

Figure 6. IPC administered prices long run pass-through.

modeling can be useful to investigate which variables can be helpful for lighting some aspects of variation of the exchange rate pass-through and to obtain some evidences about policy choices and aspects of the exchange rate pass-through.

References

[1] Betts, C. and Devereux, M.B. (2000). "Exchange Rate Dynamics in a Model of Pricing-to-Market". *Journal of International Economics*, 50(1), 215-244.

[2] Calvo, G. and Reinhart, C. (2000). "Fixing for your life". *NBER Working Paper*, n.8006, November.

[3] Campa, J.M. and Goldberg, L.S. (1995). "Exchange Rate Pass-through into Import Prices". *Federal Reserve Bank of New York*.

[4] Campa, J.M. and Goldberg, L.S. (2002). "Exchange Rate Pass-through into Import Prices: A Macro or Micro Phenomenon?". *Federal Reserve Bank of New York*.

[5] Doornick, J. A. (2001). *Ox 3.0: An Object-Oriented Matrix Programming Language*. Timberlake Consultants LTD.

[6] Durbin, J. and Koopman, S.J. (2001). *Time Series Analysis by State Space Methods*. Oxford.

[7] Fraga, A., Goldfajn, I. and Minella, A. (2003). "Inflation Targeting in Emerging Market Economies". *NBER Working Paper*, n.10.019.

[8] Goldberg, P.K. and Knetter, M.M. (1997). "Goods Prices and Exchange Rates: What Have We Learned?". *Journal of Economic Literature*, 35(3), 1243-1272.

[9] Goldfajn, I. and Valdes, R. (1999). "The Aftermath of Appreciations". *Pontifical Catholic University of Rio de Janeiro*, Working Paper n.396.

[10] Goldfajn, I. and Werlang, S.R.C. (2000). "The Pass-through from Depreciation to Inflation: A Panel Study". *Banco Central do Brasil*, Working Paper n.5.

[11] Harvey, A. C. (1989). *Forecasting, Structural Time Series Models and The Kalman Filter*. Cambridge University Press.

[12] Koopman, S. J, Shephard, N. and Doornik, J. A. (2002). "SsfPack 3.0 beta02: Statistical algorithms for models in state space". *Unpublished paper*. Department of Econometrics, Free University, Amsterdam.

[13] Menon, J. (1996). "Exchange Rate Pass-through". *Journal of Economic Surveys*, 9(2), 197-231.

[14] Olivei, G. P. (2002). "Exchange Rates and the Prices of Manufacturing Products Imported into the United States". *New England Economic Review*, First Quarter, 3-18.

[15] Pagan, A. (1980). "Some identification and estimation results for regression models with stochastically varying coefficients". *Journal of Econometrics*, 13, 341-363.

[16] Parsley, D. (1995). "Anticipated future shocks and exchange rate pass-through in the presence of reputation". *International Review of Economics*, 4(2).

[17] Pizzinga, A. (2010). "Restricted Kalman filtering: additional results". *International Statistical Review* (to appear).

[18] Pollard, P. and Coughlin, C. (2005). "Pass-through Estimates and the Choice of an Exchange Rate Index". *St. Louis FED Working Paper*.

[19] Taylor, J. (2000). "Low Inflation, Pass-through and the Pricing Power of Firms". *European Economic Review*, 44, 1389-1408.

[20] Sekine, T. (2006). "Time-Varying Exchange Rate Pass-Through: Experience of Some Industrial Countries". *BIS Working Paper*.

[21] Yang, J. (1996). "Exchange Rate Pass-through in U.S. Manufacturing Industries". *Review of Economics and Statistics*, 79(1), 95-104.

In: Kalman Filtering
Editor: Joaqun M. Gomez

ISBN: 978-1-61761-462-0
© 2011 Nova Science Publishers, Inc.

Chapter 10

QUANTIZED KALMAN FILTERING OF LINEAR STOCHASTIC SYSTEMS

Keyou You and Lihua Xie[*]
Nanyang Technological University, Singapore

PACS 05.45-a, 52.35.Mw, 96.50.Fm.

Keywords: Quantized estimation; Kalman filter; Riccati equation; Convergence analysis

1. Introduction

In recent years, networked systems such as wireless sensor networks (WSNs) have gained popularity in the research community due to their broad potential military and civilian applications. WSNs are generally composed of a large number of low-quality sensors equipped with limited communication capabilities and limited energy. However, a collective of these nodes can form a powerful network for information gathering and processing. Due to limited communication capacity and also for the sake of energy saving, the number of bits of information to be transmitted between nodes is quite restrictive. Therefore, signals such as sensor measurements are to be severely quantized before their transmissions. This introduces interesting and challenging problems such as what information is to be transmitted and how many bits are needed to represent the information in order to achieve a given performance.

Quantized estimation has been extensively investigated in literature such as [4, 6, 9, 11–13, 16] where various quantization schemes have been addressed. Luo studied the static parameter estimation under severe bandwidth constraints with each sensor's observation quantized to one or a few bits in [9]. The resulting estimators turn out to exhibit comparable variances with that of the estimator relying on un-quantized observations. Note that one of the major challenges of quantized estimation is that quantization is highly nonlinear and there lack of efficient filtering methods for nonlinear systems. By applying the particle

---

[*]This work was supported by the National Natural Science Foundation of China under grant NSFC 60828006. E-mail address: {youk0001, elhxie}@ntu.edu.sg.

filter to quantized measurements, a quantized version of particle filter is proposed by reconstructing the required probability density [6]. Unfortunately, the filtering performance is poor under a low number of quantization levels. With severe quantization, e.g., binary quantization, a dynamic quantization scheme based on feedback from the estimation center is designed for the state estimation of a hidden Markov model in [4]. The main disadvantage is that the solution involves a rather complicated on-line optimization and does not lead to a recursive filter.

A very interesting one-bit quantized innovations filter called sign-of-innovations Kalman filter (SOI-KF) was developed in [13] where a simple recursion involving time and measurement updates as in the Kalman filter is provided for state estimation. The remarkable feature of the SOI-KF lies in its simplicity which offers advantages for applications in resource-constrained sensor networks. However, the rough quantization of the SOI inevitably induces large estimation errors. A better estimate can be derived by a proper design of quantizer even with the same one bit budget. In fact, when the innovation is insignificant, by quantizing it into 1 or -1 will lead to a larger error than simply ignoring it. As such, this chapter presents a finer quantization scheme by introducing a dead zone for the quantization of innovation. To be specific, only the innovation outside the dead zone will be quantized into 1 or -1 and transmitted to the estimator. When more than one bit information can be transmitted, a multi-level quantized innovations Kalman filter (MLQ-KF) is developed. We note that an iterative sign of innovations Kalman filter allowing sequential transmission of multiple bits of information was also studied in [10], which requires a slightly higher computational load than that of MLQ-KF in this chapter.

Despite of the computational advantages of the SOI-KF and MLQ-KF, they are established under the critical assumption that the predicted density of state is Gaussian. Strictly speaking, this assumption does not hold due to the nonlinearity of quantizer. To overcome this limitation, we apply a logarithmic quantizer to the innovation and model the quantization error as a sector-bound parameter uncertainty [3]. More detailed discussion on the logarithmic quantizer is referred to [2, 3, 18].

The results of this chapter are briefly summarized as follows. First, given a multi-level quantization scheme, we derive a corresponding multi-level quantized filter that achieves the minimum mean square error (MMSE) under the assumption that the predicted probability density of state is Gaussian. The filter is given in terms of a simple Riccati difference equation (RDE) similar to that in the Kalman filter. To reveal the relationship between the number of quantization levels and the filtering performance, the optimal thresholds and the optimal multi-level quantized innovations Kalman filter are obtained. The convergence of the MLQ-KF to the Kalman filter when the number of quantization levels goes to infinity is established. We also present a max-min scheme to design a robust quantizer. The second quantized filter is based on a logarithmic quantizer acting on the innovation which does not require the Gaussian assumption on the predicted probability density of state. The associated quantized filter is again implemented recursively as the Kalman filter and are given in terms of a simple RDE. Simulation and experiment results are provided to illustrate the effectiveness and advantages of the proposed quantized filters.

The rest of this chapter is organized in the following fashion. The problem is formulated in Section 2. The MLQ-KF is derived in Section 3, where we first derive the filter based on a general quantization scheme and then proceed to seek the optimal quantization. A max-min quantization optimization is also described to address the robust quantization problem. A worst-case quantized filter via logarithmic quantization is presented in Section 4. The filter stability is analyzed in Section 5. Simulations and experiments are carried out in Section 6. Some concluding remarks are drawn in Section 7.

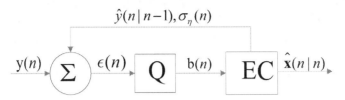

Figure 1. Network configuration.

Notations: We use $p(x|y)$ to denote the conditional probability density function of the random variable $\mathbf{x}$ given the random variable $\mathbf{y}$. $\mathcal{N}(\mathbf{x}; m, P)$ is the normal distribution with mean m and covariance matrix P. $\phi(\cdot)$ denotes the density function of a standard Gaussian random variable and $T(\cdot)$ represents its tail probability function. For the sake of simplicity, we adopt the notations that $z^N = \{z_1, \ldots, z_N\}$ and $z_*^N = \{z_1^*, \ldots, z_N^*\}$.

2. Problem Formulation

Consider the following discrete-time linear stochastic system:

$$\mathbf{x}(n) = A(n)\mathbf{x}(n-1) + \mathbf{w}(n); \qquad (1)$$
$$y(n) = \mathbf{h}(n)^T\mathbf{x}(n) + v(n), \qquad (2)$$

where $\mathbf{w}(n) \in \mathbb{R}^p$ and $v(n) \in \mathbb{R}$ are two uncorrelated white Gaussian noises with variances $W(n)$ and $\sigma_v^2(n)$, respectively. $A(n) \in \mathbb{R}^{p \times p}$ and $\mathbf{h}(n) \in \mathbb{R}^p$ are respectively a bounded time-varying matrix and a column vector. Extension to the case of vector measurement is straightforward. The initial state $\mathbf{x}(0)$ is assumed to be Gaussian and independent of the noises $\mathbf{w}(n)$ and $v(n)$.

Suppose that the sensor node can access the one-step ahead prediction of the output $\hat{y}(n|n-1)$, which is either broadcasted by the estimation center(EC) (see Fig. 1) or computed by the sensor itself. Because of the communication constraint, the sensor information has to be quantized before transmitted to the EC. Once the sensor receives an observation $y(n)$, it is able to compute the innovation $\eta(n) := y(n) - \hat{y}(n|n-1)$, which is fed to a quantizer. Note that the innovation represents new information brought by $y(n)$ about the state $\mathbf{x}(n)$ that is not conveyed by previous observations. The intuition supporting the idea of quantizing the innovation attributes to the fact that the amplitude of the innovation is generally smaller than that of the measurement $y(n)$ and thus can be represented by a fewer number of bits. The quantized innovation is then sent to a remote estimation center via a rate limited channel to produce a state estimator.

To reduce the communication cost, we first discuss the state estimation based on quantized innovations using a moderate number of bits. The main challenge of finding an optimal filter lies in the fact that quantization is highly nonlinear and the classical nonlinear state estimation tools, such as the unscented Kalman filter and the particle filter [1], typically require intensive computations, which may be prohibitive in WSNs equipped with limited computational capacity. The problem is further complicated by the fact that the conventional approach of viewing quantization error as an additive white noise is no longer justifiable due to the coarse quantization. By treating a quantizer as an information coder, there has been a new line of research on quantized feedback control. Another important approach of treating the quantization error as a bounded parameter uncertainty to the system has attracted a lot of interests as well. For example, a sector bound approach is utilized to study the effect of logarithmic quantization in [3]. We shall propose two types of filter design methods in this chapter, which result in simple recursive filters with low computational requirements.

3. MMSE Quantized Filter Design

In this section, a quantized innovations Kalman filter that has a similar form as the Kalman filter is provided.

Let $\sigma_\eta^2(n) = \mathbf{h}(n)^T P(n|n-1)\mathbf{h}(n) + \sigma_v^2(n)$ be the prediction error (innovation) covariance, where $P(n|n-1)$ is the prediction error covariance matrix of the state at time instant n. Note that $\sigma_\eta(n)$ and the one-step output prediction $\hat{y}(n|n-1)$, either broadcasted by the EC or computed by the sensor node itself, can be accessed by the sensor node. The innovation $\eta(n)$ will be normalized by $\epsilon(n) = \eta(n)/\sigma_\eta(n)$ before sent to the quantizer. Due to the symmetric property of the standard Gaussian distribution, we consider the associated symmetric quantizer (denote by 'Q' in Fig. 1) with *quantization thresholds* $\{\pm z_i\}_{i=1}^N$, $0 = z_0 < z_1 < \cdots < z_N < \infty = z_{N+1}$, namely, the output $b(n)$ of the quantizer $q(\cdot)$ is given by

$$b(n) := q(\epsilon(n)) = \begin{cases} z_N, & z_N < \epsilon(n) \\ \vdots & \vdots \\ z_1, & z_1 < \epsilon(n) \leq z_2 \\ z_0, & 0 < \epsilon(n) \leq z_1 \\ -q(-\epsilon(n)), & \epsilon(n) \leq 0. \end{cases} \quad (3)$$

Note that when $b(n) = z_0$, it will not be transmitted to the EC. That is, for a 1-bit budget, the quantizer has 3 quantization levels unlike that in [13] where innovations are quantized to 1 or -1, depending on whether they are positive or negative. Intuitively, our approach should perform better since one more quantization level is added, which will be confirmed later in theory and simulation. Assuming that there exists no error in transmitting the quantized message $b(n)$ to the estimation center, our goal is to find and analyze the MMSE estimate for the state based on the quantized innovations. To elaborate on this point, let $\mathbf{b}_{0:n}$ denote the messages received from the sensor up to the time instant n, i.e. $\mathbf{b}_{0:n} := \{b(0), b(1), \cdots, b(n)\}$ and $\hat{\mathbf{x}}(n|n)$ represent the MMSE estimate of $\mathbf{x}(n)$ given $\mathbf{b}_{0:n}$. It is well known that the MMSE estimate is obtained by computing the conditional expectation, i.e.,

$$\hat{\mathbf{x}}(n|n) := E[\mathbf{x}(n)|\mathbf{b}_{0:n}] = \int_{\mathbb{R}^p} \mathbf{x}(n) p(\mathbf{x}(n)|\mathbf{b}_{0:n}) d\mathbf{x}(n). \quad (4)$$

The following equalities can be easily obtained [13]:

$$\hat{\mathbf{x}}(n|n-1) := E[\mathbf{x}(n)|\mathbf{b}_{0:n-1}] = A(n)\hat{\mathbf{x}}(n-1|n-1), \tag{5}$$
$$\hat{y}(n|n-1) := E[y(n)|\mathbf{b}_{0:n-1}] = \mathbf{h}(n)^T\hat{\mathbf{x}}(n|n-1). \tag{6}$$

The prediction error covariance matrix is defined by

$$\begin{aligned}P(n|n-1) :&= E[(\hat{\mathbf{x}}(n|n-1) - \mathbf{x}(n))(\hat{\mathbf{x}}(n|n-1) - \mathbf{x}(n))^T] \\ &= A(n)P(n-1|n-1)A(n)^T + W(n),\end{aligned} \tag{7}$$

where the filtering error covariance matrix is given by

$$P(n|n) := E[(\hat{\mathbf{x}}(n|n) - \mathbf{x}(n))(\hat{\mathbf{x}}(n|n) - \mathbf{x}(n))^T]. \tag{8}$$

We make an assumption for the design of a recursive quantized innovations Kalman filter.
Assumption 1: The one-step prediction density of the state is Gaussian. Precisely, $p(\mathbf{x}(n)|\mathbf{b}_{0:n-1}) \sim \mathcal{N}(\mathbf{x}(n); \hat{\mathbf{x}}(n|n-1), P(n|n-1))$.

Strictly speaking, the system state conditioned on the quantized innovations cannot remain Gaussian due to the nonlinear operator of quantization. To enable the development of a simple and practically useful recursive filter, we make the Gaussian assumption as in [13]. A similar hypothesis can also be found in [5,7]. It will be demonstrated later that the performance of the quantized filter for a moderate number of bits (say 2 or more bits) is close to the Kalman filter, suggesting that the approximation is reasonable.

3.1. Multi-level Quantized Filtering

Since the probability of distribution $p(\mathbf{x}(n)|\mathbf{b}_{0:n})$ necessary for deriving the optimal estimate in (4) is not strictly Gaussian, it leads to the propagation of posterior density impractical. Consequently, it is impossible to find an optimal state estimator with a low computational load. However, under **Assumption 1**, the nonlinear quantized filtering can be nicely solved.

Theorem 1. *([21]) Consider the system described by (1) and (2). Given a multi-level quantization of normalized innovations in (3), if **Assumption 1** holds, the correction step of the MMSE estimate of the state can be computed by*

$$\hat{\mathbf{x}}(n|n) = \hat{\mathbf{x}}(n|n-1) + \frac{f(z^N, b(n))P(n|n-1)\mathbf{h}(n)}{\sqrt{\mathbf{h}(n)^T P(n|n-1)\mathbf{h}(n) + \sigma_v^2(n)}}, \tag{9}$$

$$P(n|n) = P(n|n-1) - F(z^N)\frac{P(n|n-1)\mathbf{h}(n)\mathbf{h}(n)^T P(n|n-1)}{\mathbf{h}(n)^T P(n|n-1)\mathbf{h}(n) + \sigma_v^2(n)}, \tag{10}$$

with

$$f(z^N, b(n)) = \sum_{k=-N}^{N} I_{\{z_k\}}(b(n))\frac{\phi(z_k) - \phi(z_{k+1})}{T(z_k) - T(z_{k+1})}, \tag{11}$$

$$F(z^N) = 2\sum_{k=0}^{N} \frac{[\phi(z_k) - \phi(z_{k+1})]^2}{T(z_k) - T(z_{k+1})}, \tag{12}$$

where $z_{-j} = -z_j$ and $I_A(\cdot)$ is the standard sign function, i.e., for any subset $A \subset R$,

$$I_A(x) = \begin{cases} 1, & x \in A, \\ 0, & otherwise. \end{cases}$$

Proof. It is obvious that the σ-algebra generated by $\mathbf{b}_{0:n}$ is a sub σ-algebra generated by $y(n)$ and $\mathbf{b}_{0:n-1}$. Thus, we obtain

$$\hat{\mathbf{x}}(n|n) = E[\mathbf{x}(n)|\mathbf{b}_{0:n}] = E[E[\mathbf{x}(n)|\mathbf{b}_{0:n-1}, y(n)]|\mathbf{b}_{0:n}]. \qquad (13)$$

Under **Assumption 1**, the posterior density $p(\mathbf{x}(n)|\mathbf{b}_{0:n-1}, y(n))$ is computed as

$$p(\mathbf{x}(n)|\mathbf{b}_{0:n-1}, y(n)) = \frac{p(\mathbf{x}(n)|\mathbf{b}_{0:n-1})p(y(n)|\mathbf{x}(n))}{\int_{\mathbb{R}^n} p(\mathbf{x}(n)|\mathbf{b}_{0:n-1})p(y(n)|\mathbf{x}(n))d\mathbf{x}(n)}$$
$$= \mathcal{N}(\mathbf{x}(n), \hat{\mathbf{x}}(n|n-1), P(n|n-1))\mathcal{N}(y(n), \mathbf{h}(n)^T\mathbf{x}(n), \sigma_v^2(n)). \qquad (14)$$

By following the technique of the Kalman filter, the inner conditional expectation in (13) is given by

$$\hat{\mathbf{x}}(n|n)^* := E[\mathbf{x}(n)|\mathbf{b}_{0:n-1}, y(n)] = \hat{\mathbf{x}}(n|n-1) + K(n)(y(n) - \mathbf{h}(n)^T\hat{\mathbf{x}}(n|n-1)), \quad (15)$$

where the observer gain $K(n) = \frac{P(n|n-1)\mathbf{h}(n)}{\mathbf{h}(n)^T P(n|n-1)\mathbf{h}(n) + \sigma_v^2(n)}$.

Due to that $E[\epsilon(n)|\mathbf{b}_{0:n}] = f(z^N, b(n))$ and the fact that $\hat{\mathbf{x}}(n|n-1)$ is adapted to the σ-algebra generated by $\mathbf{b}_{0:n}$, it follows that

$$\begin{aligned}\hat{\mathbf{x}}(n|n) &= E[\hat{\mathbf{x}}(n|n)^*|\mathbf{b}_{0:n}] \\ &= \hat{\mathbf{x}}(n|n-1) + \frac{f(z^N, b(n))P(n|n-1)\mathbf{h}(n)}{\sqrt{\mathbf{h}(n)^T P(n|n-1)\mathbf{h}(n) + \sigma_v^2(n)}}. \end{aligned} \qquad (16)$$

In addition, it clearly holds that

$$P(n|n) = E[E[(\mathbf{x}(n) - \hat{\mathbf{x}}(n|n))(\mathbf{x}(n) - \hat{\mathbf{x}}(n|n))^T|\mathbf{b}_{0:n-1}, y(n)]]. \qquad (17)$$

Again, adopting the technique of deriving the Kalman filter leads to that

$$\begin{aligned}&E[(\mathbf{x}(n) - \hat{\mathbf{x}}(n|n))(\mathbf{x}(n) - \hat{\mathbf{x}}(n|n))^T|\mathbf{b}_{0:n-1}, y(n)] \\ &= E[(\mathbf{x}(n) - \hat{\mathbf{x}}(n|n)^*)(\mathbf{x}(n) - \hat{\mathbf{x}}(n|n)^*)^T|\mathbf{b}_{0:n-1}, y(n)] + \\ &\quad E[(\hat{\mathbf{x}}(n|n) - \hat{\mathbf{x}}(n|n)^*)(\hat{\mathbf{x}}(n|n) - \hat{\mathbf{x}}(n|n)^*)^T|\mathbf{b}_{0:n-1}, y(n)] \\ &= P(n|n)^* + E[(\hat{\mathbf{x}}(n|n) - \hat{\mathbf{x}}(n|n)^*)(\hat{\mathbf{x}}(n|n) - \hat{\mathbf{x}}(n|n)^*)^T|\mathbf{b}_{0:n-1}, y(n)] \\ &= P(n|n)^* + \frac{P(n|n-1)\mathbf{h}(n)\mathbf{h}(n)^T P(n|n-1)}{\mathbf{h}(n)^T P(n|n-1)\mathbf{h}(n) + \sigma_v^2(n)}(\epsilon(n) - f(z^N, b(n)))^2, \quad (18)\end{aligned}$$

where $P(n|n)^* = P(n|n-1) - \frac{P(n|n-1)\mathbf{h}(n)\mathbf{h}(n)^T P(n|n-1)}{\mathbf{h}(n)^T P(n|n-1)\mathbf{h}(n) + \sigma_v^2(n)}$. The second equality follows from (14) and (15) while the third equality follows from (15) and (16). The last equality is due to that $f(z^N, b(n))$, $y(n)$ and the predictor $\hat{\mathbf{x}}(n|n-1)$ are adapted to σ-filed generated by $\mathbf{b}_{0:n-1}$ and $y(n)$. Under **Assumption 1**, we further obtain that

$$\begin{aligned} E\left[(\epsilon(n) - f(z^N, b(n)))^2\right] &= E\left[E[(\epsilon(n) - f(z^N, b(n)))^2 | \mathbf{b}_{0:n-1}]\right] \\ &= E\left[E[\epsilon^2(n) | \mathbf{b}_{0:n-1}]\right] - E\left[E[f^2(z^N, b(n)) | \mathbf{b}_{0:n-1}]\right] \\ &= 1 - F(z^N). \end{aligned}$$

In view of (17) and (18), the estimation error covariance is computed as follows:

$$P(n|n) = P(n|n-1) - F(z^N)\frac{P(n|n-1)\mathbf{h}(n)\mathbf{h}(n)^T P(n|n-1)}{\mathbf{h}(n)^T P(n|n-1)\mathbf{h}(n) + \sigma_v^2(n)}. \tag{19}$$

The proof is thus completed. ∎

In comparison with the Kalman filter, the MLQ-KF is a recursive application of (5), (7), (9) and (10), which is similar to the Kalman filter. Both require only a few algebraic operations per iteration. However, using the quantized innovation induces a penalty on the optimal uncertainty reduction by a factor $1 - F(z^N)$ due to the loss of information in quantization of innovation, as observed in (10). This penalty can be minimized by choosing optimal quantization thresholds. In the sequel, we call $F(z^N)$ *performance recovery factor*. The implementation of the MLQ-KF is summarized in **Algorithm 1**.

Algorithm 1: Implementation of MLQ-KF

- Initialization:
  Prior estimate $\hat{\mathbf{x}}(-1|-1)$ and covariance matrix $P(-1|-1)$.

- Predict (time update):

$$\begin{aligned} \hat{\mathbf{x}}(n|n-1) &= A(n)\hat{\mathbf{x}}(n-1|n-1); \\ P(n|n-1) &= A(n)P(n-1|n-1)A^T(n) + W(n). \end{aligned}$$

- Correction (quantized measurement update):

$$\begin{aligned} \hat{\mathbf{x}}(n|n) &= \hat{\mathbf{x}}(n|n-1) + \frac{f(z^N, b(n))P(n|n-1)\mathbf{h}(n)}{\sqrt{\mathbf{h}(n)^T P(n|n-1)\mathbf{h}(n) + \sigma_v^2(n)}}, \\ P(n|n) &= P(n|n-1) - F(z^N)\frac{P(n|n-1)\mathbf{h}(n)\mathbf{h}(n)^T P(n|n-1)}{\mathbf{h}(n)^T P(n|n-1)\mathbf{h}(n) + \sigma_v^2(n)}. \end{aligned}$$

It is interesting to note that if the quantized message is sent through a lossy digital channel and the packet dropout process follows an i.i.d. binary Bernoulli process $\{\gamma(n)\}_{n\geq 0}$ with packet dropout rate r, i.e., $P(\gamma(n) = 0) = r$, the corresponding MLQ-KF can be similarly derived by defining

$$f(z^N, b(n)) = \gamma(n) \sum_{k=-N}^{N} I_{\{z_k\}}(b(n)) \frac{\phi(z_k) - \phi(z_{k+1})}{T(z_k) - T(z_{k+1})},$$

$$F(z^N) = 2(1-r) \sum_{k=0}^{N} \frac{[\phi(z_k) - \phi(z_{k+1})]^2}{T(z_k) - T(z_{k+1})},$$

where $\gamma(n) = 1$ indicates that the packet is successfully received by the estimator while $\gamma(n) = 0$ corresponds to packet dropout. This implies that for the i.i.d. packet dropout, the *performance recovery factor* is further reduced by the successful transmission probability $1 - r$. Note that both quantization and packet dropout result in information losses. We mention that the joint effects of packet dropout and finite communication data rate on mean square stabilizability of a linear system are exactly quantified in [19, 20].

3.2. Optimal Quantization Thresholds

Note that the performance of the MLQ-KF is approximately measured by the covariance matrix $P(n|n)$. According to Theorem 1, it is clear that the optimal thresholds z_*^N of the quantizer in (3) can be obtained by the maximization of $F(z^N)$, i.e.,

$$z_*^N = \arg \max_{z^N} F(z^N). \tag{20}$$

We interpret the optimization problem in (20) as the optimal quantization to minimize the quadratic distortion of a standard Gaussian random variable so that it can be numerically solved by the Lloyd's algorithm [8]. Let

$$\alpha_k = \frac{\int_{z_k}^{z_{k+1}} x\phi(x)dx}{\int_{z_k}^{z_{k+1}} \phi(x)dx}, \forall k \in \{0, \ldots, N\}.$$

It can be easily verified that

$$F(z^N) = 1 - 2 \sum_{k=0}^{N} \int_{z_k}^{z_{k+1}} (x - \alpha_k)^2 \phi(x) dx \triangleq 1 - 2D(z^N).$$

Hence, we obtain that

$$z_*^N = \arg \min_{z^N} D(z^N).$$

Using the Lloyd's algorithm, the optimal quantization thresholds can be numerically solved. Alternatively, they can be obtained by evoking the Matlab command,

$$[x, y] = \text{fmincon}(fun, x_0, A, b, Aeq, beq, lb, ub).$$

See Table 1 for the optimal quantization thresholds and the optimal performance recovery factor, respectively for $N = 1, 2, 3$ and 4. Note that the above optimization problem does not have a closed form solution. However, in our case the optimal quantizer is fixed for a

Table 1. Solutions to (20) for optimal quantization thresholds

N=1	N=2	N=3	N=4
$z_1^* = 0.612$	$z_1^* = 0.382$ $z_2^* = 1.244$	$z_1^* = 0.280$ $z_2^* = 0.874$ $z_3^* = 1.611$	$z_1^* = 0.221$ $z_2^* = 0.681$ $z_3^* = 1.198$ $z_4^* = 1.866$
$F(z_*^1) = 0.810$	$F(z_*^2) = 0.920$	$F(z_*^3) = 0.956$	$F(z_*^4) = 0.972$

given bit number and is designed off-line. Thus, a numerical optimization does not cause any problem.

As mentioned above, $b(n) = 0$ will not be transmitted to the EC as it will not improve the predicted estimate. Thus, $[-z_1, z_1]$ can be viewed as a dead zone. In the SOI-KF of [13]), there does not exist a dead zone, i.e., $z_1 = 0$. The comparison of the performance recovery factor for the cases of with and without dead zone is showed in Fig. 2. For single bit quantization, the optimal quantized filter with a dead zone gives a much improved performance recovery factor of 0.810 as compared to the performance recovery factor of $(2/\pi \approx 0.637)$ of SOI-KF. Fig. 2 also indicates that the higher of a bit rate, the larger of the performance recovery factor. In fact, with $N = 4$, the performance recovery factor is already very close to 1. The relationship given in Fig. 2 will be useful in estimating the required bits number for desirable filtering performance and the stability of filter.

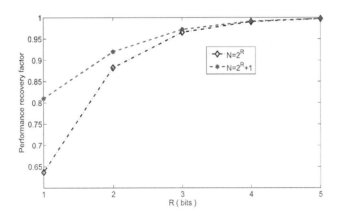

Figure 2. $F(z_*^N)$ versus R bits

3.3. Convergence Analysis

The following result establishes the convergence of the MLQ-KF to the Kalman filter when $N \to \infty$.

Theorem 2. ([21]) Let $\triangle = sup_{k \in \mathbb{N}} \triangle_k$, where $\triangle_k = |z_k - z_{k+1}|$, consider the symmetric quantization thresholds of quantizer (3) satisfying that

1. $\triangle_k \leq \triangle \to 0$;
2. $S(N) = \sum_{k=1}^{N-1} \triangle_k \to \infty$ as $N \to \infty$.

Then, $z_k \to z_{k+1}$ implies that

$$\frac{\phi(z_k) - \phi(z_{k+1})}{T(z_k) - T(z_{k+1})} \to z_{k+1} \qquad (21)$$

and

$$F(z^N) \to 1 \text{ as } N \to \infty. \qquad (22)$$

Proof. For $\triangle \to 0$, $\triangle z_k \to 0$ as well. Hence

$$\begin{aligned}
\lim_{z_k \to z_{k+1}} \frac{\phi(z_k) - \phi(z_{k+1})}{T(z_k) - T(z_{k+1})} &= \lim_{\triangle z_k \to 0} \frac{\phi(z_{k+1} - \triangle z_k) - \phi(z_{k+1})}{\int_{z_{k+1} - \triangle z_k}^{z_{k+1}} \phi(t) dt} \\
&= \lim_{\triangle z_k \to 0} \frac{\phi'(z_{k+1} - \triangle z_k)}{\phi(z_{k+1} - \triangle z_k)} \\
&= z_{k+1}.
\end{aligned} \qquad (23)$$

In addition, the following can be established.

$$\begin{aligned}
S : &= \sum_{k=1}^{\infty} \frac{[\phi(z_k) - \phi(z_{k+1})]^2}{T(z_k) - T(z_{k+1})} \\
&= \frac{1}{\sqrt{2\pi}} \sum_{k=1}^{\infty} \frac{[e^{(-\frac{z_k^2}{2})} - e^{(-\frac{z_{k+1}^2}{2})}]^2}{\int_{z_k}^{z_{k+1}} e^{(-\frac{t^2}{2})} dt} \\
&= \frac{1}{\sqrt{2\pi}} \sum_{k=1}^{\infty} \frac{[e^{(-\frac{z_k^2}{2})} - e^{[-\frac{(z_k + \triangle z_k)^2}{2}]}]^2}{e^{-\frac{(z_k + \theta_k \triangle z_k)^2}{2}} \triangle z_k} \\
&= \frac{1}{\sqrt{2\pi}} \sum_{k=1}^{\infty} e^{-\frac{z_k^2}{2}} \frac{[1 - e^{-(\frac{(\triangle z_k)^2}{2} - z_k \triangle z_k)}]^2}{\triangle z_k e^{-\frac{(\theta_k \triangle z_k)^2}{2} - \theta_k z_k \triangle z_k}}
\end{aligned} \qquad (24)$$

where $0 \leq \theta_k \leq 1$. Using the Taylor expansion for the exponential function in (24), we have $S := S_0 + S_1$

$$S_1 = \frac{1}{\sqrt{2\pi}} \sum_{k=1}^{\infty} z_k^2 e^{-\frac{z_k^2}{2}} \triangle z_k \xrightarrow{(\triangle \to 0)} \frac{1}{\sqrt{2\pi}} \int_0^{\infty} t^2 e^{-t^2/2} dt = \frac{1}{2} \qquad (25)$$

and there exist nonnegative integers $0 \leq i, j, u, v \leq n < \infty$ and $|c_{i,j,v,u}| < \infty$ such that

$$\begin{aligned}
S_0 &= \frac{1}{\sqrt{2\pi}} \sum_{i,j,u,v} \sum_{k=1}^{\infty} [c_{i,j,v,u} e^{(-\frac{z_k^2}{2})} (\triangle z_k)^{2+i} z_k^j \theta_k^u o((\triangle z_k)^v)] \\
&< \triangle \sum_{i,j,u,v} c_{i,j,v,u} \frac{1}{\sqrt{2\pi}} \times \sum_{k=1}^{\infty} z_k^j e^{(-\frac{z_k^2}{2})} \triangle z_k := C\triangle \to 0 \text{ (as } \triangle \to 0\text{)}. \quad (26)
\end{aligned}$$

where C is a finite constant because for $\triangle \to 0$,

$$\frac{1}{\sqrt{2\pi}} \sum_{k=1}^{\infty} z_k^j e^{-\frac{z_k^2}{2}} \triangle z_k \to \frac{1}{\sqrt{2\pi}} \int_0^{\infty} t^j e^{-t^2/2} dt < E[\mathbf{x}^j] < \infty$$

for standard Gaussian random variable $\mathbf{x}$ and $\forall j \in \mathbb{N}$. This completes the proof. ∎

Remark 1. As $z_k \to z_{k+1}$, then $\epsilon(n) = \frac{\eta(n)}{\sigma_\eta(n)} \to z_{k+1}$ and thus $f(z^N, b(n)) \to \epsilon(n)$ by (21). In light of (9),(10) and (22), the filter becomes

$$\hat{\mathbf{x}}(n|n) \to \hat{\mathbf{x}}(n|n-1) + \frac{P(n|n-1)\mathbf{h}(n)\eta(n)}{\mathbf{h}(n)^T P(n|n-1)\mathbf{h}(n) + \sigma_v^2(n)},$$

$$P(n|n) \to P(n|n-1) - \frac{P(n|n-1)\mathbf{h}(n)\mathbf{h}(n)^T P(n|n-1)}{\mathbf{h}(n)^T P(n|n-1)\mathbf{h}(n) + \sigma_v^2(n)},$$

which implies that the MLQ-KF converges to the Kalman filter.

3.4. Robust Quantization

This section will examine the design of a robust quantizer when the prediction error covariance is not known exactly. In practice, the error covariance may not be accurate due to quantization error or we may not be able to transmit the covariance every time instant due to limited communication capacity.

Let $\delta(n) = \sigma_\eta(n)/\sigma_{e,\eta}(n) \in [\underline{\delta}, \bar{\delta}]$, where $\sigma_\eta(n)$ is the actual prediction error covariance and $\sigma_{e,\eta}(n)$ is an estimated one. $\underline{\delta}, \bar{\delta}$ are known lower and upper bounds of $\delta(n)$. Denote $\hat{\epsilon}(n) = \eta(n)/\sigma_{e,\eta}(n)$, the quantization scheme (3) is modified as:

$$b(n) := q(\hat{\epsilon}(n)) = \begin{cases} z_N, & \delta z_N < \hat{\epsilon}(n) \\ \vdots & \vdots \\ z_1, & \delta z_1 < \hat{\epsilon}(n) \leq \delta z_2 \\ z_0, & 0 < \hat{\epsilon}(n) \leq \delta z_1 \\ -q(\hat{\epsilon}(n)), & \hat{\epsilon}(n) \leq 0. \end{cases} \quad (27)$$

By some minor modifications, the performance recovery factor is revised as $F(\delta z^N)$. Since δ is not exactly known, we design a robust max-min quantizer by the following optimization:

$$\bar{z}_*^N = \arg\max_{z^N} \min_{\delta \in [\underline{\delta}, \bar{\delta}]} F(\delta z^N). \quad (28)$$

Examining the properties of function $F(z^N)$ reveals the fact that given z^N, the minimum is attained at the boundaries $\underline{\delta}$ or $\bar{\delta}$. Hence, the optimization problem is reduced to

$$\bar{z}_*^N = \arg\max_{z^N} \min\{F(\underline{\delta} z^N), F(\bar{\delta} z^N)\}.$$

It is easy to check that

$$\frac{1}{\max_{z^N} \min\{F(\underline{\delta} z^N), F(\bar{\delta} z^N)\}} = \min_{z^N} \max\left\{\frac{1}{F(\underline{\delta} z^N)}, \frac{1}{F(\bar{\delta} z^N)}\right\}. \quad (29)$$

We recall the command $x = \text{fminimax}(fun, x_0)$ in the Optimization Toolbox of Matlab to get numerical solutions. For instance, let $\underline{\delta} = 1$ and $\bar{\delta} = 2$. For $N = 1$, $\bar{z}_1^* = -0.4086$ and $F(\delta^* \bar{z}_*^1) = 0.7890$ with $\delta^* = 1$ or 2; for $N = 2$, $\bar{z}_1^* = 0.2498$, $\bar{z}_2^* = 0.8638$, and $F(\delta^* \bar{z}_*^2) = 0.8994$ with $\delta^* = 1$ or 2. For $N = 1$ and $\bar{\delta} \to \infty$, the robust quantized filter will reduce to the SOI-KF.

4. Worst-case Quantized Filter

The quantized innovations Kalman filter derived in the last section is of a similar form as the standard Kalman filter and is of practical importance. However, recall that the derivation of the quantized innovations filter requires **Assumption 1** which does not hold theoretically. To overcome this problem, we present a worst-case quantized filter in this section by applying a logarithmic quantizer to the innovations. The estimator is constructed as follows:

$$\hat{\mathbf{x}}(n) = A(n)\hat{\mathbf{x}}(n-1) + L(n)Q(y(n) - \mathbf{h}(n)^T \hat{\mathbf{x}}(n-1)), \tag{30}$$

where $Q(\cdot)$ is a logarithmic quantizer and $L(n)$ is the estimation gain to be designed. We now recall the definition of a logarithmic quantizer [3]. A quantizer is called a *logarithmic quantizer* if it has the form:

$$Q(v) = \begin{cases} u^{(i)}, & \text{if } \frac{1}{1+\delta} u^{(i)} < v \le \frac{1}{1-\delta} u^{(i)}, v > 0; \\ 0, & \text{if } v = 0; \\ -Q(-v), & \text{if } v < 0 \end{cases} \tag{31}$$

where $u^{(i)}$, $i \in \mathbb{N}$, are from the set:

$$\begin{aligned} \mathcal{U} &= \{\pm u^{(i)} : u^{(i)} = \rho^i u^{(0)}, i = \pm 1, \pm 2, \ldots\} \\ &\cup \{\pm u^{(0)}\} \cup \{0\}, \quad 0 < \rho < 1, u^{(0)} > 0, \end{aligned} \tag{32}$$

ρ represents the quantizer *density* and the initial level $u^{(0)}$ is given in advance. The logarithmic quantizer is illustrated in Fig. 3, where the quantization error can be characterized by a sector-bound uncertainty of size $\delta = \frac{1-\rho}{1+\rho}$. One fundamental property of logarithmic quantization is that the relative quantization error $\Delta(v)$ is uniformly bound by δ, i.e., $Q(v) - v = v\Delta(v), \Delta(v) \in [-\delta, \delta]$. This multiplicative representation of quantization error is more appropriate than the additive one which has been considered in many existing works of signal processing and control since it makes sense when the signal to be quantized is large, the quantization error will be large. It should be noted that the floating-point roundoff in scientific calculations may be treated as logarithmic quantization as well [15]. In General, an exact description of the uncertainty $\Delta(v)$ is difficult. As such, we model the quantization error $\Delta(v)$ as a bounded parameter uncertainty which is deterministic and satisfies $|\Delta(v)| \le \delta$. It is worthy pointing out that the deterministic description of $\Delta(v)$ may be conservative since the quantization error of the logarithmic quantizer is a subset of $\{e \in \mathbb{R} : |e| \le \delta\}$. Our objective here is to design a filter of the form (30) such that the upper bound of the filtering error covariance is minimized, i.e.,

$$\min_{L(n)} \text{Tr}(P(n)). \tag{33}$$

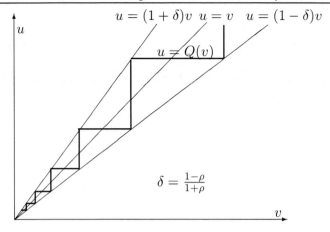

Figure 3. logarithmic quantizer.

where $P(n)$ is an upper bound of the estimation covariance matrix as follows:

$$E[\mathbf{e}(n)\mathbf{e}(n)^T] \leq P(n).$$

This filter design problem is referred to as *worst-case quantized filter design*.

Theorem 3. *([17]) Consider the system described by (1) and (2). Given the logarithmic quantizer with density ρ, the optimal robust quantized filter gain of (30) is given by*

$$L(n) = \frac{P(n|n-1)\mathbf{h}(n)}{\mathbf{h}(n)^T P(n|n-1)\mathbf{h}(n) + \sigma_v^2(n)} \tag{34}$$

where $P(n|n-1)$ is given by

$$P(n|n-1) = A(n)P(n-1)A(n)^T + W(n). \tag{35}$$

with $P(n)$ satisfying the following Riccati difference equation:

$$P(n) = P(n|n-1) + W(n) - (1-\delta^2)\frac{P(n|n-1)\mathbf{h}(n)\mathbf{h}(n)^T P(n|n-1)}{\mathbf{h}(n)^T P(n|n-1)\mathbf{h}(n) + \sigma_v^2(n)}. \tag{36}$$

Proof. In the light of (1), (2) and (30), it follows that

$$\mathbf{e}(n) = \left[I - (1+\Delta(n))L(n)\mathbf{h}(n)^T\right]\mathbf{e}(n|n-1) - (1+\Delta(n))L(n)v(n) \tag{37}$$

where $\mathbf{e}(n|n-1) = A(n)\mathbf{e}(n-1) + w(n-1)$ and the relative quantization error $\Delta(n) \in [-\delta, \delta]$. Then, for a given $\Delta(n)$, the filtering error covariance can be obtained as

$$P_{\Delta,n} = F(n)P_{\Delta,n|n-1}F(n)^T + \sigma_v^2(n)(1+\Delta(n))^2 L(n)L(n)^T,$$

where $F(n) = I - (1+\Delta(n)L(n)\mathbf{h}(n)^T$, $P_{\Delta,n} = E[\mathbf{e}(n)\mathbf{e}(n)^T]$, and $P_{\Delta,n|n-1} = E[\mathbf{e}(n|n-1)\mathbf{e}(n|n-1)^T]$. We now introduce the following Riccati inequality:

$$P(n) \geq F(n)P(n|n-1)F(n)^T + \sigma_v^2(n)(1+\Delta(n))^2 L(n)L(n)^T. \tag{38}$$

If a fixed $P(n) > 0$ satisfies the above inequality for any $\Delta(n) \in [-\delta, \delta]$, it can be easily argued that it provides an upper bound for $P_{\Delta,n}$. By Schur complement, (38) can be rewritten as

$$\overline{M}(n) + \overline{L}(n)\Delta(n)E(n) + E(n)^T\Delta(n)\overline{L}(n)^T \leq 0, \tag{39}$$

where

$$\overline{M}(n) = diag\{M(n), -r_{1,n}^{-1}\}, \overline{L}(n) = \begin{bmatrix} L(n) \\ 0 \end{bmatrix},$$

$$\overline{E}(n) = [-\mathbf{h}(n)^T P(n|n-1) + r_{1,n}L(n)^T \quad 1],$$

$$M(n) = -P(n) + [I - L(n)\mathbf{h}(n)^T]P(n|n-1)[I - L(n)\mathbf{h}(n)^T]^T + \sigma_v^2(n)L(n)L(n)^T,$$

$$r_{1,n} = \mathbf{h}(n)^T P(n|n-1)\mathbf{h}(n) + \sigma_v^2(n).$$

It follows that (39) holds for any $\Delta(n) \in [-\delta, \delta]$ if and only if there exists an $\beta(n) > 0$ such that [?]

$$\overline{M}(n) + \beta(n)\overline{L}(n)\overline{L}(n)^T + \beta(n)^{-1}E(n)^T E(n) \leq 0.$$

By applying Schur complement again, we have that

$$(\beta(n) - r_{1,n})^{-1}\left[-P(n|n-1)\mathbf{h}(n) + r_{1,n}L(n)\right]\left[-P(n|n-1)\mathbf{h}(n) + r_{1,n}L(n)\right]^T$$
$$+ M(n) + \beta(n)\delta^2 L(n)L(n)^T \leq 0 \tag{40}$$

with the constraint

$$r_{1,n} < \beta(n). \tag{41}$$

In the above, we have applied the equality

$$\beta(n)^{-1} + \beta(n)^{-2}(r_{1,n}^{-1} - \beta(n)^{-1})^{-1} = (\beta(n) - r_{1,n})^{-1}.$$

By some straightforward manipulations and completion of square, (40) leads to

$$P(n) \geq P(n|n-1) - (1-\delta^2)r_{2,n}^{-1}P(n|n-1)\mathbf{h}(n)\mathbf{h}(n)^T P(n|n-1) \tag{42}$$
$$+ \beta(n)(\beta(n) - r_{1,n})^{-1}r_{2,n}\left[L(n) - P(n|n-1)\mathbf{h}(n)r_{2,n}^{-1}\right] \tag{43}$$
$$\times \left[L(n) - P(n|n-1)\mathbf{h}(n)r_{2,n}^{-1}\right]^T \tag{44}$$

where $r_{2,n} = \beta(n)\delta^2 + (1-\delta^2)r_{1,n}$. In view of (41) and the fact that $r_{1,n} > 0$, the optimal filter gain is obtained as

$$L(n) = P(n|n-1)hr_{2,n}^{-1}.$$

In this case, (44) is reduced to

$$P(n) \geq P(n|n-1) - (1-\delta^2)r_{2,n}^{-1}P(n|n-1)\mathbf{h}(n)\mathbf{h}(n)^T P(n|n-1).$$

Further, noting the constraint (41) and taking the limit, $\beta(n) \to r_{1,n}$, in order to minimize the trace of the right hand side matrix, then $r_{2,n} \to r_{1,n}$ and the optimal filter gain is given by (34) and the Riccati difference inequality becomes

$$P(n) \geq P(n|n-1) - (1-\delta^2)r_{1,n}^{-1}P(n|n-1)\mathbf{h}(n)\mathbf{h}(n)^T P(n|n-1).$$

By setting the above inequality to equality in order to give the minimum upper bound, the Riccati difference equation (36) follows. ∎

5. Filter Stability Analysis

We now investigate the stability of the MLQ-KF and the worst-case quantized filter when the plant matrices A, $\mathbf{h}$ and the noise variances W, σ_v^2 are time-invariant. Observe that the RDE (10) and (36) can be written in the form:

$$P_{n+1} = AP_nA' + W - \beta^* AP_n\mathbf{h}(\mathbf{h}'P_n\mathbf{h} + \sigma_v^2)^{-1}\mathbf{h}'P_nA' \qquad (45)$$

with $\beta^* = F(z_*^N)$ for the MLQ-KF and $\beta^* = 1 - \delta^2$ for the worst-case quantized filter. The corresponding algebraic Riccati equation (ARE) is

$$P = APA' + W - \beta^* AP\mathbf{h}(\mathbf{h}'P\mathbf{h} + \sigma_v^2)^{-1}\mathbf{h}'PA'. \qquad (46)$$

Theorem 4. *([3]) Consider the RDE (45) and assume that $(A, Q^{1/2})$ and $(\mathbf{h}', A)$ are respectively controllable and detectable. Then for an unstable A, the ARE (46) has a positive definite solution P and the RDE (45) admits a unique positive definite solution P_n satisfying $P_n \to P$ for any $P_0 \geq 0$ if and only if*

$$\beta^* > 1 - \frac{1}{\prod_i |\lambda_i^u(A)|^2},$$

where $\lambda_i^u(A)$ are unstable eigenvalues of A.

For the MLQ-KF, the above gives a clear relationship between the system dynamics and the number of bits required for achieving the convergence of the RDE.

6. Simulations & Experiments

6.1. Simulations

Consider the following discrete time equations of motion for target tracking [14]:

$$\mathbf{x}(n+1) = \begin{bmatrix} 1 & \tau & \tau^2/2 \\ 0 & 1 & \tau \\ 0 & 0 & 1 \end{bmatrix} \mathbf{x}(n) + \mathbf{w}(n) \qquad (47)$$

$$y(n) = [1\ 0\ 0]\mathbf{x}(n) + v(n) \qquad (48)$$

where $\mathbf{x}(n) = [x_1(n)\ x_2(n)\ x_3(n)]^T$ with $x_i(n)$, $i = 1, 2, 3$ respectively the target position, speed and acceleration at time t. The input random signal $\mathbf{w}(n)$ is a discrete time white noise sequence and independent on the additive white noise $v(n)$ with variance σ_R^2. When the sampling interval τ is sufficiently small, the covariance matrix of $\mathbf{w}(n)$ is given by

$$W = 2\alpha\sigma_m^2 \begin{bmatrix} \tau^5/20 & \tau^4/8 & \tau^3/6 \\ \tau^4/8 & \tau^3/3 & \tau^2/2 \\ \tau^3/6 & \tau^2/2 & \tau \end{bmatrix},$$

where σ_m^2 is the variance of the target acceleration and α is the reciprocal of the maneuver time constant.

We use the following data: $\tau = 0.1s, \alpha = 0.1, \sigma_m^2 = 1, \sigma_R^2 = 10^{-2}$. It can be easily verified that $(A, W^{1/2})$ and $(\mathbf{h}', A)$ are controllable and observable pairs. Corollary 4 implies the stability of the 1-bit and 2-bit quantized innovation Kalman filters, which are denoted by 1-LQ-KF and 2-LQ-KF, respectively. The initial state is $\mathbf{x}(0)$ is a random variable with zero mean and covariance matrix [14]:

$$P(0|0) = \begin{bmatrix} \sigma_R^2 & \sigma_R^2/\tau & 0 \\ \sigma_R^2/\tau & 2\sigma_R^2/\tau^2 & 0 \\ 0 & 0 & 0 \end{bmatrix}.$$

The filtering error variances of the position for the 1-LQ-KF and SOI-KF are compared in Fig. 4. The variances are obtained by Monte Carlo simulations based on 500 samples. As expected, the 1-LQ-KF outperforms the SOI-KF. Fig. 5 shows that for the 2-bit case, the computed filtering error variance calculated by the MRDE (45) is close to the estimated one obtained by Monte Carlo simulations based on 500 samples, indicating that the variance computed by the MRDE gives a good approximation of the true variance. The results for the speed and acceleration estimates are similar and omitted. Next, we evaluate the robust minimax quantizer where it is assumed that the prediction error variance is randomly perturbed by δ, which is uniformly distributed over $[0.8, 1.2]$. We design a 2-bit robust minimax quantizer (2-LQ-RKF) according to (27) with $\bar{z}_*^2 = [0.3035\ 1.0094]$. For comparison, a non-robust quantizer (nominal) quantizer given in (3) is implemented as well. Fig. 6 shows that the 2-LQ-RKF has a smaller filtering error variance for the position than the non-robust one based on Monte-Carlo simulations of 500 samples.

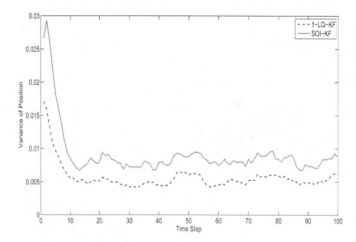

Figure 4. Comparison of the filtering error variance of the position estimate given by the 1-LQ-KF and SOI-KF based on 500 samples.

6.2. Experiments On WSN Based Target Tracking

We now implement the proposed optimal 1-bit quantized innovation Kalman filter and the SOI-KF of [13] in a distributed wireless sensor network based target tracking platform. In

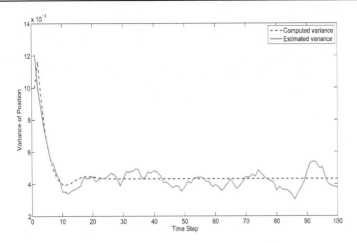

Figure 5. Comparison of the position error variance computed by the MRDE (45) and the estimated variance by Monte Carlo simulations based on 500 samples.

the platform, 9 sonar sensors are uniformly deployed in a square region of about $9m^2$ and the position (x_j, y_j) of the j-th sensor is known. Our purpose is to track a moving target (uncooperative robot). Under this setting, we have the following state equation:

$$\mathbf{x}(n+1) = G(\triangle t_n)\mathbf{x}(n) + \boldsymbol{\omega}(n, \triangle t_n)$$

where $\mathbf{x}(n) = (x(n), \dot{x}(n), y(n), \dot{y}(n))^T$ is the state of the target at the n-th measurement time t_n. $\triangle t_n = t_{n+1} - t_n$ is the difference between t_{n+1} and t_n, i.e. the n-th sampling interval. $G(\triangle t_n)$ is the transition matrix depending on $\triangle t_n$, defined by

$$G(\triangle t_n) = \begin{bmatrix} 1 & \triangle t_n & 0 & 0 \\ 0 & 1 & 0 & 0 \\ 0 & 0 & 1 & \triangle t_n \\ 0 & 0 & 0 & 1 \end{bmatrix}$$

$\boldsymbol{\omega}(n, \triangle t_n)$ is the Gaussian process noise with zero mean and the covariance matrix

$$Q(\triangle t_n) = q \begin{bmatrix} \frac{1}{3}\triangle t_n^3 & \frac{1}{2}\triangle t_n^2 & 0 & 0 \\ \frac{1}{2}\triangle t_n^2 & \triangle t_n & 0 & 0 \\ 0 & 0 & \frac{1}{3}\triangle t_n^3 & \frac{1}{2}\triangle t_n^2 \\ 0 & 0 & \frac{1}{2}\triangle t_n^2 & \triangle t_n \end{bmatrix}$$

where q is a known scalar, determining the intensity of the process noise.

The sensors collect information about their distance to the target. At each time instant, only one sensor takes its measurement. The measurement model given by

$$z_j(n) = h_j(\mathbf{x}(n)) + v_j(n) \qquad (49)$$

where j is the sensor index, $v_j(n)$ is a white Gaussian noise with covariance matrix $R_j(n) = p$, h_j is a nonlinear measurement function depending on $\mathbf{x}(n)$, given by

$$h_j(\mathbf{x}(n)) = \frac{a}{\sqrt{(x(n) - x_j)^2 + (y(n) - y_j)^2}}$$

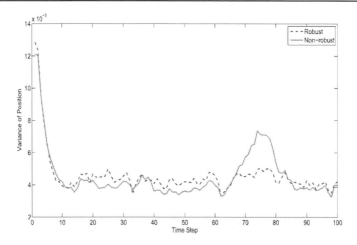

Figure 6. Comparison of the position error variances by the robust and non-robust 2-bit quantized filters obtained from Monte Carlo simulations based on 500 samples.

In our experiment, the sampling interval is $\triangle t_n = 0.1s$, $q = 50$, $p = 0.0001$, $a = 40$. To implement the algorithms, we linearize (49) in a neighborhood of $\hat{\mathbf{x}}(n|n-1)$ to obtain

$$\gamma_j(n) = z_j(n) - \hat{z}(n|n-1) = \mathbf{h}_j(n)(\mathbf{x}(k) - \hat{\mathbf{x}}(n|n-1))$$

Here $\mathbf{h}_j(n)$ is the Jacobian matrix of $h_j(\cdot)$ evaluated at $\hat{\mathbf{x}}(n|n-1)$ and is given by

$$\mathbf{h}_j(n) = \frac{[-a(\hat{x}(n|n-1) - x_j)\ 0\ -a(\hat{y}(n|n-1) - y_j)\ 0]}{[(\hat{x}(n|n-1) - x_j)^2 + (\hat{y}(n|n-1) - y_j)^2]^{3/2}}$$

Experimental results of the SOI-KF [13] and the proposed optimal 1-bit MLQ-KF are shown in Fig 7. It shows that the SOI-KF [13] failed to track the target over the entire course where our 1-bit MLQ-KF successfully fulfilled the tracking duty.

7. Conclusions

Extending the existing work of the SOI-KF, we have developed a general multi-level quantized innovation filter with a dead zone. Our result is derived under the assumption that the system state conditioned on the quantized innovations is Gaussian which is reasonable for quantizer of a moderate number of bits. The distinct feature of the developed quantized innovation filter lies its simplicity and efficiency. The convergence of the filter to the Kalman filter when the number of quantization levels goes to ∞ has also been established. The result will be useful in applications such as wireless sensor networks where communication capacity is limited. To remove the Gaussian assumption, a worst-case quantized filter based on a logarithmic quantizer has been presented which minimizes the upper bound of the mean square estimation error covariance among all admissible uncertainties. Simulations and experiments demonstrated the respective theoretic results.

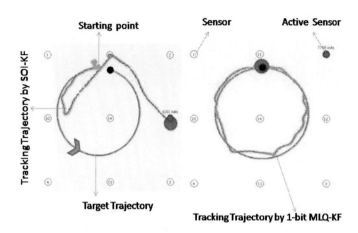

Figure 7. The true target trajectory is the circle. The left figure shows the tracking result of SOI-KF [13] and the right one is that of our 1-bit MLQ-KF.

References

[1] M. Arulampalam, S. Maskell, N. Gordon, and T. Clapp, "A tutorial on particle filters for online nonlinear/non-Gaussian Bayesian tracking," *IEEE Transactions on signal processing*, vol. 50, no. 2, 2002.

[2] N. Elia and S. Mitter, "Stabilization of linear systems with limited information," *IEEE Transactions on Automatic Control*, vol. 46, no. 9, pp. 1384–1400, 2001.

[3] M. Fu and L. Xie, "The sector bound approach to quantized feedback control," *IEEE Transactions on Automatic Control*, vol. 50, no. 11, pp. 1698–1711, 2005.

[4] M. Huang and S. Dey, "Dynamic quantization for multisensor estimation over bandlimited fading channels," *IEEE Transactions on Signal Processing*, vol. 55, no. 9, pp. 4696–4702, 2007.

[5] K. Ito and K. Xiong, "Gaussian filters for nonlinear filtering problems," *IEEE Transactions on Automatic Control*, vol. 45, no. 5, pp. 910–927, 2000.

[6] R. Karlsson and F. Gustafsson, "Particle filtering for quantized sensor information," ser. *Proceedings of the 13th European Signal Processing Conference*, Antalya, Turkey, September 2005.

[7] J. Kotecha and P. Djuric, "Gaussian particle filtering," *IEEE Transactions on Signal Processing*, vol. 51, no. 10, pp. 2592–2601, 2003.

[8] S. Lloyd, "Least squares quantization in PCM," *IEEE transactions on information theory*, vol. 28, no. 2, pp. 129–137, 1982.

[9] Z. Luo, "Universal decentralized estimation in a bandwidth constrained sensor network," *IEEE Transactions on Information Theory*, vol. 51, no. 6, 2005.

[10] E. Msechu, S. Roumeliotis, A. Ribeiro, and G. Giannakis, "Decentralized quantized Kalman filtering with scalable communication cost," *IEEE Transactions on Signal Processing*, vol. 56, no. 8, pp. 3727–3741, 2008.

[11] N. Newton, "Observations preprocessing and quantization for nonlinear filters," *SIAM Journal on Control and Optimization*, vol. 38, no. 2, pp. 482–502, 2000.

[12] G. Pagès and H. Pham, "Optimal quantization methods for nonlinear filtering with discrete-time observations," *Bernoulli*, vol. 11, no. 5, pp. 893–932, 2005.

[13] A. Ribeiro, G. Giannakis, and S. Roumeliotis, "SOI-KF: Distributed Kalman Filtering with Low-Cost Communications using the Sign Of Innovations," *IEEE Transactions on Signal Processing*, vol. 54, no. 12, 2006.

[14] R. Singer, "Estimating Optimal Filter Tracking Performance for Manned Maneuvering Targets," *IEEE Transactions on Aerospace and Electronic Systems*, vol. 6, no. 4, pp. 473–483, 1970.

[15] B. Widrow, I. Kollar, and M. Liu, "Statistical theory of quantization," *IEEE Transactions on Instrumentation and Measurement*, vol. 45, no. 2, pp. 353–361, 1996.

[16] W. Wong and R. Brockett, "Systems with finite communication bandwidth constraints-Part I: State estimation problems," *IEEE Transactions on Automatic Control*, vol. 42, no. 9, pp. 1294–1299, 1997.

[17] L. Xie and L. Shi, "State Estimation over Unreliable Network," *Proc. 7th Asian Control Conference*, pp. 453–458, 2009.

[18] K. You, W. Su, M. Fu, and L. Xie, "Attainability of the minimum data rate for stabilization of linear systems via logarithmic quantization," *to appear in Automatica*, August 2010.

[19] K. You and L. Xie, "Minimum data rate for mean square stabilizability of linear systems with Markovian packet losses," *to appear in IEEE Transactions on Automatic Control*, 2010.

[20] ——, "Minimum data rate for mean square stabilization of discrete LTI systems over lossy channels," *to appear in IEEE Transactions on Automatic Control*, 2010.

[21] K. You, L. Xie, S. Sun, and W. Xiao, "Multiple-level quantized innovation Kalman filter," *Proc. 17th IFAC World Congress, also to appear in the Transactions of the Institute of Measurement & Control*, pp. 1420–1425, 2008.

In: Kalman Filtering
Editor: Joaquín M. Gomez

ISBN: 978-1-61761-462-0
© 2011 Nova Science Publishers, Inc.

Chapter 11

KALMAN FILTER TO ESTIMATE DYNAMIC AND IMPORTANT PATTERNS OF INTERACTION BETWEEN MULTIPLE VARIABLES

Harya Widiputra,[*] Russel Pears and Nikola Kasabov
The Knowledge Engineering and Discovery Research Institute, New Zealand

Abstract

Estimating the state of nature processes has been a very challenging task for human beings. These processes include not only those which occur in the biological field, i.e., interaction between genes; or those which emerge in the ecological field, i.e., how humidity is related to the level of sun ray, wind speed and rain; but also processes which exist in the global financial area, i.e., how a stock market in a specific country influences or is being influenced by the other stock markets in different countries directly or indirectly, and how the macro economic factors in a single country are related to each other from time to time. Being able to model past and present states of these processes would lead us to the possibility of understanding how various variables or things in this world interact, which in the end would offer us the knowledge to estimate future states of the systems in which these processes take place.

1. Introduction

As we may have already known and agreed, there are "things" in this world which are closely connected to each other, influencing the others, causing changes in the behaviour of other things, being affected by activities of others, and so forth. In general, we believe that there are things which are referred to as variables in this world that interact with each other, and this interaction controls, in one way or the other, how expression or behaviour of these variables changes dynamically over time. For instance, people in the finance area (and most probably people everywhere) take into account that movement of a stock market index in a specific country is being affected by movement of other stock market indexes across the world [1], or people in the biomedical area understand that there exists a Gene Regulatory

---
[*]E-mail address: harya.widiputra@aut.ac.nz

Network (GRN) which governs movement of the genes expression level in the human body [26].

The task of extracting and modelling interdependencies or patterns of interaction between variables has become a challenging task, in particular for research in the information science studies. Being able to accomplish such missions is expected to lead us to the ability of understanding how observed variables in a specific environment, i.e. finance, biology, weather, etc., move together, inhibit each other, are connected to each other and how their interaction profile changes dynamically over time. It will also help us identify important variables which have the most influencing power in governing the state of a system and predicting the observed variables expression values at future time points.

In 1960, a paper describing a recursive solution to discrete-data linear filtering problem was published by R.E. Kalman [33], this proposed method is known as the Kalman filter (named after him). The Kalman filter is a set of mathematical equations that provides an efficient computational (recursive) means to estimate the state of a process, in a way that it minimises the mean of the squared error. Different studies have revealed that the filter is very powerful in several aspects: it supports estimations of past, present, and even future states, and it can do so even when the precise nature of the modelled system is unknown.

This chapter presents studies which have used first-order differential equations and the Kalman filter to extract and model a pattern of interactions (past, current and future states) between variables from a set of multiple time-series data. It will describe how to implement the Kalman filter to identify a set of first-order differential equations that describe the dynamics of the representative trajectories and use these equations to discover important variable interactions, model them (as interaction networks), and use the discovered model to predict values at future time points. Unlike the well-known approach Multivariate Data Analysis (which models relationship between multiple observed variables in the structure of a single dependent variable being influenced by several independent variables, as in Equation 1)

$$y = a_1 x_1 + a_2 x_2 + \ldots + a_n x_n + b, \qquad (1)$$

where y is the dependent variable; $x_1, x_2, .., x_n$ are the independent variables; $a_1, a_2, ..., a_n$ are the coefficient of influence; and b is a constant value, this method would extract and model interactions between observed variables in a complete, fully-connected network.

The chapter will also present results of using this methodology to extract dynamic and important interaction patterns between: stock markets in the Asia Pacific region; stock markets in Asia, Europe and the United States; Indonesia's macroeconomic factors (i.e. inflation rate, interest rate, and stock market index and central bank interest rate); gene expression dataset; and air pressure dataset from observation stations in New Zealand. Additionally, this chapter would also exhibit prediction results of multiple time-series data using modelled interaction networks which were extracted through the Kalman filtering process.

2. Stochastic Models and Estimation

Provided a physical system, for instance a moving aircraft, a chemical process, or the national economy, an engineer or a researcher would try to develop a mathematical model that adequately represents some aspects of the behaviour of the observed system. Through physical insights, fundamental "laws", and empirical experiments, he tries to model the interrelationships among observed variables of interests, inputs to the system and outputs from the system.

With such mathematical models and the tools provided by system and control theories, one should be able to investigate the structure of the system. Furthermore, they are also capable of designing a process which will act as a compensator to alter the state of the system so that it would generate desired responses. Estimating the state of relationship between various observed variables deals with building a mathematical model that is able to capture changes in this state of relationship when they are vibrantly changing in the future [37].

In probability theory, a stochastic process, sometimes called a random process, is the counterpart to a deterministic process (or deterministic system) [31]. Instead of dealing with only one possible reality of how the process might evolve over time (as is the case, for example, for solutions of an ordinary differential equation), in a stochastic or random process there is some indeterminacy in its future evolution described by probability distributions [7]. This means that even if the initial condition (or starting point) is known, there are many possibilities the process might go through, but some paths may be more probable and others less. In the simplest possible case (discrete time), a stochastic process amounts to a sequence of random variables known as a time series [17].

Familiar examples of processes modelled as stochastic time series include stock market and exchange rate fluctuations, signals such as speech, audio and video, medical data such as a patient's EKG, EEG, blood pressure or temperature, and random movement such as Brownian motion or random walks.

In the methodology that will be described in the following sections of this chapter, we will use multiple time-series dataset sampled regularly or irregularly, to extract interdependencies between the observed variables. Stochastic model and estimation is utilised, since we would like to be able to capture the dynamics of interdependencies between the observed variables, as the observed systems can be categorised complex dynamic systems (i.e. stock market indexes, macroeconomics factors, gene expression data and weather dataset) [13],[16].

3. Modelling Interaction Patterns in a Dynamic System with First-Order Differential Equations

3.1. Discrete-Time Approximation of First-Order Differential Equations

In the upcoming explained methodology, we modelled the state of interdependencies between the observed variables as discrete-time approximation of first-order differential equations, given by:

$$x_{t+1} = Fx_t + \varepsilon_t, \qquad (2)$$

where $x_t = (x_1, x_2, ..., x_n)'$ is the observed measurements at the t-th time interval and n is the number of observed variables modelled, ε_t is a noise component with covariance

$E = cov(\varepsilon_t)$, and $F = (f_{ij}); i = 1$ to n, $j = 1$ to n is the transition matrix relating x_t to x_{t+1}. This equation is related to the continuous first-order differential equation $dx/dt = \Psi x + e$ by $F = \tau\Psi + I$ and $\varepsilon_t = \tau e$ where τ is the time interval. We implement here a discrete approximation instead of a continuous model for the ease of modelling [22] and processing the irregular time-course data (this is due to the use of Kalman filter). Besides the fact that this approach is a tool widely used for modelling biological processes, there are two advantages in using the first-order differential equations.

First, relations of the observed variables can be elucidated from the transition matrix F through choosing a threshold value $(\zeta; 1 > \zeta > 0)$. If $|f_{ij}|$ is larger than the threshold value ζ, $x_{t,j}$ is considered to have significant influence on $x_{t+1,i}$. A positive value of f_{ij} indicates a positive influence and vice-versa. Second, they can be easily manipulated with Kalman filter to handle irregularly sampled data, which allow parameter estimation, likelihood evaluation, model simulation and prediction.

Yet, there is a main drawback in using differential equations. It requires the estimation of n^2 parameters for the transition matrix F and $n(n-1)/2$ parameters for the noise covariance E. To minimise the number of model parameters, we estimate only F and set E to a small value. For instance, if we are observing two series, and both consist of only 4 samples, then we avoid over-parameterisation by setting n to 4, (which is the maximum number of samples) before the number of parameters exceeds the number of training data {It matches the number of model parameters (size of F is n^2) to the number of training data ($n \times 4$ samples)}.

To cope with irregularly sampled data, we implement the state-space methodology and the Kalman filter. We treat the actual trajectories as a set of unobserved or hidden variables called the *state variable*, and then apply the Kalman filter to compute their optimal estimates based on the observed measurements. The state variables that are regular or complete can now be applied to perform model functions like prediction, parameter estimations instead of the observed measurement that are irregular or incomplete. This approach is better than the interpolation methods as it prevents false modelling by trusting a fixed set of interpolated points that may be erroneous.

3.2. State-Space Representation

In previous section the term *state variable* was introduced. The state variables can be considered as the smallest possible subset of the system variables that can represent the entire states of the system in any given time. The phrase *system state transition function* would be the transfer function representing the interaction between observed variables occurring in any given time that maps present values of the state variables (system state) to their future values.

The minimum number of state variables required to represent a given system, n, is usually equal to the order of the system's defining differential equation. If the system is represented in transfer function form, the minimum number of state variables is equal to the order of the transfer function's denominator after it has been reduced to a proper fraction. It is important to understand that converting a state space realisation to a transfer function form may lose some internal information about the system, and may provide a description of a system which is stable, when the state-space realisation is unstable at certain points.

A mathematical model of a system as a set of input, output and state variables related by the first-order differential equations is called as *state-space representation*. State-space models are a flexible family of models which fits the modelling of many scenarios. The strongest feature of state-space models is the existence of very general algorithm for filtering, smoothing and predicting [18].

Observed variables are expressed as vectors and the differential equations are written in matrix form (which is most suitable to be used when the dynamical system is linear and time invariant) to map the number of inputs, outputs and state variables. The state-space representation (which is also known as the "time-domain approach") provides a convenient and compact way to model and analyse systems with multiple inputs and outputs (suitable to extract patterns of interaction in multiple time-series data). With p inputs and q outputs, we would otherwise have to write down $p \times q$ Laplace transforms to encode all the information about a system. Unlike the frequency domain approach, the use of the state-space representation is not limited to systems with linear components and zero initial conditions. State-space refers to the space whose axes are the state variables, in which the state of the system can be represented as a vector within that space.

To apply the state-space methodology, a model must be expressed in the following format called the *discrete-time state-space representation*,

$$x_{t+1} = \Phi x_t + w_t \tag{3}$$

$$y_t = A x_t + v_t \tag{4}$$

$$cov(w_t) = Q, cov(v_t) = R \tag{5}$$

where x_t is the system state; y_t is the observed data; Φ is the state transition matrix that relates x_t to x_{t+1} ; A is the linear connection or interaction matrix that relates x_t to y_t; w_t and v_t are uncorrelated white noise sequences whose covariance matrices are Q and R respectively. The first equation, called the *state equation*, describes the dynamic behaviour of the state variables. The other equation is called the *observation equation* and it relates the system states to the observation.

In order to represent the discrete-time model in the state-space form, we simply substitute the discrete-time equation,

$$x_{t+1} = F x_t + \varepsilon_t, \tag{6}$$

into the state equation,

$$x_{t+1} = \Phi x_t + w_t. \tag{7}$$

by setting $\Phi = F$, $w_t = \varepsilon_t$ and $Q = E$. We then form a direct mapping between the system states and the observations by setting $A = I$. The state transition matrix Φ (functional equivalent to F) is the parameter of interest as it relates the future response of the system to the present state and governs the dynamics of the entire system. However, in this methodology, the covariance matrices Q and R are of secondary interest and are fixed to small values to reduce the number of model's parameters.

3.3. The Discrete Kalman Filter [14]

As an introductory section to the Kalman filter, in this part of the chapter we will first give a brief explanation about the Kalman filter and then how it can be used to compute the optimal estimates of our state-space model. A Kalman filter is simply an optimal recursive data processing algorithm which is also an optimal estimator for a dynamic linear system [28].

The optimality aspect of Kalman filter in this case is of course in respect to virtually any criterion that makes sense. One aspect of this optimality shows that the Kalman filter incorporates all information that can be provided to it (measurement of observed variables) [30]. It processes all information, regardless of their precision, to estimate the current values of the observed variables, with use of (1) knowledge of the system and measurement device dynamics, (2) the statistical description of the system noises, measurement errors and uncertainty in the dynamics models, and (3) any available information about initial conditions of the observed variables.

The word *recursive* mentioned in the previous paragraph simply means, unlike other data processing methods, Kalman filter does not require previous data to be kept in a memory, therefore it will not reprocess again when new measurement of observed variables becomes available. This is a very important characteristic of a filter implementation.

Figure 1 shows how Kalman filter can be used to find the optimal estimate of the system state. By associating the system state to how observed variables interact with each other in a specific system (e.g. stock market system, weather system, biological system, etc.), the idea is to use Kalman filter to estimate or model how these observed variables connect (in respect of how they interact with each other) at a specific time point and how this connection is changing dynamically over time. The Kalman filter combines all available measurement data (i.e. stock prices, stock market indexes, gene expressions value, etc.) plus prior knowledge about the system, to produce an estimate of the system state in such a manner that error is minimised statistically.

One may ask, if we know that the system we are observing is a non-linear dynamic system then what the point of implementing the Kalman filter to estimate the system state is. To answer this, Maybeck [28] in his book has pointed out some basic assumptions in the Kalman filter. One of them is about the justifiability of Kalman filter for linear system modelling. Even though in real world most of the systems are non-linear, it is a typical engineering approach to linearise some nominal points or trajectory to achieve a perturbation model or error model. Linear systems are desirable in that they are easier to be manipulated with engineering tools and linear system (or differential equation) theory is much more complete and practical than non-linear ones.

Supporting this, upcoming sections will show that by considering the observed systems (which are complex dynamic system) as linear systems, we are able to extract important and useful knowledge about interactive patterns between observed variables.

The fact is that, there are means of extending the Kalman filter concept to some non-linear applications or developing non-linear filters directly (which has been done through the development of the extended Kalman filter [14]). Nevertheless, these are considered only if linear models prove inadequate which is not the case here.

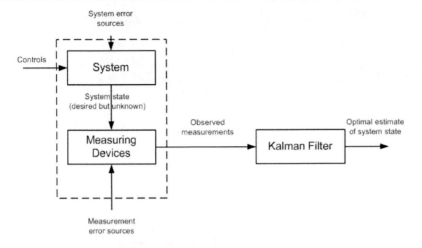

Figure 1. Typical Kalman filter application [29].

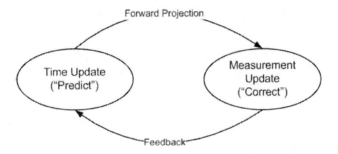

Figure 2. The ongoing discrete Kalman filter cycle. The *time update* projects the current state estimate ahead in time. The *measurement update* adjusts the projected estimate by an actual measurement at that time [14].

3.4. The State-Space Model Estimation through Kalman Filtering

Kalman filter estimates the states of a state-space model by using a form of a feedback control, this means that the filter works by estimating the state at some point in time and then obtains feedback in the form of (noisy) measurement. Therefore, basically there are two important groups of equations in the Kalman filter formulation. The first one is the *time update* equations. This equation are responsible for projecting forward (in time) the current system state and error covariance estimates to obtain *priori* estimates for the next step. In our study, we use these forward recursions to compute the state estimates.

$$x_{t+1} = Ax_t + Bu_t + w_t, \qquad (8)$$

where the w_t being the process noise is waived in this methodology.

As for likelihood evaluation and parameter estimation, we use the other group of equations which is the *measurement update* equations. These equations will also act as *corrector* equations. Indeed, the final algorithm resembles that of a *predictor-corrector* algorithm for solving numerical problems. A general concept of this algorithm can be seen in Figure 2,

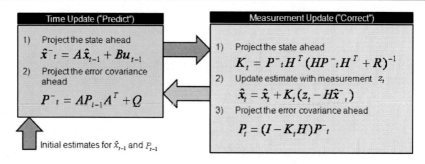

Figure 3. A complete picture of the operation of the Kalman filter, combining the high-level diagram of Figure 2 with Equations 9,10,11,12 and 13 [14].

while the complete picture of the operation of Kalman filter is shown in Figure 3.

The specific equations for the time updates are presented in these equations below,

$$\hat{x}_{t+1}^- = A\hat{x}_t + Bu_t \tag{9}$$

$$P_{t+1}^- = AP_t A^T + Q \tag{10}$$

while the equations for measurement updates are described in these equations below,

$$K_{t+1} = P_{t+1}^- H^T \left(HP_{t+1}^- H^T + R \right)^{-1} \tag{11}$$

$$\hat{x}_{t+1} = \hat{x}_{t+1}^- + K_{t+1} \left(z_{t+1} - H\hat{x}_{t+1}^- \right) \tag{12}$$

$$P_{t+1} = (I - K_{t+1} H) P_{t+1}^- \tag{13}$$

From the equations we can observe how the time update equations project the state and covariance estimates forward from time step t to step $t+1$. The $n \times n$ matrix A (where n is the number of observed variables) in the difference Equation 9 relates the state at current time step t to the state at future step $t+1$. This formula can be associated to the form of *discrete-time state-space representation* as in Equation 3. The $n \times l$ matrix B relates the optional control input $u \in \Re^l$ (where in the *discrete-time state-space representation* is waived) to the state x. The $m \times n$ matrix H in the measurement equations (as in Equation 11) relates the state to the observed data (measurement) z_t. Here, this formula is also being associated to the *discrete-time state-space representation* as in Equation 3.

The first task during the measurement update is to compute the Kalman gain, K_{t+1}, and the next step is to actually measure the process to obtain z_{t+1}, and then go to generate a *posteriori* state estimate by incorporating the measurement as in Equation 12. The final step is to obtain a *posteriori* error covariance estimate through Equation 13. After each measurement update, the process is repeated with the previous *posteriori* estimates used to project or predict the new a *priori* estimates. This recursive characteristic is one of the very appealing features of the Kalman filter, where the filter recursively conditions the current estimate on all of the past measurements.

4. Gene Regulatory Network Discovery from Time-Series Gene Expression Data [26]

The dynamics interactions between DNA, RNA and protein lead to a genetic regulatory networks (GRN) and in turn control the gene regulation. Directly or indirectly in a cell, such molecules interact either in a positive or in a repressive way, therefore it is hard to obtain the accurate computational models through which the final state of a cell can be predicted with certain accuracy.

Gene regulatory network discovery and modelling is an important target in researches related to the biological systems as they are considered to be the systems controlling the fundamental mechanism that govern biological systems. A single gene (an observed variable) interacts with many other genes (other observed variables) in the cell, inhibiting or promoting directly or indirectly, the expression of some of them at the same time [20].

The discovery of GRN from time-series of gene expression observations can be used to: (1) identify important genes in relation to a disease or a bilogical function, (2) gain an understanding on the dynamic interaction between genes, and (3) predict gene expression values at future time points. One of the major approaches that have been conducted to deal with the discovery and modelling of GRN involves the use of differential equations [26]. Yet, we can also perceive the problem of GRN discovery as an attempt to solve and model dynamics interaction between the multiple observed variables whose measurements are made in timely basis (multiple time-series).

In their work, Kasabov et al. [26], have introduced a distinctive method that integrates Kalman filter and Genetic Algorithm (GA) [12] for GRN discovery and modelling from a time-series gene expression data. The GA itself is used to select a small number of genes (feature selection to reduce the possibility of huge dimension of observed variables) and the Kalman filter is used to derive the GRN of this selected observed genes. In this section of the chapter we will focus only on the part in which Kalman filter has been used to extract GRN, while the use of GA to select a small number of genes from the whole observed genes will not be analysed.

The GRN extraction can be seen as the process of finding important patterns of interaction from multiple observed variables. In this case, we can associate the GRN as the equation or function that maps system state at time point t to system state at time point $t + 1$. Using the time-series gene expression data we can map this to the state-space representation as in Equation 3 and Equation 4, and then apply the Kalman filter to find the optimal estimate of the observed system.

In their experiment, Kasabov et al. [26] have tried to identify most significant genes (4 genes) out of the 32 pre-selected genes which are considered to regulate telomerase {Telomerase is an enzyme that adds DNA sequence repeats ("TTAGGG" in all vertebrates) to the 3' end of DNA strands in the telomere regions, which are found at the ends of eukaryotic chromosomes [26]} in the GRN. They also apply the joint normalisation with the interval of $[-1, 1]$ to the plus and minus series, the purpose of this is to preserve the information on the difference between the series in the mean.

To select the four most significant genes, Kasabov et al. [26] have applied the GA and the exhaustive search which interestingly give similar results. These selected genes can be seen in Table 1, where they have put the rank for top ten genes out of the 32 pre-selected

Table 1. Significant genes extracted by GA and through an exhaustive search from 32 selected genes [26].

Rank	Indices of significant genes found by GA (Freq. of occurance in Minus GRNs, Freq. of occurance in Plus GRNs) and their accession numbers in Genbank	Indices of significant genes found by exhaustive search (gene Index)
1	27(179, 185) X59871	20 M98833
2	21(261, 0) U15655	27 X59871
3	12(146, 48) J04101	33 X79067
4	33(64, 118) X79067	12 J04101
5	20(0, 159) M98833	6 AL021154
6	22(118, 24) U25435	29 X66867
7	11(0, 126) HG3523-HT4899	5 D50692
8	5(111, 0) D50692	22 U25435
9	18(0, 105) D89667	10 HG3521-HT3715
10	6(75, 0) AL021154	13 J04102

genes. Using the top for genes found by GA and exhaustive search (33, 8, 27, 21), GRN can be discovered by feeding the expression values (as observed measurement) to the first-order differential equation (as in Equation 2) and use Kalman filter to compute the optimal estimates of the past, present and future state of the state-space model (as in Equation 3). The identified GRN can also be used for model simulation and prediction. The GRN dynamics can also be visualised with a network diagram using the influential information extracted from the state transition matrix F.

Results of extracted GRNs are illustrated in Figure 4 and Figure 5. Based on these figures we can observe that for the plus series, the network diagram in Figure 4.a shows that gene 27 has the most significant role regulating all other genes (note that gene 27 has all its arrows out-going). The network simulation in Figure 4.b fits the true observations well and the predicted values appear to be stable, suggesting that the model is considerably accurate and robust. As for the minus series, the network diagram is shown in Figure 5.a, it is quite clear that the extracted GRN is different from that of the plus series. Here the role of gene 27 is not as prominent. The relationship between genes is no more casual but interdependent, with genes 27, 32 and 21 simultaneously affecting each other. Again, the network simulation result (generated using the extracted GRN or transition matrix) in Figure 5.b shows that the model fits the data well and the prediction appears reasonable.

Results show that Kalman filter deals effectively with irregular and scarce data collected from a large number of variables (the actual observations contain 12,625 genes, each of which sampled 4 time at irregular time intervals, however only 32 genes have been selected due to the specific objective of the work). GRNs are modelled as discrete-time approximations of first-order differential equations and Kalman filter is applied to estimate the true

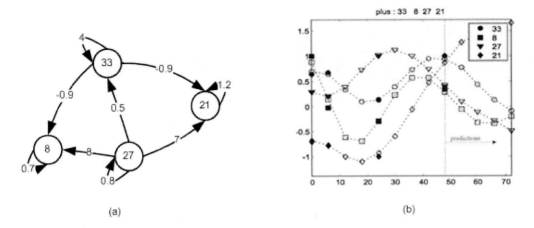

Figure 4. The identified best GRN of top-four genes for the plus series: (a) network diagram; (b) the network simulation and gene expression prediction over future time. Solid markers represent observations [26].

gene trajectories from the irregular observations (time update equations) and to evaluate the likelihood of the GRN models (measurement update equations). The extracted GRNs are considered to be accurate and robust as the network simulation results fit the observed data well and that it provides us with a reasonably stable prediction value.

5. Extraction and Modelling of Dynamic Interaction between Economic Variables

In the previous section, we have outlined the use of Kalman filter to extract gene regulatory network (GRN) from a given time-series gene expression data. It has also been discovered that the extracted GRNs are able to generate a good network simulation and rationally excellent prediction. In this section of the chapter we are going to introduce the use of Kalman filter to extract and model patterns of interaction between observed economic variables (as the observed measurement). Additionally, we will also integrate the incremental learning concept with the Kalman filter in order to capture and reveal dynamic patterns of interaction between the observed economic variables which are expected to change over time.

5.1. Modelling and Predicting the Behaviour of Multiple Interactive Stock Markets in the Asia Pacific Region [16]

The globalised security markets of today are characterised with interdependencies, and often demonstrate contagious behaviour in periods of crisis [25]. An increasing number of studies are addressing the effects of such interrelationships, along with the challenge of relationship identification and modelling within a globalised environment. Chiang and Doong [36] consider stock returns and volatility and find that four out of seven Asian markets present a significant relationship between stock returns and unexpected volatility. In their research on the French, German and UK indexes, and the corresponding stock index future

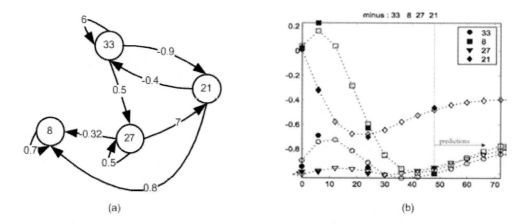

Figure 5. The identified best GRN of top-four genes for the minus series: (a) network diagram; (b) the network simulation and gene expression prediction over future time. Solid markers represent observations [26].

markets, Antoniou et al. [1] signify that the behaviour of a domestic market is influenced by the foreign markets. Collin and Biekpe [9] further study contagion and interdependence in the African markets, find evidence for contagion from the most traded market, while Serguieva et al. in [5],[6] have focused their study in Asia during the time of crisis.

However, these studies do not report the ability to simultaneously capture multiple dynamic relationships between interactive markets. This task serves as the main reason to why we are looking at the possibility of using the Kalman filter to extract such relationship by modelling the interactions of the financial variables dataset as discrete-time approximations of first-order differential equations (as in Equation 2). Furthermore, to meet the challenge of modelling and predicting multiple markets, it is essential to capture their interactive behaviour in a dynamic fashion. In this section we are going to describe our proposed method called as the Dynamic Interaction Network (DIN) that analyses a set of multiple time-series data in the form of stock market indexes and reveals important interrelationships between them.

In our proposed method, we use the Kalman filter to compute the optimal estimates of the relationships between stock market indexes (as the observed variables). Here, we model the interdependencies between multiple time-series data of stock market indexes as discrete-time approximations of first-order differential equations which will lead to the extraction of dynamic interaction network (DIN) which is considered to govern movement of these observed stock market indexes altogether.

The extracted DINs model is expected to reveal dynamic interactions among stock markets. This learned information will be useful to detect influences between stock markets (directly or indirectly) and to evaluate their degree of dependency as well. Through capturing dynamic influences among markets, we will also be able to foresee how these interactive stock markets will behave in the future.

Figure 6 exemplifies such behaviour with the trajectories of the markets indexes over an 72-week period spanning from February 2005 to November 2007. The 10 markets selected in the Asia Pacific region include Australia (AORD), China (SSE), Hong Kong (HSIX), In-

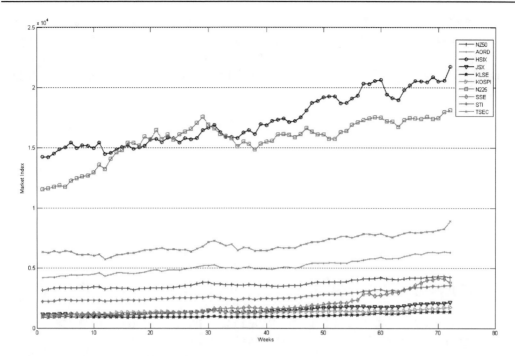

Figure 6. February 2005 to November 2007 Stock market indexes trajectories in the Asia Pacific Region [16].

donesia (JSX), Japan-Nikkei (N225), Malaysia (KLSE), New Zealand (NZ50), Singapore (STI), South Korea (KOSPI) and Taiwan (TSEC). This is the dataset that we use in conducting the experiments to extract patterns of DIN from multiple interactive stock markets.

As in Equation 2, we would like to find the transition matrix F which maps system state at time point t to system state at time point $t + 1$. This transition matrix F is considered to be the interaction model among the observed variables that governs movement of the ten observed stock market indexes in the Asia Pacific region over time.

In our work, we split the 72-week dataset to 50 weeks of training dataset (to extract the initial DIN) and 22 testing dataset (to test the reliability of extracted DIN). To reveal a reliable model, we run 10 different trials on the training dataset and average each of the interaction coefficient values in the transition matrix before constructing the DIN. Results of the modelling process are: (1) the transition matrix, (2) the DIN diagram of interactive stock market and (3) a graph comparing actual and simulated stock indexes trajectories. Figure 7 presents the results of the DIN modelling process. Part (c) of the figure clearly shows that the estimated trajectories, based on the extracted DIN, closely track the actual trajectories. Therefore, we can conclude Kalman filter is a good approximator for future values of multiple stock indexes when it is being modelled as discrete-time approximations of first-order differential equations.

When we take a look at Figure 7.b., we can see DIN model reveals that Hong Kong (HSIX), Taiwan (TSEC), Singapore (STI), South Korea (KOSPI) and Japan (N225) are the influential markets in the region. This is in agreement with those markets being based in the economic powerhouse of the Asia Pacific region and they tend to influence other countries which are economically related to them. For instance, Figure 7.b recognises the influence Hong Kong exerts on China (Shanghai stock market), Japan, Korea, Malaysia and Singapore. DIN model also shows that China (SSE) is affected by most of the other markets. To analyse the result, we need to consider China's strategic approach to its Asian neighbours. In this case, China:

1. has become one of the largest traders and investors with many Asian countries;

2. exports primarily consumer goods to most countries in Asia Pacific;

3. is more than just a trading partner, as it also invests extensively in the region (e.g. China recently became one of the largest investors in Indonesia by buying into oil and gas interest);

4. is the largest foreign investor in some of the smaller economies in Southeast Asia.

By proposing to negotiate a free trade agreement with the ASEAN countries, China offered to share the benefits of its economic growth, while reminding the region of their growing reliance on China. All these contribute to the unique position of China, and to the number of vertices involved in the DIN model as in Figure 7.b.

Next, we also find that the DIN model identifies interactions between Taiwan (TSEC), Hong Kong (HSIX), Singapore (STI) South Korea (KOSPI) and Australia (AORD), which are in agreement with previous findings by Masih and Masih [4] in their research on the dynamics of stock market interdependency. Though their results are based on the period from 1982 to 1994, we conclude that the relationships among these countries are preserved in the more recent period. Therefore, for some leading economies with relatively stable economic infrastructure, consistent interactions exist in a form of long-term stock market relationships. In conclusion, the transition matrix in Figure 7.a represents meaningful relationships between the observed variables (stock market indexes in the Asia Pacific region) and can be used to build a reliable and stable DIN model.

Analysing how the interactions model change when new measurements of the system become available would be of interest. We definitely would like to know if the proposed method is capable to capture changes in the relationship between observed variables when new data is available. To accommodate this, we implement the incremental learning process to our proposed method. The incremental learning process is simply a process of updating the knowledge that we already have by taking into account new experiences when the environment changes. In this case, it means that we would like to update our knowledge about the interactive patterns between the observed variables by updating (creating new connections, removing existing connections or updating existing connections between nodes) the DIN model when new information arrives. In our work we implement the incremental learning process when we do the testing process (using the 22-week dataset).

Now we will be analysing how the DIN model changes when new data becomes available. Here, we extract a new DIN model using the 10-week dataset after the 50-week dataset that we have used to extract DIN model in Figure 7.b. Additionally, another DIN model is also extracted again after a further 10 weeks. These extracted DIN models are illustrated in Figure 8.a.

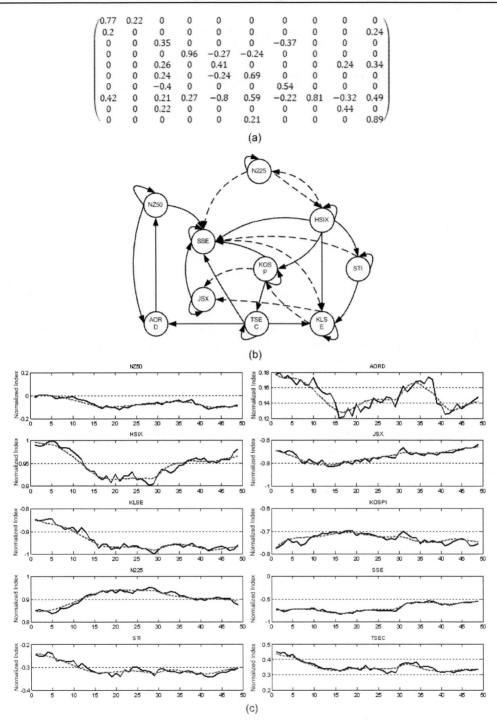

Figure 7. Results of the DIN modelling process: (a) transition matrix; (b) DIN model (note: dashed edges in the interaction network diagram represent negative relationships between vertices); (c) actual (solid line) and DIN estimated trajectories (dashed line) during training process using 50 weeks' data [16].

Table 2. Error rates for prediction of 10 stock market indexes 22 weeks ahead in the Asia Pacific region [16].

Market	RMSE
NZ50	0.0117
AORD	0.0108
HSIX	0.0066
JSX	0.0123
KLSE	0.0143
KOSPI	0.0140
N225	0.0076
SSE	0.0375
STI	0.0097
TSEC	0.0142

Feeding 10 weeks of new data leads to significant changes in the DIN model. The most significant change is the disappearance of interaction between New Zealand (NZ50) and Australia (AORD). A closer look at the trajectories (as in Figure 6) reveals that, for the 10 weeks of new data, NZ50 and AORD move quite independently from each other, in contrast to their behaviour in the previous 50 weeks. Another noteworthy change in Figure 8.b is that after feeding 20 weeks of new data, the interaction between NZ50 and AORD reappears. Again Figure 6f is relevant here, as it shows that starting from week 62, upward movement in NZ50 index is accompanied by corresponding increases in the AORD index and vice versa, restoring the linkage between the two markets. Therefore, it can be concluded that the DIN model adapts to changes in stock market conditions faithfully represents the interactions among markets.

The Asia Pacific Stock Market Indexes Prediction with DIN

We test the extracted DIN model to predict future values of multiple stock indexes. The test dataset covers 20-week data, following chronologically the 50 weeks of training dataset. First, we consider the appropriateness of fit of the DIN model at all 20 points of the test dataset, plotting actual test trajectories and the simulated trajectories together. Second, we calculate the prediction error using the root-mean-squared error, to measure how good the DIN prediction performance is. The trajectories generated using DIN model are presented in Figure 9, showing that it provides good accuracy of prediction, and this is confirmed by the error measurements as in Table 2, which are considerably small. We can conclude that by using 50-week data as training dataset and by integrating the incremental learning concept to update knowledge about relationship between stock markets, we have been able to extract reliable DINs model which are sufficiently accurate to be used for predicting future values.

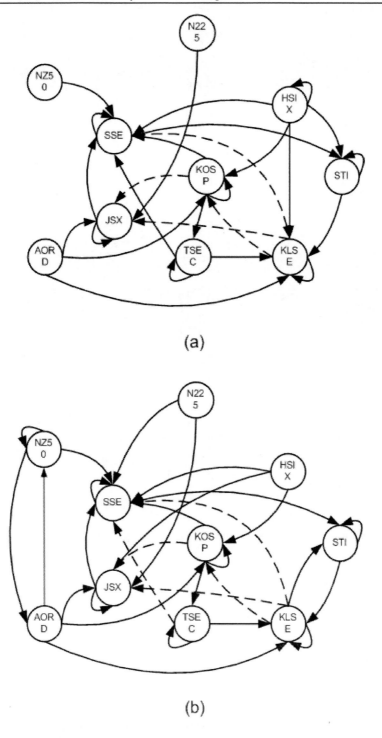

Figure 8. Feeding new data and updating the DIN model: (a) DIN model after 10 weeks; (b) DIN model after 20 weeks (note: dashed edges show inverse relationship between vertices) [16].

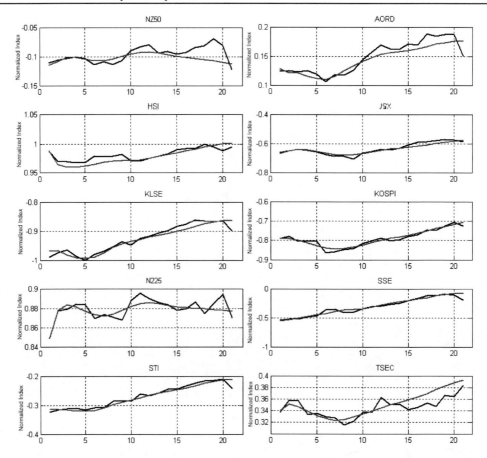

Figure 9. Comparison between actual (in black) and predicted (in grey) trajectories using DIN model over the 22-weeks test dataset [16].

5.2. Modelling Vibrant Relationship between Stock Market and Macroeconomic Factors

Researchers have used various methods and observed several stock market related indicators to speculate the trend of the stock market index. A number of researchers stated that macroeconomics condition is a good indicator to forecast stock market index. Some studies conducted in Ghana [8], China [38] and Singapore [32] have confirmed this observation. In addition, the study carried out in Thailand [35] showed that macroeconomics condition is more relevant to stock market condition before the financial crisis in 1998. Furthermore, other study suggests that macroeconomics factor offer better insight in predicting recession in stock market than predicting stock returns [34],[3]. The method used to find the correlation between stock market and macroeconomics in some of the previous researches is the co-integration analysis [35].

In this section, we are going to describe the employment of DIN model extraction as a distinctive approach to find important and dynamic relationships between stock market index and macroeconomics factors in a specific country. Through DIN realisation, we intend to acquire:

1. dynamic and complex relationship between stock market index and macroeconomics factors;

2. relationship among macroeconomics factors them self;

3. ultimately, the prediction of future values of each observed variables.

Economics data from Indonesia including: (1) the Indeks Harga Saham Gabungan (IHSG) as the stock market index, (2) the exchange rates {Indonesian Rupiah (IDR) to the United States Dollar (USD)}, (3) bank interest rates, and (4) Indonesian Central Bank certificate index (SBI) are taken into account.

As in the previous sections, we also model the interaction network between the observed stock market index and the macroeconomics factors as discrete-time approximations of first-order differential equations, and use Kalman filter to compute the optimal estimate of relationship between these observed variables based on the observed dataset. Data sample in this work spans from June 2000 to April 2009. They have undergone preprocessing step, which are data trend removal and linear normalisation. Figure 10 illustrates trajectories of the economic variables data which are used.

From the transition matrix and the DIN model as in Figure 11, it can be seen that IHSG is mainly being influenced by itself and slightly influenced by other macroeconomics factors. This finding is resulted from a research conducted in Thailand [35], where it was concluded that after the financial crisis which attacked Asia in 1998, macroeconomics factor started to lose their significant influence to movement of stock market index. It also illustrates that the stock market index in Indonesia in period of June 2000 to January 2007 (we have used data from this period as training dataset) has insignificant relation to the country's economic condition. This discovery conforms to the outcome of research conducted by Liu and Sun [38], where they concluded that in developing countries there is no strong relationship between stock market and macroeconomics factors. Moreover, extracted DIN model also verifies a statement by Chen [34], which deliberates that macroeconomics factors are of better use to predict bear market than stock returns.

When the dataset spanning from February 2007 to April 2009 is employed, we have the transition matrix and DIN model in Figure 12 as the result. The figure shows that after 108 weeks, relationship among economic variables has changed compared to the one extracted in earlier time, as in Figure 11. Extracted DIN model shows that IHSG is now being influenced negatively by bank interest rates, even though only at a low level of stimulus. This finding complies with a theory supported by Maysami, Howe and Hamzah [32], which says that low interest rates reduce the costs of borrowing and motivate corporate to expand their business which will lead to the increase of corporate stock price. Moreover, as stocks are most likely bought with borrowed money, high interest rates may prevent people from borrowing money and buying stocks.

Prediction of Stock Market Index and Macroeconomics Factors with DIN

We also tested the extracted DIN model from the observed economic variables to predict their future values. As the result, Figure 13 and Table 3 show the comparison of the

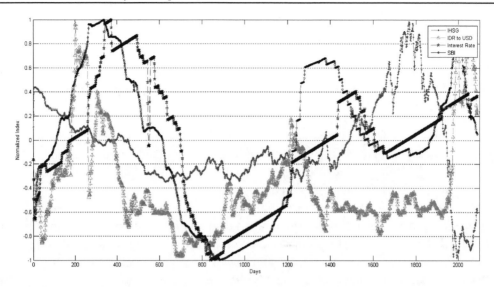

Figure 10. Trajectories of IHSG, IDR to USD exchange rate, bank interest rate and central bank certificate index on weekly basis, spanning from June 2000 to April 2009.

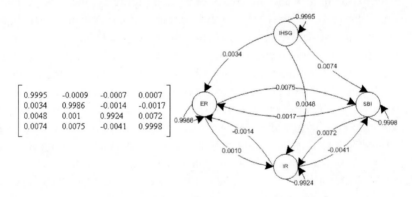

Figure 11. Transition matrix and DIN model generated from data spanning from June 2000 to January 2007.

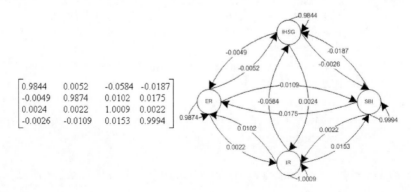

Figure 12. Transition matrix and DIN model generated using data spanning from February 2007 to April 2009, approximately 2 years after previous DIN model (as in Figure 11) was extracted.

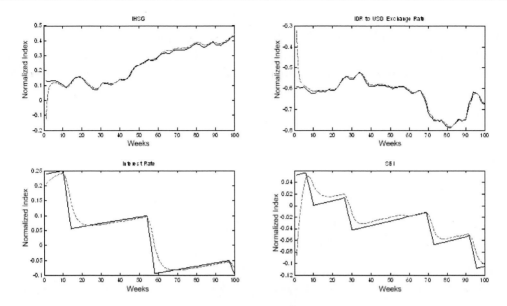

Figure 13. Actual (solid line) and predicted (dashed line) trajectories of macroeconomics factors in Indonesia generated with extracted DIN model.

Table 3. Error rates for prediction of the Indonesian macroeconomics factors.

Macroeconomics Factor	Period	RMSE
IHSG	Feb07-Apr09	0.041541
Exchange Rate (IDR to USD)	Feb07-Apr09	0.042522
Bank Interest Rate	Feb07-Apr09	0.019131
Indonesian Central Bank certificate index (SBI)	Feb07-Apr09	0.021105

actual and predicted trajectories of four economic variables and the root-mean-squared error to justify the prediction accuracy respectively. Again, we can observe that not only the extracted DINs model reveals important and dynamic changes in patterns of interaction among the observed economic variables, it has also been proven that the extracted DINs model can be used to predict movement of multiple economic variables with a reasonable degree of accuracy.

5.3. Modelling the Interactive Patterns of International Stock Markets [21]

Interactions between stock markets have been researched in the past few years. Globalisation has brought interdependence among stock markets around the world, in which a change in one stock market affects other stock markets. A study by [19] stated that in 1990, the relationship between international stock markets is stronger than the preceding years, due to relaxation of foreign ownership restrictions. Some other researchers [10],[23],[24] and [27] also found that local stock markets are generally influenced by major stock markets in the world (i.e. U.S., United Kingdom and Japan). Several studies applied the co-integration analysis method to find the correlation between the stock markets. Unlike these previous studies which have used the correlation analysis to find existence of correlation between pair of markets, we use the Kalman filter to extract complete, fully-connected network showing relationship between multiple variables (not just pair of markets) by formulating the interdependency between the observed stock markets as discrete-time approximation of first-order differential equations.

To model patterns of interaction between global stock markets, we take data of stock indexes from eight countries in the world, which are: (1) United States (DJA), (2) United Kingdom (FTSE), (3) Germany (DAX), (4) South Africa (JSE), (5) China (SSE), (6) Japan (N225), (7) Indonesia (JKSE) and (8) Australia (AORD). Data samples span from September 2006 to June 2009. In the conducted experiment the dataset has undergone data preprocessing steps which are trend removal and linear normalisation.

As a result of the DIN model extraction process using dataset spanning from September 2006 to September 2008, we acquired the transition matrix and the DIN model as illustrated in Figure 14. Here, it can be seen that DJA of the United States is mainly being influenced by itself and significantly by some other countries (United Kingdom, Germany, China and South Africa). This finding falls in with results from research conducted by Bessler [11], where it was concluded that the United States stock market is strongly influenced by itself and significantly by United Kingdom and Germany. It also illustrates that the other stock markets are significantly influenced by the United States and United Kingdom stock markets.

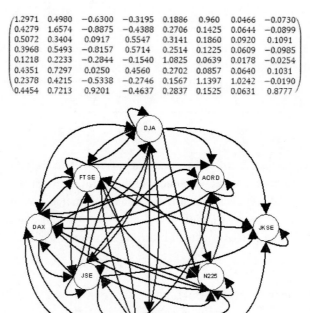

Figure 14. Transition matrix and DIN model extracted from international stock market indexes, spanning from September 2006 to September 2008 [21].

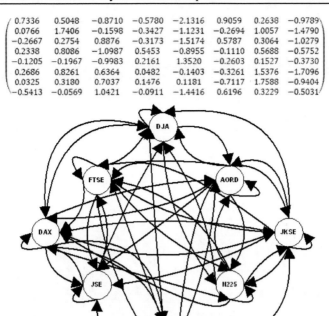

Figure 15. Transition matrix and DIN model extracted from international stock market indexes, spanning from September 2008 to June 2009 (10 months after the DIN model in Figure 14 is extracted) [21].

The Indonesian stock market (JKSE) appears to have the weakest influence compared to the other stock markets, since it only significantly influences itself. This result is also expected since Indonesia is considered to be the "weakest" country compared to the other seven countries from an economy perspective. As we can observe the extracted DIN model, instead of influencing the other countries, Indonesia is being influenced by the other leading countries, for instance United States, Germany, United Kingdom, Japan and China.

By feeding the dataset which spans from September 2008 to June 2009 we are able to extract a new DIN model, as shown in Figure 15. The DIN model shows that after 10 months, relationship among the observed stock markets has changed. The new extracted DIN model shows that DJA is now being influenced strongly by SSE, the stock market of China, and significantly by the rest of the other countries. We can also see that SSE of China has become the most influencing stock market among the other countries, while AORD of Australia is shown to have interdependency with the other countries (except with United Kingdom and South Africa). This finding complies with theory supported by Drew and Chong [23]. The result is also in line with finding by Beelders [27], since there is also interaction between the South Africa stock market (JSE) with the other observed countries.

This section describes (yet again), that by modelling relationship or interactions or interdependencies between observed multiple time-series data as discrete-time approximations of first-order differential equations, and by using a good estimator to compute the optimal estimates of the state-space model of the observed multiple time-series data, we are capable

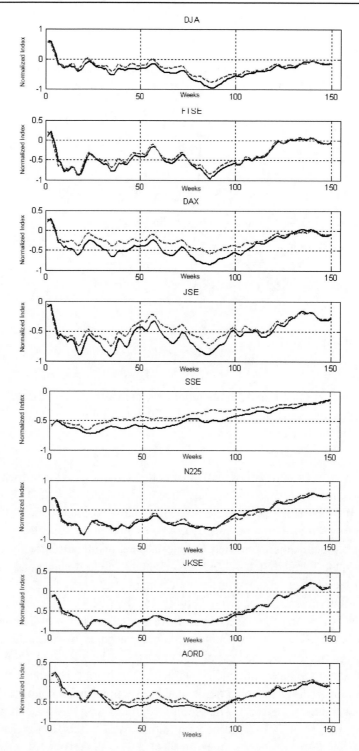

Figure 16. Actual (solid line) and predicted (dashed line) trajectories of eight international stock markets generated with extracted DIN model [21].

Table 4. Error rates between predicted and actual trajectories for eight international stock markets [21].

Stock Market	Period	RMSE
DJA	Sept 2008-June 2009	0.11366
FTSE	Sept 2008-June 2009	0.064351
DAX	Sept 2008-June 2009	0.18052
JSE	Sept 2008-June 2009	0.11763
SSE	Sept 2008-June 2009	0.1055
N225	Sept 2008-June 2009	0.074479
JKSE	Sept 2008-June 2009	0.034345
AORD	Sept 2008-June 2009	0.093822

of extracting important and dynamic patterns of interaction between the observed variables (series). Additionally, these extracted network models (DIN models) can be employed to perform multiple time-series prediction in a considerably good accuracy. Results of predicting movement of international stock markets is illustrated in Figure 16.

6. Dynamic Interaction Network to Model Interactive Patterns in Weather Dataset

The word "weather" can be defined as the state of the atmosphere when measured on a scale of hot or cold, wet or dry, calm or stormy, clear or cloudy. Different from climate, which is the term for the average atmospheric conditions over longer periods of time, weather generally refers to day-to-day temperature, air pressure, wind movement (speed and direction) and precipitation activity.

Weather occurs due to density (temperature and moisture) differences between one place and another. These differences can occur due to the sun angle at any particular spot, which varies by latitude from the tropics. The strong temperature contrast between polar and tropical air gives rise to the jet stream.

One of the most important factors that governs or has the greatest influence of circulating the weather systems is the wind, as it directs movement of the air in a certain velocity. Wind is caused by differences in pressure. When a difference in pressure exists, the air is accelerated from higher to lower pressure. On a rotating planet the air will be deflected by the Coriolis effect, except exactly on the equator. Globally, the two major driving factors of large scale winds (the atmospheric circulation) are the differential heating between the equator and the poles (difference in absorption of solar energy leading to buoyancy forces) and the rotation of the planet.

The atmosphere or the weather system is a chaotic system, so small changes to one part of the system can grow to have large effects on the system as a whole. This makes it difficult to accurately predict weather more than a few days in advance, though weather forecasters are continually working to extend this limit through the scientific study of weather, meteo-

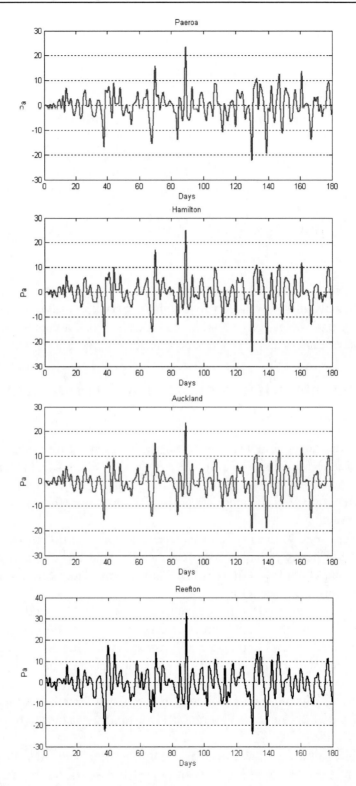

Figure 17. Trajectories of air pressures data from 4 observation stations in New Zealand, spanning from October 2009 to Decemeber 2009.

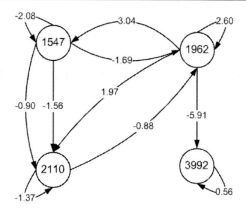

Figure 18. DIN model for pattern of interactions between air pressures in Auckland, Hamilton, Poriroa and Reefton, extracted using data from September 2009 to December 2009.

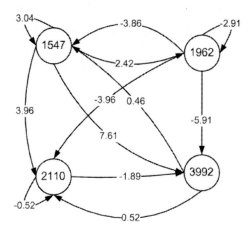

Figure 19. DIN model for pattern of interactions between air pressures in Auckland, Hamilton, Poriroa and Reefton, extracted using data from January 2010 to February 2010.

rology. It is theoretically impossible to make useful day-to-day predictions more than about two weeks ahead, imposing an upper limit to potential for improved prediction skill.

Based on the fact that a small change to one part of the system can lead to a complete change in the weather system as a whole, it would be of interest to observe variables of the atmosphere in order to extract and model how they are related to each other and how they will behave as a system.

Analysing the air pressure sampled irregularly in different places by modelling their interdependencies as discrete-time approximation of first-order differential equations would help us forecast movement of the wind, which in the end would lead us to the possibility of developing a new method for weather system prediction.

In this section of the chapter we are going to outline results of extracting patterns of interaction between air pressures observed in different areas in New Zealand starting from October 2009 until end of December 2010. Trajectories of observed air pressure data is illustrated in Figure 17. Please notice that trajectories of data collected from Auckland,

Poriroa and Hamilton expose similar shape. This is expected as these three observation stations are located in the same region (within a radius of 100 kilometers, part of the North Island of New Zealand). In contrast to these three trajectories, the last trajectory of air pressure data collected in Reefton (part of the South Island of New Zealand), moves in different fashion.

Only by examining trajectories of the air pressure data from the four observation stations we could recognise that there exists a strong relationship or interaction between movement of air pressure in Auckland, Poriroa and Hamilton. And, there should be a noteworthy interaction among these three areas with Reefton which is located more than 1000 kilometers down south. Extracted interaction network model based on the observed air pressure data (spans from October 2009 until end of December 2009) is illustrated in Figure 18.

As it has been expected, the extracted interaction network model shows strong connections between Auckland, Poriroa and Hamilton, while the interaction with Reefton can be classifed weak. Interestingly, when we feed more data to the system (20 weeks observation data) the interaction network model changes significantly (as in Figure 19). Interaction between Auckland, Porioa and Hamilton resides in a stable manner, but their interaction with Reefton has increased significantly. This finding suits the actuality of the weather system being a chaotic system, in which its state changes dynamically from time to time.

The results of being able to extract dynamic patterns of interaction between air pressure in different observation areas using the Kalman filter opens the possibility of extracting complex interaction patterns between variables in our weather system (including: the temperature, air pressure, wind movement, precipitation level, etc.). Eventually being able to overcome such knowledge would help us understand more how our weather system works and how it transforms its state vibrantly.

7. Conclusion

Given a set of measurements of observed multiple variables sampled on a regular or irregular timely basis (multiple time-series dataset), we are capable of extracting the dynamic interaction models between these observed variables by modelling the governing interdependencies as discrete-time approximation of first-order differential equations. The methodology is to treat the actual trajectories (actual observed measurements) as a set of unobserved or hidden variables (using the state-space representation), and then to apply a reliable and stable estimator algorithm to compute the transfer function that maps past system state to present system state and present system state to future system state based on the observed measurements.

Kalman filter, being a set of mathematical equations that provides vigorous and efficient computational (recursive) means to estimate the state of a process, in a way that it minimises the mean-squared-error, has confirmed its capability to solve the problem of estimating the past, present and future states of the state-space representation of the observed measurements. The filter has also exposed the ability to succeed even when the observed variables are being sampled irregularly or when the precise nature of the system is unknown.

Through conducted experiments, using data from different fields, for instance: (1) biomedical (GRN expression data), (2) financial (stock market indexes, macroeconomics factors and currency exchange rates) and (3) weather system (air pressures), it is found that we are capable of revealing and modelling important and dynamic patterns of interaction between the observed variables sampled on a timely basis. Furthermore, the methodology that has been explained in this chapter gives us insights about the applications of the Kalman filter which can be described as follows,

1. Kalman filter is a robust computing method that can be used to estimate the actual trajectories of multiple time-series data (even when they are being observed irregularly);

2. by being able to estimate the actual trajectories of observed multiple time-series data, Kalman filter is also capable of computing the past, present and future state of the system which leads to the ability of extracting the transition matrix which maps system-state at current time-point to system-state at future time-point;

3. the extracted transition matrix represents important and complex interdependencies between the observed variables in the form of an interaction network;

4. not only the interaction network exposes essential relationships; it also can be employed to predict future values of observed variables with an acceptable degree of accuracy.

An interesting topic is to explore the possibilities of implementing the extended Kalman filter which was developed to estimate systems with non-linear relationships between the contributing variables, and to analyse whether it could provide us with a better model and understanding of how variables (things) in this world work together, influencing one another directly or indirectly.

References

[1] A. Antoniou, G. Pescetto and A. Violaris. "Modelling International Price Relationships and Interdependencies between the Stock Index and Stock Index Future Markets of Three EU Countries: A Multivariate Analysis." Journal of Business, Finance and Accounting, vol. 30, pp. 645-667, 2003.

[2] A. Ben-Dor and Z. Yakhini. "Clustering Gene Expression Patterns." In Proceedings of the Third Annual International Conference on Computational Molecular Biology, Lyon, France, 1999.

[3] A. Humpe and P. Macmillan. "Can Macroeconomics Variables Explain Long Term Stock Movement? A Comparison of the US and Japan." CDMA Working Paper No. 07/20, October 2007. Available at SSRN: http://ssrn.com/abstract=1026219.

[4] A. Masih and R. Masih. "Dynamic Modelling of Stock Market Interdependencies: An Empirical Investigation of Australia and the Asian NICs." Working Papers 98-18: 1323-9244, University of Western Australia, 1998.

[5] A. Serguieva and H. Wu. "Computational Intelligent in Financial Contagion Analysis." International Journal on Complex Systems, vol. 2229, pp. 1-12, 2008.

[6] A. Serguieva, T. Kalganova and T. Khan. "An Intelligent System for Risk Classification of Stock Investment Projects." Journal of Applied Systems Studies, vol. 4(2), pp. 236-261, 2003.

[7] A.H. Jazwinski. Stochastic Processes and Filtering Theory. New York: Academic Press, 1970.

[8] A.M. Adam and G. Tweneboah. "Macroeconomics Factors and Stock Market Movement: Evidence from Ghana." MPRA Paper No. 11256, 2008.

[9] D. Collins and N. Biekpe. "Contagion and Interdependence in African Stock Markets." The South African Journal of Economics, vol. 71(1), pp. 181-194, 2003.

[10] D. Isakov and C. Perignon. "On the Dynamic Interdependence of International Stock Markets: A Swiss Perspective." Swiss Journal of Economics and Statistics, vol. 136, pp. 123-146, 2000.

[11] D.A. Bessler. "The Structure of Interdependence in International Stock Markets." Journal of International Money and Finance, vol. 22, pp. 261-287, April 2003.

[12] D.E. Goldberg. Genetic Algorithms in Search, Optimization and Machine Learning Reading. MA: Addison-Wesley, 1989.

[13] F.C. Schweppe. Uncertain Dynamic Systems. Englewood Cliffs, New Jersey: Prentice-Hall, 1973.

[14] G. Welch and G. Bishop. "An Introduction to the Kalman Filter." Computer Science Working Papers TR95-041, University of North Carolina at Chapel Hill, 2006.

[15] G.M. Caporate, A. Serguieva and H. Wu. "A Mixed-game Agent-based Model for Simulating Financial Contagion."In Proceedings of the 2008 Congress on Evolutionary Computation, Piscataway, NJ: IEEE Press, pp. 3420-3425, 2008.

[16] H. Widiputra, R. Pears, A. Serguieva and N. Kasabov. "Dynamic Interaction Networks in Modelling and Predicting the Behaviour of Multiple Interactive Stock Markets." International Journal of Intelligent System in Accounting, Finance and Management, Special Issue, vol. 16, pp. 189-205, 2009.

[17] J.J. Deyst. "Estimation and Control of Stochastic Processes." Unpublished course notes. M.I.T. Dept. of Aeronautics and Astronautics, Cambridge, Massachusetts, 1970.

[18] J.S. Bay. Fundamental of Linear State Space Systems. Boston: WCB/McGraw-Hill, 1999.

[19] K. Phylaktis and F. Ravazzolo. "Stock Market Linkages in Emerging Markets: Implications for International Portfolio Diversification."Journal of International Financial Markets, Institutions and Money, vol. 15, Issue 2, pp. 91-106, April 2005.

[20] L. Friedman, Nachman and Pe'er. "Using Bayesian Networks to Analyze Expression Data." Journal of Computational Biology, vol. 7, pp. 601-620, 2000.

[21] L. Lukmanto, H. Widiputra and Lukas. "Dynamic Interaction Network to Model the Interactive Patterns of International Stock Markets." World Academy of Science, Engineering and Technology, vol. 59, pp. 257-261, 2009.

[22] M. Aoki. Optimization of Stochastic Systems - Topics in Discrete-Time Systems. New York: Academic Press, 1967.

[23] M. Drew and L. Chong. "Stock Market Interdependence: Evidence from Australia." Discussion Papers in Economic, Finance and International Competitiveness, School of Economics and Finance, Queensland University of Technology, Discussion Paper No. 106, February 2002.

[24] M. Glezakos, A. Merika and H. Kaligosfyris. "Interdependence of Major World Stock Exchanges: How is the Athens Stock Exchange Affected?." International Research Journal of Finance and Economics, Issue 7, pp. 24-39, January 2007.

[25] M. Psillaki and D. Margaritis. "Long-Run Interdependence and Dynamic Linkages in International Stock Markets: Evidence from France, Germany and the U.S." Journal of Money, Investment and Banking, Issue 4, 2008.

[26] N. Kasabov, Z. Chan, V. Jain, I. Sidorov and D. Dimitrov. "Gene Regulatory Network Discovery from Time-series Gene Expression Data: A Computational Intelligence Approach." Lecture Note in Computer Science, vol. 3316, pp. 1344-1353, 2004.

[27] O. Beelders. "International Stock Market Interdependence: A South African Perspective.", March 2002. Available at SSRN: http://ssrn.com/abstract=304323 or doi:10.2139/ssrn.304323.

[28] P.S. Maybeck. "The Kalman Filter - An Introduction for Potential Users." TM-72-3, Air Force Flight Dynamics Laboratory, Wright-Patterson AFB, Ohio, June 1972.

[29] P.S. Maybeck. Stochastic Models, Estimation, and Control, Volume 1. New York: Academic Press, 1979.

[30] R. Brown. Introduction to Random Signal Analysis and Kalman Filtering. New York: Wiley, 1983.

[31] R. Lewis. Optimal Estimation with an Introduction to Stochasic Control Theory. New York: Wiley, 1986.

[32] R.C. Maysami, L.C. Howe and M.A. Hamzah. "Relationship between Macroeconomics Variables and Stock Market Indices: Cointegration Evidence from Stock Exchange of Singapore's All-S Sector Indices. Jurnal Pengurusan 24, pp. 47-77, 2004.

[33] R.E. Kalman. "A New Approach to Linear Filtering and Prediction Problems." Transaction of The ASME - Journal

[34] S. Chen. "Predicting the Bear Stock Market: Macroeconomic Variables as Leading Indicators," Journal of Banking & Finance, Elsevier, vol. 33(2), pp. 211-223, February 2009.

[35] T. Brahmasrene and K. Jiranyakul. "Cointegration and Causality between Stock Index and Macroeconomics Variables in an Emerging Market." Academy of Accounting and Financial Studies Journal, September 2007.

[36] T.C. Chiang and S. Doong. "Empirical Analysis of Stock Returns and Volatility: Evidence from Seven Asian Stock Markets based on TAR-GARCH." Review of Quantitative Finance and Accounting, vol. 17, pp. 301-318, 2001.

[37] T.P. McGarty. Stochastic Systems and State Estimation. New York: Wiley, 1974.

[38] Y. Liu and L. Sun. "Analysis of Cointegration between Macroeconomics Variables and Stock Index." In Proceedings of the 2008 Fourth International Conference on Natural Computation, vol. 05, pp. 318-322, 2008.

[39] Z. Chan, N. Kasabov and L. Collins. "A Two-stage Methodology for Gene Regulatory Network Extraction from Time-course Gene Expression Data." Expert System with Applications, vol. 30, pp. 59-63, 2006.

In: Kalman Filtering
Editor: Joaquín M. Gomez

ISBN 978-1-61761-462-0
© 2011 Nova Science Publishers, Inc.

Chapter 12

KALMAN FILTERS FAMILY IN GEOSCIENCE AND BEYOND

Olivier Pannekoucke[1,*] *and Christophe Baehr*[1,2]
[1]Météo-France/CNRS, CNRM/GAME (URA 1357)
[2]Université de Toulouse Paul Sabatier, Institut de Mathématiques, France

Keywords: Data assimilation technics, non-linear filter, Kalman filter, Ensemble methods, particle filter.

1. Introduction

Being able to predict the weather is one of the greatest challenges of mankind. This success relies on the Kalman filter equations, and its various generalization or approximations. The aims of the chapter is to see why Kalman equations are needed and also to provide various generalization and approximation of information dynamics.

At a theoretical level, the atmosphere lies in a particular phase space whose state at time q is denoted by X_q. All the physical process imply a time evolution of this state from q to $q+1$ and it is formally written by

$$X_q = \tilde{\mathcal{M}}_q(X_{q-1}), \tag{1}$$

where $\tilde{\mathcal{M}}_q$ corresponds to the propagator underlying to the nature. A numerical weather prediction model is a dynamical system that incorporate all pertinent physical process to provide a worth information toward the forecaster. This corresponds to a set of partial derivative equation that have to be time-integrated from a known state. Of course, this procedure assumes that numerical weather prediction is a deterministic process: one state leads to a one and only one time evolution of the flow (we hope it is). The numerical model can be viewed as a simple non-linear equation that makes evolving the numerical representation of the atmosphere X_q from the time q to the time $q+1$ according to

$$X_q = \mathcal{M}_q(X_{q-1}) + \mathcal{W}_q, \tag{2}$$

---
[*]E-mail address: olivier.pannekoucke@meteo.fr

where $\mathcal{M}_q$ corresponds to the propagator of the numerical model that differs from the nature propagator $\tilde{\mathcal{M}}_q$ implying to introduce a correction W_q representing a model error due to various approximations of the real physics, for instance the parametrization of the turbulence or of the diphasic process in clouds. W_q is assumed to be a centered Gaussian random variable with covariance matrix Q_q. Of course, despite of the complexity of the model and all the very clever things you put inside, to be useful you need to provide the right initial state at time q so to obtain the right time evolution of the real atmosphere (and also before it happens to be expandable, that is the constraint we face).

Unfortunately, the atmosphere is known to be chaotic, by this word, we mean that the atmospheric flow is sensitive to the initial conditions. This crucial aspect has been highlighted by Lorenz in the 60's from numerical study using only 3 freedom degrees [53]. The equations of Lorenz's model describes, at the simplest level, the convection dynamic through a Galerkin projection on the first three Fourier modes. The amplitude of the three modes are denoted by (X, Y, Z) and their dynamic associated to $\tilde{\mathcal{M}}$ is the non-linear dynamical system

$$\begin{cases} \frac{dX}{dt} = \sigma(Y - X), \\ \frac{dY}{dt} = (r - Z)X - Y, \\ \frac{dZ}{dt} = XY - bZ, \end{cases} \quad (3)$$

where the classical parameter set $(\sigma, b, r) = (10, 8/3, 28)$ is used to obtain a chaotic dynamic. For this non-linear dynamic, it is possible to show that the system is deterministic: it satisfies the non-linear Cauchy-Lipschitz theorem. But, despite on its simplicity, no numerical solution is able to produce a real solution. This system is intrinsically unstable, and computation in finite precision leads to errors that increase exponentially with time. Instead, the numerical solution live in a portion of the state space called a strange attractor. We cannot compute a real solution but we can document its "climate".

If we extrapolate a little Lorenz's results to real applications. We have to retain that long time numerical predictions fail to reproduce the real atmospheric flow. There exists a duration from which the numerical solution diverges from the real weather. Even if you provide the exact numerical state of the atmosphere, the use of a numerical approximation of the real dynamic leads irreparably to another solution after a while that depends on the weather situation.

All is not lost. Fortunately, the atmosphere is observed that inquire about its real state. These observations, denotes Y_q, are related to the real numerical state by

$$Y_q = \mathcal{H}_q(X_q), \quad (4)$$

where $\mathcal{H}_q$ is the observation operator mapping the phase space into the observations space. This relationship can not be inversed directly. One reason, is that data are not free from noise and thus we retain the relation

$$Y_q = \mathcal{H}_q(X_q) + V_q, \quad (5)$$

where V_q stands for error associated to the measures (noise) but also to the use of $\mathcal{H}_q$ (interpolations, approximations of physical process like radiative transfer). V_q is assumed to be a centered Gaussian random variable with covariance matrix R_q. The only thing one can construct is an estimation $\widehat{X}_q$ the truth X_q knowing the observation Y_q. Of course, this estimation is then time integrated to provide an estimation of the further state $\overline{X}_q$ thanks to the dynamic

$$\overline{X}_q = \mathcal{M}_q(\widehat{X}_{q-1}). \qquad (6)$$

Mathematically, if $Y_{0:q}$ denotes the list of observations from a zero time, $\overline{X}_q$ and $\widehat{X}_q$ are defined as the conditionnal expectation $\overline{X}_q = \mathbb{E}(X_q/Y_{0:q-1})$, the best information knowing the observation from time 0 to $q-1$; and $\widehat{X}_q = \mathbb{E}(X_q/Y_{0:q})$, the best information knowing the observation from time 0 to $q-1$ and knowing the last observations given at the same time q. Note that we have also that $\overline{X}_q = \mathbb{E}(X_q/\widehat{X}_{q-1})$ and $\widehat{X}_q = \mathbb{E}(X_q/\overline{X}_q, Y_q)$. Thereafter, the nonlinear operator $\mathcal{M}_q$ and $\mathcal{H}_q$ are assumed linear for facilitate the demonstrations. These are denoted by M_q and H_q.

The basic issue is to estimate the most likely state of the atmosphere $\widehat{X}_q$ knowing the observation. This step is called data assimilation in geosciences, and the most likely state is called the analysis state. A natural way to solve the problem is the use of the Kalman filtering, where the analysis appears as a corrected state using the observation of a prior information or background. This chapter introduces various forms and applications of linear/nonlinear filtering with some pedagogical experiments. First, a theoretical presentation examines the concepts while a second part is more dedicated to practical use inspired by examples encountered in geophysical applications, but that could be applied in other engineering sciences.

The Kalman equations can be set out as the restriction of a general nonlinear filter. The nonlinear filter is entirely determined by the Feynman-Kac probability measures, deduced from the formalism of Markov processes using the conjunction of probability distributions. Then the filtering process is seen as a posterior probability time transport and a prior probability update transport. In particular, the update Markov kernel is shown non-unique and various example are discussed. The Feynman-Kac distribution of the non-linear filter may be approached using particle systems, detailed here. Kalman equations are deduced from this formalism under Gaussian and linear assumptions. Another demonstration is detailed where the Kalman equations are view as a linear regression onto a subspace spanned by the observations.

The Ensemble Kalman filter method (EnKF) is introduced as a combination of ensemble techniques and Kalman engineering. We put forward the strength of the EnKF and a awkward limitation : it convergences to a stochastic process which is not the non-linear filter under non-linear dynamics or non-Gaussian probability distribution.

In the angle of data assimilation, a practical variational implementation of Kalman filter is presented with emphasis on the 4D-Var method: a temporal extension of the Kalman filter applied in order to adjust a temporal forecast to the observations measured at different time steps. In oder to extend the 4D-Var method with full non-linear dynamics and non-Gaussian distributions, a temporal extension of the genetic particle filter is introduced: the 4D-PF.

Due to the huge size of the control vector occurring in geophysical applications, the necessary modeling of the background error covariance matrix is reminded, and some covariance models are introduced: the wavelet and the diffusion formulation for covariances. Both models are detailed and illustrated within simple applications on the sphere: a synthetic ensemble data obtained from non-linear time integrations of perturbed initial conditions evolving with a non-divergent barotropic dynamic. These formulations correspond to some parametric model of covariance matrix. The estimation of their parameters from ensemble methods is explained with emphasis on perturbed analysis ensemble. The diagnostic of the local length-scale is employed as a practical tool for estimating the heterogeneity of covariance functions.

We start now by the theoretical presentation of information dynamics provided by the non-linear filtering.

2. Non-linear Filtering and the Kalman Filter

The non-linear filtering is the general framework to combine known informations about a system and its evolution. The fusion of informations relies on the Bayes' rule, that can written under a binary operator. The term "information" is a quite idle notion but it has to be understood within this context as a probability distribution that quantify a certain amount of knowledge (our knowledge). This way to interpret probabilities is highly powerful and made easy to use by introducing the conjunction operator. Then the Kalman filter equations appear as a particular case of the general theory, where distributions are assume Gaussian and the dynamic linear. We start with a change on perspective about the probability interpretation and introduce the conjunction formalism.

2.1. Non-linear Filter Formalism

2.1.1. The Conjunction Operator

Probability theory can be interpreted from two different manners [47]: the "objective" versus the "subjective" school of thought. The former school considers the probability of an events as its frequency observed from random experiments, while the latter regards probability as expressions of human ignorance. The subjective school provides a powerful insight for the quest for information, highlighting the understanding of data assimilation process. Among all possible states provided by the forecast process, only certain of them match the observations or any other posterior knowledges. This restricted set appears as consistent with both prior and posterior informations.

A merely way to formally represent this constraint on information is to introduce the probability conjunction [82, 81]. It corresponds to the equivalent within probability theory of the logical operation 'and'. Considering X_1 and X_2 as being two random variables associated to probabilities P_1 and P_2 standing for information on a system, and assuming that these probability are featured by their density of probability p_i, so that $P_i(d\mathbf{x}_i) = p_i(\mathbf{x_i})d\mathbf{x}_i$, then we can construct a third random variable $X_{1\wedge 2}$ whose probability distribution is the conjunction $P_1 \wedge P_2$, defined from its density, by $(p_1 \wedge p_2)(\mathbf{x}_{1\wedge 2}) = \frac{1}{Z}p_1(\mathbf{x}_{1\wedge 2})p_2(\mathbf{x}_{1\wedge 2})$, where $Z = \int p_1(\mathbf{x}_{1\wedge 2})p_2(\mathbf{x}_{1\wedge 2})d\mathbf{x}_{1\wedge 2}$ is the normalization term. Z is also called the *partition function* by analogy with the statistical mechanic.

This formalism is now applied in the particular case where P_1 is a prior information and where P_2 is a posterior information. It is not a necessary tool to express the filtering process, but offers a powerful vision of it by reducing the formalism to a simple binary operation onto probabilities.

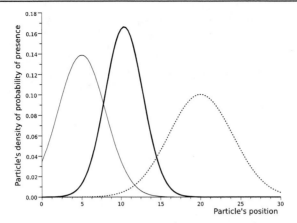

Figure 1. Illustration of the conjunction operator applied in the retrieval position of a particle. Two information p_1 (thin solid line) and p_2 (dashed line) are merged together to form a third information $p_1 \wedge p_2$ (bold solid line) the conjunction distribution.

2.1.2. Reminders on Stochastic Process

Thereafter, the system (atmosphere, ocean, or surface parameters) is assumed to be represented, at time t, by a state vector $\mathbf{x}_t$ living in $\mathbb{R}^n$ where n denotes the number of freedom degrees. Due to the limited knowledge available about true state $\mathbf{x}_t$, this quantity is assume to be a random vector $\mathbf{X}_t$ represent a stochastic process where $t \in [0, T]$. Given a discretization $0 \leq t_0 < t_1 < \cdots < t_q \leq T$, the state $\mathbf{X}_{t_k}$ is simply denoted by $\mathbf{X}_k$. The stochastic process is then entirely defined given the marginal law

$$P(\mathbf{X}_0 \in d\mathbf{x}_0, \cdots, \mathbf{X}_q \in d\mathbf{x}_q),$$

where $d\mathbf{x}_k$ denotes an elementary volume in $\mathbb{R}^n$ surrounding the point $\mathbf{x}_k$ of coordinate $(x_k^i)_{i \in [1,n]}$ and corresponding to Cartesian product

$$d\mathbf{x}_k \equiv [x_k^1, x_k^1 + dx_k^1] \times \cdots \times [x_k^n, x_k^n + dx_k^n].$$

Such a trajectory is denoted by $\mathbf{X}_{0:q} = (\mathbf{X}_0, \cdots, \mathbf{X}_q)$, and its probability is denoted $P_{0:q}(\mathbf{X}_{0:q} \in d\mathbf{x}_{0:q})$ with the event $\mathbf{X}_{0:q} \in d\mathbf{x}_{0:q} \Leftrightarrow (\mathbf{X}_0 \in d\mathbf{x}_0, \cdots, \mathbf{X}_q \in d\mathbf{x}_q)$. The expectation operator applied on a function f arbitrary chosen [1] is defined by $\mathbb{E}[f(\mathbf{X}_{0:q})] = \int f(\mathbf{x}_{0:q}) P_{0:q}(X_{0:q} \in d\mathbf{x}_{0:q})$.

Under assumption of a Markov process, since conditional probability only depends on the last known state, the probability

$$P_{0:q}(\mathbf{X}_{0:q} \in d\mathbf{x}_{0:q}) = \left[\prod_{k=1}^{q} P(\mathbf{X}_k / \mathbf{X}_{0:k-1} = \mathbf{x}_{0:k-1})\right] P(\mathbf{X}_0 \in d\mathbf{x}_0),$$

is reduced to

$$P_{0:q}(\mathbf{X}_{0:q} \in d\mathbf{x}_{0:q}) = \left[\prod_{k=1}^{q} M_k(\mathbf{x}_{k-1}, d\mathbf{x}_k)\right] P(\mathbf{X}_0 \in d\mathbf{x}_0),$$

---

[1] f is assumed to ensure the convergence of integrals.

where $M_k(\mathbf{x}_{k-1}, d\mathbf{x}_k)$ is the Markov kernel transition defined by

$$M_k(\mathbf{x}_{k-1}, d\mathbf{x}_k) = P(\mathbf{X}_k \in d\mathbf{x}_k / \mathbf{X}_{k-1} = \mathbf{x}_{k-1}).$$

The Markov kernel $M_k(\mathbf{X}_{k-1}, d\mathbf{x}_k)$ is the stochastic time evolution from the state $\mathbf{X}_{k-1}$ to the state $\mathbf{X}_k$. Such process can be view as the dynamics of information similarly to the Liouville equation [29] or the Fokker-Planck equation [48]. The time evolution of probability clearly appears when introducing by P_q the probability $P(\mathbf{X}_q \in d\mathbf{x}_q)$, in that case,

$$P_{q+1} = M_{q+1} \bullet P_q,$$

- can be viewed as a matrix vector product where M_{q+1} (respectively P_q) stands for the matrix of the transition (respectively the state vector of information) in the discrete case.

For a Markov process, the operation $\bullet$ can be extended to the trajectory case by $P_{0:q} = M_q \bullet P_{0:q-1}$ and recursively with

$$P_{0:q} = M_q \bullet M_{q-1} \bullet \cdots \bullet M_1 \bullet P_0.$$

The dynamics of the real system appears as a particular sample of the random process $\mathbf{X}_q$. The aim of data assimilation is to provide the best estimates of $\mathbf{X}_q$, knowing the observation over a temporal window. It can be shown that the optimal solution is reached by conditional expectation $\widehat{\mathbf{X}}_q = \mathbb{E}[\mathbf{X}_q/\mathbf{Y}_{0:q}]$ where $\mathbf{Y}_{0:q}$ denotes the random vector of observations times from 0 to q. But observations appear as a particular kind of information and can be generalized with other sources of *a priori* informations. Then en estimated state is the made evolving trough the time thanks to the numerical model leading to another process $\overline{\mathbf{X}}_{q+1} = \mathbb{E}[\mathbf{X}_{q+1}/\mathbf{Y}_{0:q}]$. At least, there is four process: the real process $\mathbf{X}_q$, its observations $\mathbf{Y}_q$, its estimation knowing the observation until time $q - 1$ given by $\overline{\mathbf{X}}_q$, and its estimation knowing the observation until time $q - 1$ plus the new observation at time q given by $\widehat{\mathbf{X}}_q$. This is now detailed thanks to the notion of conjunction of two probabilities.

2.1.3. Feynman-Kac Probability Measures

As described in section (2.1.2.), the prior information of the state vector is provided by the marginal probability $P_{0:q}(\mathbf{X}_{0:q} \in d\mathbf{x}_{0:q})$. Now, if a posterior source of information is known about $\mathbf{X}_q$, its constrains our knowledge of the state $\mathbf{X}_q$ and taking the form of a probability $G_q(\mathbf{X}_q)$. The resulting distribution that respects both constraint is given by the conjunction of probability $G_q \wedge P_{0:q}$ defined in the previous paragraph. In this framework, G_q is called the *potential function* [2]. It can be seen as a weighted Markov chain, where the potential is given by a random medium.

The probability $\widehat{\eta}_{0:q}$ corresponding to the information constrained at each time by a potential function is defined as the conjunction of probability

$$\widehat{\eta}_{0:q} = G_q \wedge G_{q-1} \wedge \cdots \wedge G_0 \wedge P_{0:q},$$

thereafter we only consider the marginal distribution at time q $\widehat{\eta}_q$.

---

[2] Actually, G_q is not necessarily associated to a probability distribution but at least a function with values in in $[0, 1]$ and with $\eta(G_q) > 0$.

The time evolution $\widehat{\eta}_q$, with the Markov transition $\mathbf{M}_{q+1}$, leads to the information $\overline{\eta}_{q+1}$ defined by

$$\overline{\eta}_{q+1} = \mathbf{M}_{q+1} \bullet \widehat{\eta}_q. \tag{7}$$

It is easy to deduce that $\widehat{\eta}_q$ is the so-called Boltzmann-Gibbs transformation of $\overline{\eta}_q$ by the potential G_q

$$\widehat{\eta}_q = G_q \wedge \overline{\eta}_q. \tag{8}$$

The measures $\overline{\eta}_q$ and $\widehat{\eta}_q$ are the two Feynman-Kac probability measures. A cycling process relates the two measures: $\overline{\eta}_q$ is deduced from the time evolution of $\widehat{\eta}_{q-1}$, while $\widehat{\eta}_q$ is deduced from the Bolzmann-Gibbs transform of $\overline{\eta}_q$ with posterior information, and so on.

The Boltzmann-Gibbs transformation can be written under a transport from

$$\widehat{\eta}_q = \mathbf{S}_q \bullet \overline{\eta}_q, \tag{9}$$

where $\mathbf{S}_q$ is a non-unique selection kernel, this is the Markov representation of the transition. For instance, the following kernels

$$\mathbf{S}^1_{q,\overline{\eta}_q}(\overline{x}_q, d\widehat{x}_q) = G_q \wedge \overline{\eta}_q(d\widehat{x}_q), \tag{10}$$

$$\mathbf{S}^2_{q,\overline{\eta}_q}(\overline{x}_q, d\widehat{x}_q) = G_q(\overline{x}_q)\delta_{\overline{x}_q}(d\widehat{x}_q) + [1 - G_q(\overline{x}_q)]\, G_q \wedge \overline{\eta}_q(d\widehat{x}_q), \tag{11}$$

are admissible, where $\delta_x(dy)$ denotes the Dirac measure. At this level, there is no differences among the two kernels. Since the probability $\overline{\eta}_q$ is present in $\mathbf{S}^1_{q,\overline{\eta}_q}$ and $\mathbf{S}^2_{q,\overline{\eta}_q}$, they are mean-field operator. Moreover $\mathbf{S}^2_{q,\overline{\eta}_q}$ is a genetic selection kernel since the adaptivity of a particle is taken into account at the sampling level.

It follows that the cycling process relating the two Feynman-Kac measures can be sum up into the following diagram

$$\overline{\eta}_q \xrightarrow[\mathbf{S}_{q,\overline{\eta}_q}]{\text{selection}} \widehat{\eta}_q \xrightarrow[\mathbf{M}_{q+1}]{\text{prediction}} \overline{\eta}_{q+1}. \tag{12}$$

2.1.4. Particle Approximation of Feynman-Kac Measures

In practice, time evolution of Feynman-Kac measures is not feasible and a particle approximation is employed. This strategy is common in data assimilation where the equations of the Kalman filter are resolved thanks to an ensemble method. It consists of the following steps.

Let $\left(\overline{\mathbf{X}}_q^{k,N}\right)_{k \in [1,N]}$ be a collection of N independent random variables sampling the law $\overline{\eta}_q$, then the empirical distribution law

$$\overline{\eta}_q^N = \frac{1}{N} \sum_k \delta_{\overline{\mathbf{X}}_q^{k,N}}. \tag{13}$$

converges [3] toward the distribution $\overline{\eta}_q$ because of the law of large numbers. The equivalent of the best estimator $\overline{X}_q = \mathbb{E}(X_q/Y_{0:q-1})$, under the discrete approximation is thus

$$\overline{X}_q^N = \frac{1}{N} \sum_k \overline{X}_q^{k,N}. \tag{14}$$

The distribution $\widehat{\eta}_q$ is also approximated by a discretized version $\widehat{\eta}_q^N$ with the dynamic

$$\widehat{\eta}_q^N = G_q \wedge \overline{\eta}_q^N. \tag{15}$$

The definition of the conjunction implies that the density $\widehat{\eta}_q^N$ is a multinomial law

$$\widehat{\eta}_q^N = \sum_k p_k \delta_{\overline{\mathbf{X}}_q^{k,N}}, \tag{16}$$

where

$$p_k = \frac{G_q(\overline{\mathbf{X}}_q^{k,N})}{\sum_j G_q(\overline{\mathbf{X}}_q^{j,N})}. \tag{17}$$

In term of Markov process, the distribution $\widehat{\eta}_q^N$ corresponding to the Gibbs transform of $\overline{\eta}_q^N$ can be formulated as a linear transport

$$\widehat{\eta}_q^N = \mathbf{S}_q^N \bullet \overline{\eta}_q^N, \tag{18}$$

where $\mathbf{S}_q^N$ is a selection kernel. This transition serves to produce a new collection $(\widehat{\mathbf{X}}_q^{k,N})$ of particles so that

$$\widehat{\eta}_q^N = \frac{1}{N} \sum_k \delta_{\widehat{\mathbf{X}}_q^{k,N}}. \tag{19}$$

The equivalent of the best estimator $\widehat{X}_q = \mathbb{E}(X_q/Y_{0:q})$, under the discrete approximation is thus

$$\widehat{X}_q^N = \sum_k p_k \overline{X}_q^{k,N} = \frac{1}{N} \sum_k \widehat{X}_q^{k,N}. \tag{20}$$

This collection is then evolved in time thanks to the prediction kernel $\mathbf{M}_{q+1}$ and

$$\overline{\eta}_{q+1}^N = \mathbf{M}_{q+1}^N \bullet \widehat{\eta}_q^N, \tag{21}$$

which produces a new collection $\left(\overline{\mathbf{X}}_{q+1}^{k,N}\right)$. This dynamic is sum up with the diagram

$$\left\{ \left(\overline{\mathbf{X}}_q^{k,N}\right)_{k\in[1,N]}^{\overline{\eta}_q^N} \right\} \xrightarrow[\mathbf{S}_q]{\text{selection}} \left\{ \left(\widehat{\mathbf{X}}_q^{k,N}\right)_{k\in[1,N]}^{\widehat{\eta}_q^N} \right\} \xrightarrow[\mathbf{M}_{q+1}]{\text{prediction}} \left\{ \left(\overline{\mathbf{X}}_{q+1}^{k,N}\right)_{k\in[1,N]}^{\overline{\eta}_{q+1}^N} \right\} \tag{22}$$

---

[3] This convergence has to be understood as follows. Let $\mathbf{X}$ be a random variable of law η and $(\mathbf{X}^{k,N})$ be a collection of N independent random variables sampling the law η. The empirical distribution is defined by $\eta^N = \frac{1}{N} \sum_k \delta_{\mathbf{X}^{k,N}}$. The empirical expectation applied on a test function f is defined by $I^N(f) = \frac{1}{N} \sum_k f(\mathbf{X}^{k,N})$. Then, convergence of the empirical expectation applied on a test function f toward the expectation $I(f) = \mathbb{E}[f(\mathbf{X})]$ means that the law of the random variable $\sqrt{N}\left[I^N(f) - I(f)\right]$ is closed to the Gaussian $\mathcal{N}(0, \sigma^2)$ with zeros mean and variance σ^2, indicating a convergence in $O(1/\sqrt{N})$.

Examples of selection kernel can be given by the discretization of selection kernels introduced in the previous section. For instance, if the selection kernel is given by the discretized version $\mathbf{S}^{1,N}_{q,\overline{\eta}^N_q}$ of the kernel $\mathbf{S}^1_{q,\overline{\eta}_q}$, then the collection $\left(\widehat{\mathbf{X}}^{k,N}_q\right)$ is defined as N sample distributed according to the multinomial law Eq. (16). For the discrete kernel $\mathbf{S}^{2,N}_{q,\overline{\eta}^N_q}$, then the previous sample is preceded by a Bernoulli selection with parameter $G_q\left(\overline{\mathbf{X}}^{k,N}_q\right)$ leading to either conserving the particle $\overline{\mathbf{X}}^{k,N}_q$ with a probability $G_q\left(\overline{\mathbf{X}}^{k,N}_q\right)$, or sampling according to the multinomial law Eq. (16) with a probability $1 - G_q\left(\overline{\mathbf{X}}^{k,N}_q\right)$. It is easy to understand why this kernel is called a genetic selection kernel, since the selection consists in to reinforced the well adapted states.

If a given selection kernel has no impact on the exact dynamics, the choice of a discretized selection kernel constrains the convergence rate of $\overline{\eta}^N_q$ (respectively $\widehat{\eta}^N_q$) toward $\overline{\eta}_q$ (respectively $\widehat{\eta}_q$). Hence, it can be shown that the variance error associated to the kernel $S^{2,N}_{q,\overline{\eta}^N_q}$ is smaller than the variance error associated to the kernel $S^{1,N}_{q,\overline{\eta}^N_q}$ [59]. This is justified since the Bernoulli step can lead to keep the particle even is its weight $G_q\left(\overline{\mathbf{X}}^{k,N}_q\right)$ is very low. Thus compared to $\mathbf{S}^{1,N}_{q,\overline{\eta}^N_q}$, $\mathbf{S}^{2,N}_{q,\overline{\eta}^N_q}$ is better appropriate to reproduce large tails of probability distributions. The convergence can be improved following [59, 4] by introducing the kernel

$$\mathbf{S}^3_{q,\overline{\eta}_q}(\overline{x}_q, d\widehat{x}_q) = \varepsilon_q G_q(\overline{x}_q)\delta_{\overline{x}_q}(d\widehat{x}_q) + [1 - \varepsilon_q G_q(\overline{x}_q)] G_q \wedge \overline{\eta}_q(d\widehat{x}_q), \qquad (23)$$

where $\varepsilon_q = 1/EssSup_{\overline{\eta}_q}(G_q)$ with $EssSup_{\overline{\eta}_q}$ the essential supremum associated to the distribution $\overline{\eta}_q$ applied on the potential G_q. The discretized version $\mathbf{S}^{3,N}_{q,\overline{\eta}^N_q}$ of this kernel is obtained similarly to $\mathbf{S}^{2,N}_{q,\overline{\eta}^N_q}$ and replacing the essential supremum by $EssSup_{\overline{\eta}_q}(G_q) \approx Max\{G_q(\overline{X}^{k,N}_q)\}$. Comparing to $\mathbf{S}^{2,N}_{q,\overline{\eta}^N_q}$ the genetic kernel $\mathbf{S}^{3,N}_{q,\overline{\eta}^N_q}$ conserve the particle with the maximum potential since its Bernoulli parameter is now of 1.

Few remarks need to be mention. Firstly, it has to be noted that particles $\widehat{\mathbf{X}}^{k,N}_q$ are in interaction with particles $\overline{\mathbf{X}}^{k,N}_q$ leading to a potentially break down of the independence of the collection $\left(\overline{\mathbf{X}}^{k,N}_q\right)$. However, it can be shown that if the ensemble is independent at time $q = 0$, then this independence is preserve along the dynamic. This property is called the chaos propagation [59, 78].

Secondly, it is important to not confuse the convergence of discreted laws toward exact laws and the convergence of the best estimator toward the true state, where the error is bigger to the Cramer-Rao bound [71, 54].

2.1.5. Illustration on a Simple Model

To an application, the model we consider now, is given by the following recurrence

$$\begin{cases} X_q = \widetilde{\mathcal{M}}_q(X_{q-1}) = X_{q-1} + \mathcal{N}_q \mathcal{P}_q, \\ X_0, \end{cases} \qquad (24)$$

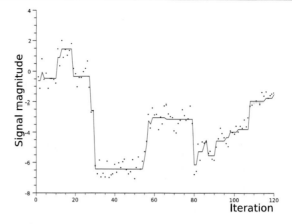

Figure 2. Example of synthetic data obtained from the model Eq.(24) with the synthetic reference trajectory X_q (solid line) and the synthetic observation set Y_q (dots).

where $\mathcal{N}_q$ (respectively $\mathcal{P}_q$) are random variables following a Gaussian law (resp. a Poisson law) variable with zero mean and a variance 1 (resp. of parameter [4] $\mu = 0.1$, with mean and variance equal to μ). Hence the corresponds to a non-Gaussian linear dynamic.

A synthetic observation set Y_q is obtained from a reference trajectory X_q with starting point $X_0 = 0$, that corresponds to a sample of the dynamic Eq.(24), where observation at time q is constructed by

$$Y_q = X_q + \sqrt{R_q}\zeta_q, \qquad (25)$$

where $\sqrt{R_q} = 0.4$ and ζ_q is a sample of a reduced centered Gaussian random variable of covariance matrix the identity. An illustration of this synthetic data based is reported on figure 2. The reference X_q (solid line) is a piecewise constant curve, surrounded by the halo of synthetic observations Y_q (dots).

A particle approximation of the non-linear filter has been implemented using the discrete version of the selection kernel $\mathbf{S}^3$ Eq.(23) and under the assumption that the dynamics is perfectly known, given by Eq.(24). Results obtained with this approach are given in Fig. 3 for the different particle ensemble size $Ne = \{25, 100, 400, 1600\}$ and using the synthetic data of Fig. 2. This corresponds to four estimated process $\widehat{X}_q^N = \frac{1}{N}\sum_k \widehat{X}_q^{k,N}$, one for each ensemble size N. All the panels represent the signal of reference (solid line) surrounded by the noisy observations (dots), the optimal estimation (dashed line) surrounded by the uncertainty range provided by the standard deviation of the ensemble (dotted lines). From these panels, it can be observe that the convergence of the filter toward the signal depends on the ensemble size. The convergence is even faster than the ensemble size is large. For instance, some defects can be observed at each step of the signal of reference, see e.g. at the transition near the iteration 30 on panel (a) for $Ne = 25$. For large ensemble, the Cramer-Rao bound seems to be reached: on panels (c-d) the signal is included within the uncertainty tube, see e.g. the tube between iterations 30 to 50.

---

[4] Whe have the probability $P(\mathcal{P}_q = n) = e^{-\mu}\frac{\mu^n}{n!}$.

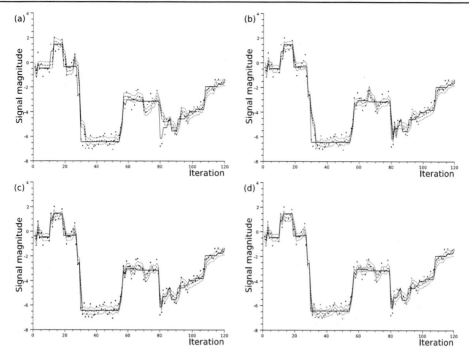

Figure 3. Filtering by using a particle approximation of the non-linear filter with different ensemble size: (a) $Ne = 25$, (b) $Ne = 100$, (c) $Ne = 400$ and (d) $Ne = 1600$.

2.2. Derivation of the Kalman Filter Equation from the Non-linear Framework

2.2.1. Feynman-Kac Measures under Gaussian and Linear Dynamics

Now, the particular case of a Gaussian linear dynamic is considered. It means that the process X_q has the form

$$X_q = M_q X_{q-1} + W_q, \qquad (26)$$

where M_q is the linear forecast operator and W_q is a random Gaussian variable $\mathcal{N}(0, Q_{q+1})$ with zero mean and covariance matrix Q_{q+1}. The observations are related to the state X_q thanks to

$$Y_q = H_q X_q + V_q, \qquad (27)$$

where H_q is the observation operator mapping the state model into the observation space, and V_q is a random Gaussian variable $\mathcal{N}(0, R_q)$ with zero mean and covariance matrix R_q. In this framework, it is assumed that W_q and V_q are independent, and that $\overline{\eta}_0$ is a Gaussian measure. It follows that $\overline{\eta}_q$ and $\widehat{\eta}_q$ are also Gaussian, respectively $\mathcal{N}(\overline{X}_q, \overline{P}_q)$ and $\mathcal{N}(\widehat{X}_q, \widehat{P}_q)$. Hence, these law are completely featured by their mean $\overline{X}_q = \mathbb{E}(X_q/\widehat{X}_{q-1})$ and $\widehat{X}_q = \mathbb{E}(X_q/\overline{X}_q, Y_q)$; and their covariance matrix de covariance $\overline{P}_q = Var(X_q/\widehat{X}_{q-1})$ and $\widehat{P}_q = Var(X_q/\overline{X}_q, Y_q)$. It is easy to prove that means and covariances of $\overline{\eta}_q$ and $\widehat{\eta}_q$ are given by the Kalman filter equations. This can be illustrate as follow.

The Feynman-Kac formalism provides that $\widehat{\eta}_q = G_q \wedge \overline{\eta}_q$ leading in the framework of Gaussian density to a log-likelihood

$$\mathcal{J}(\widehat{x}_q) = -\log \widehat{\eta}_q(\widehat{x}_q) = \frac{1}{2}||\widehat{x}_q - \overline{X}_q||^2_{\overline{P}_q^{-1}} + \frac{1}{2}||Y_q - H\widehat{x}_q||^2_{R_q^{-1}}, \qquad (28)$$

where $||\cdot||_F$ stands for the norm associated to the Mahalanobis distance, i.e. $||x||^2_F = x^T F x$. $\mathcal{J}$ is a cost function. Since we are concerned by a log-likelihood function, that is unique within to a constant of normalization, and because $\mathcal{J}$ is quadratic it follows that it can be written

$$\mathcal{J}(\widehat{x}_q) = \frac{1}{2}||\widehat{x}_q - \widehat{X}_q||_{\widehat{P}_q^{-1}}, \qquad (29)$$

and one has to determine $\widehat{X}_q$ and $\widehat{P}_q$. The equivalence of Eq.(28) and (29) implies that their Hessian matrix are equal, leading to

$$\mathcal{J}'' = \widehat{P}_q^{-1} = \overline{P}_q^{-1} + H_q^T R_q^{-1} H_q, \qquad (30)$$

that fixes the matrix $\widehat{P}_q$. Otherwise, the optimum state should ensure that $\widehat{X}_q = ArgMin\mathcal{J}$, or $\nabla \mathcal{J}_{\widehat{X}_q} = 0$. With the gradient of $\mathcal{J}$, Eq.(28), given by

$$\nabla \mathcal{J}_{\widehat{x}_q} = \overline{P}_q^{-1}(\widehat{x}_q - \overline{X}_q) - H_q^T R_q^{-1}(Y_q - H_q \widehat{x}_q),$$

the solution is obtained from the linear equation system

$$\left(\overline{P}_q^{-1} + H_q^T R_q^{-1} H_q\right)(\widehat{X}_q - \overline{X}_q) = (Y_q - H_q \overline{X}_q),$$

so

$$\widehat{X}_q = \overline{X}_q + \left(\overline{P}_q^{-1} + H_q^T R_q^{-1} H_q\right)^{-1} H_q^T R_q^{-1}(Y_q - H_q \overline{X}_q).$$

By using the Sherman-Morrison-Woodbury formula it can be shown that,

$$K_q = \left(\overline{P}_q^{-1} + H_q^T R_q^{-1} H_q\right)^{-1} H_q^T R_q^{-1} = \overline{P}_q H_q^T (H_q \overline{P}_q H_q^T + R)^{-1}.$$

Hence, the correction equation of the Kalman filter is recovered

$$\widehat{X}_q = \overline{X}_q + K_q(Y_q - H\overline{X}_q). \qquad (31)$$

Once the equations are written, the Kalman filter holds in five equations, two equations for the prediction step

$$\overline{X}_q = M_q \widehat{X}_{q-1}, \qquad (32)$$
$$\overline{P}_q = M_q \widehat{P}_{q-1} M_q^T + Q_q, \qquad (33)$$
$$\qquad (34)$$

and three equations for the correction step

$$K_q = \overline{P}_q H_q (H_q \overline{P}_q H_q^T + R_q)^{-1}, \qquad (35)$$
$$\widehat{X}_q = \overline{X}_q + K_q(Y_q - H_q \overline{X}_q), \qquad (36)$$
$$\widehat{P}_q = (I - K_q H_q) \overline{P}_q. \qquad (37)$$

By this way, the selection step corresponding to the Boltzmann-Gibbs transformation of $\overline{\eta}_q$ to provide $\widehat{\eta}_q$ is simply replaced by a correction step. This is summarized thanks to the following diagram.

$$\overline{\eta}_q = \mathcal{N}(\overline{X}_q, \overline{P}_q) \xrightarrow{\text{correction}} \widehat{\eta}_q = \mathcal{N}(\widehat{X}_q, \widehat{P}_q) \xrightarrow{\text{prediction}} \overline{\eta}_{q+1} = \mathcal{N}(\overline{X}_{q+1}, \overline{P}_{q+1}).$$

Note that the Cramer-Rao bound is reached and given by $\widehat{P}_q$.

2.2.2. Illustration in a Simple Model

We illustrate now the Kalman equations applied within the simple example introduced in section (2.1.5.). Comparison of the particle filter with a Kalman filter approach requires to introduce a linear dynamic and a Gaussian assumption for the probabilities. The simple Gaussian model that can be deduced from Eq.(24) is

$$\begin{cases} X_q = \mathcal{M}_q(X_{q-1}) + W_q = X_{q-1} + W_q, \\ X_0, \end{cases} \quad (38)$$

where W_q is Gaussian random variable with zero mean and covariance matrix Q_q constant here and equal to Q. This random variable corresponds to the model error that should model the discrepancy between the stationary dynamic and the full non Gaussian dynamic Eq.(24). It means that we replace the discrepancy by a Gaussian random variable of same mean and same variance. In this particular case, the magnitude of Q is known, it corresponds to the variance of $\mathcal{N}_q \mathcal{P}_q$. It can be shown [5] that the mean is $\mathbb{E}(\mathcal{N}_q \mathcal{P}_q) = 0$ and the variance is $V(\mathcal{N}_q \mathcal{P}_q) = \mu + \mu^2$. It results that $Q \approx 0.1$.

The figure 4 illustrates the filtering by using the Kalman filter with variable values for the model error covariance $Q \in \{0, 0.001, 0.01, 0.1\}$. This mimic an error in the estimation of the real Q. When the model error is low (panels a-b) the convergence of the filter toward the signal is of exponential type. The signal is too smooth compared to the reference with a large error to reconstruct the time series. When the model error is large (panel d) the estimated signal follows the observations. Hence, the observation error is no really filtered.

To better understand the covariance dynamics and how to experiment it, a 1D example is presented.

2.3. Illustration of the Kalman Filter Covariance Dynamic

We propose now a simple illustration of the Kalman filter covariance dynamics on a circular testbed. The circle is discretized into $N_g = 301$ grid points, the circle is view as being an Earth great circle of radius $a = 6400\,km$. The dynamic is a simple advection transport of a passive tracer h by a wind field $U = 20\,m.s^{-1}$. The time integration is set to $T = 6\,H$.

---

[5] Let $W = \mathcal{G}\mathcal{P}$, where $\mathcal{G}$ is a centered Gaussian random variable of variance σ^2 and $\mathcal{P}$ is a Poisson random variable of parameter μ. Then $P(W \in [w, w+dw]) = \sum_{k=0}^{\infty} P(k\mathcal{G} \in [w, w+dw]/\mathcal{P}=k) P(\mathcal{P}=k)$. But, $k\mathcal{G}$ is a centered Gaussian variable of variance k^2, and because of the independence between $\mathcal{G}$ and $\mathcal{P}$ we obtain thus $P(W \in [w, w+dw]) = \left(\sum_{k=0}^{\infty} \frac{1}{\sqrt{2\pi}k} e^{-w^2/2k^2} e^{-\mu}\frac{\mu^k}{k!}\right) dw$. Then, the expectation of W is thus $\mathbb{E}(W) = \sum_{k=0}^{\infty} \mathbb{E}(k\mathcal{G}) e^{-\mu}\frac{\mu^k}{k!} = 0$ and its variance is $V(W) = \sum_{k=0}^{\infty} V(k\mathcal{G}) e^{-\mu}\frac{\mu^k}{k!} = \sum_{k=0}^{\infty} k^2\sigma^2 e^{-\mu}\frac{\mu^k}{k!}$. With $k^2 = k(k-1) + k$ we obtain that $V(W) = \sigma^2 e^{-\mu}\left(\sum_{k=0}^{\infty} k(k-1)\frac{\mu^k}{k!} + \sum_{k=0}^{\infty} k\frac{\mu^k}{k!}\right) = \sigma^2(\mu^2 + \mu)$.

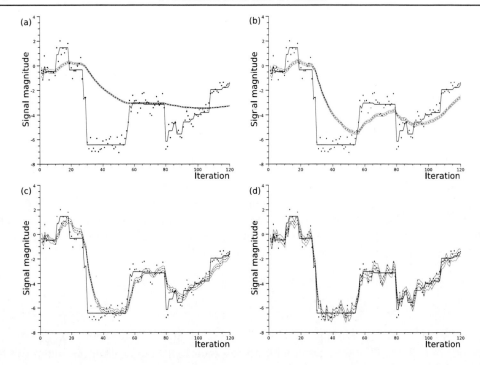

Figure 4. Filtering by using a Kalman filter (see Eq.(38)) for various value of Q: (a) $Q = 0$, (b) $Q = 0.001$, (c) $Q = 0.01$ and (d) $Q = 0.1$.

Hence, the solution of the dynamic is a simple translation of length $\mathcal{L}_{U,T} = 430\, km$. In order to represent the inflation of error over time integration an inflation rate $\alpha = 1.009$. Thus, the model can be formulated as the following simple partial derivative equation

$$\partial_t h + U \partial_x h = \alpha h, \qquad (39)$$

where x denotes the space coordinate, linked with the angular coordinate θ by $x = \theta a$. The explicit solution of the linear Eq.(39) takes the form

$$\boldsymbol{M}(h) = \alpha\, h(x - \mathcal{L}_{U,T}), \qquad (40)$$

where $\mathcal{L}_{U,T} = 430\, km$. Because of the linear dynamic, the equation of the Kalman filter occur. Latter, the model is assumed to be perfect so that $\boldsymbol{Q}_q = 0$. It results that the inflation of the variance is in α^2. The observation network considered here, is heterogeneous with one observation per grid point between $90°$ and $270°$. The observation covariance error is defined by $\boldsymbol{R}_q = \sigma_o^2 \boldsymbol{I}$, where $\sigma_o = 0.95$ is the observation standard deviation and $\boldsymbol{I}$ is the unit matrix. The initial condition for the prior covariance matrix $\boldsymbol{P}_0$ is an homogeneous Gaussian covariance matrix defined by $\rho(r) = e^{-\frac{r^2}{2L_h^2}}$, where r is the separation between two point on the circle and $L_h = 500\, km$.

The figure 5 illustrates the dynamic of the covariance matrix $\boldsymbol{P}_q$ as described by equation (33) for the time evolution of the posterior information into prior information, and equations (35-37) for the correction step. The different panels represent some covariance

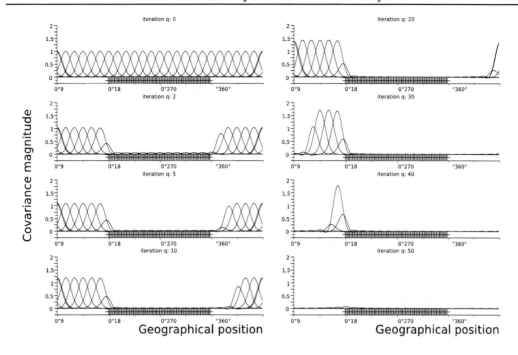

Figure 5. Dynamic of the prior information as described within the simple dynamic over the circle for the various time iterations $q \in \{0, 2, 5, 10, 20, 30, 40, 50\}$.

functions (all covariance functions are not reported here), where each Gaussian-like curve is a covariance function. The variance field corresponds in this example to the magnitude of the covariance function. The inflation of variances is visible comparing the covariance magnitude at $q = 0$ (top panel of left column) and the one at $q = 20$ (top panel of the right column). The discussion onto the variance field will be presented latter in section (4.1.). The advection of uncertainty reduction is clearly visible here, since the reduction of covariance magnitude is the fingerprint of the displacement of the information. This will be detailed in section (4.1.) and (4.2.3.).

Another to obtain the Kalman equation is now proposed. This second way to construct these equation is interesting since it provide new tools for validations of operational implementation of optimal analysis.

2.4. The Kalman Filter Is a Linear Regression

The Kalman filter is the optimal estimator in case of linear dynamics and observations subject to Gaussian perturbations or errors. The main purpose of the filter is to retrieve the Gaussian laws of the state vector according to the observation series. There are two probability laws to compute: one is the prediction law using the linear Gaussian dynamics, the other is the update law using the current observations. As linear combinations of Gaussian probabilities, the two filter laws are still Gaussian. The calculations of the Kalman estimator may be seen from many different angles: it is at the same time, the minimum of variance and the maximum of likelihood estimator, the least square solution of a linear quadratic problem, the solution of a linear variational problem. But it is also, in a Hilbertian context, the projection on the subspace generated by the observations corresponds to a linear regression. This is the presentation we make now.

2.4.1. The Conditioning Given a Series of Observations Is a Projection

The conditional expectation of a random variable knowing another one can be seen as a projection. This manner of thinking the conditioning returns the conditional mean as a linear regression. Here we describe this method in a formal framework while its application to the Kalman filter equations will be given in the next subsection.

We consider the space of random variable with finite energy $\mathbb{L}^2 = \mathcal{L}^2(\Omega, \mathcal{F}, \mathbb{P})$ where Ω is the random event space, $\mathcal{F}$ is a σ-algebra and $\mathbb{P}$ a probability measure. We define an inner product or scalar product on $\mathbb{L}^2$ by: $(X, Y) \in \mathbb{L}^2 \times \mathbb{L}^2 :< X, Y >= \mathbb{E}(XY)$ and a norm on $\mathbb{L}^2$ by $\|X\|^2 =< X, X >$.

A random variable $Y \in \mathbb{L}^2$ is a function of a random event $\omega \in \Omega$. The set of borelian square integrable function f, considering the composite functions $f(Y)$, generates a vector subspace of $\mathbb{L}^2$ denoted $span(Y)$. The subspace spanned by the σ-algebra $\sigma(Y)$, denoted $\mathbb{Y}$ is contained in $span(Y)$ and is a subspace of $\mathbb{L}^2$ with a natural isometry.

With this inner product, $\mathbb{L}^2$ becomes a Hilbertian space and the notion of projection appears. But before we have to refer to the orthogonality of two random variables in $\mathbb{L}^2$. Considering the inner product of (X, Y), we have : $X \perp Y \Leftrightarrow \mathbb{E}(XY) = 0$. Using this, we point out that for any random variable X and Y in $\mathbb{L}^2$, $[X - \mathbb{E}(X/Y)] \perp f(Y)$ is true for any square integrable function f.

The orthogonal projection of a random variable X onto the subspace $\mathbb{Y}$ is defined by the random variable $Z \in \mathbb{Y}$ which realizes the infimum distance $\|X - Z\| = \inf_{z \in \mathbb{Y}} \|X - z\|$. We denote $proj_{\mathbb{Y}}(X)$ the projection of X onto $\mathbb{Y}$. Two results are immediate : $[X - proj_{\mathbb{Y}}(X)] \perp \mathbb{Y}$ and the projection is a linear operator.

The properties of conditional expectations ensure that for any random variables X and Y, $\mathbb{E}(X/\sigma(Y)) \in \mathbb{Y}$. Obviously for any function $f \in \mathbb{L}^2$, $\mathbb{E}\big(f(Y)\mathbb{E}[X - \mathbb{E}(X - \sigma(Y))]\big) = 0$. Therefore we have the important result $\mathbb{E}(X/\sigma(Y)) = proj_{\mathbb{Y}}(X)$.

In conclusion, if $(Y_q)_{q \geq 0}$ is a series of observations, the vector space $span(1, Y_0, \ldots, Y_q)$ is a subspace of $\mathbb{L}^2(\Omega, \sigma(Y_{0:q}), \mathbb{P})$ and the conditional mean $\mathbb{E}(X/Y_{0:q})$, which will be the filter estimation, is an orthogonal projection of X onto the vector subspace $span(1, Y_0, \ldots, Y_q)$.

2.4.2. The Kalman Filter Equations

Thanks to the projection tool, we present now how the Kalman filter equation can be recovered. For an unreachable state X_q seen for any time step $q \geq 0$ by the observations Y_q, the purpose of the filter is to compute the parameters of two Gaussian laws, the predictor $Law(X_q \,/\, Y_{0:q-1})$ and the update $Law(X_q \,/\, Y_{0:q})$.

Before establishing the Kalman filter equations, we have to extend the result of the previous paragraph about the expression of the conditional expectation $\mathbb{E}(X/Y)$ as a linear regression where X and Y are two square integrable random variable. To prove that, we write $\mathbb{E}(X/Y)$ as an affine function of Y. We know that the slope of the regression is given by $a = \dfrac{cov(\mathbb{E}(X/Y), Y)}{cov(Y,Y)} = \dfrac{cov(X,Y)}{cov(Y,Y)}$. To evaluate the constant term b of the regression, taking the expectation of $\mathbb{E}(X/Y) = aY + b$ and using a property of the conditional mean, we find $b = \mathbb{E}(X) - a\mathbb{E}(Y)$. The linear regression is therefore: $\mathbb{E}(X/Y) = \mathbb{E}(X) - \dfrac{cov(X,Y)}{cov(Y,Y)}[Y - \mathbb{E}(Y)]$.

If we have to learn in real time with a filter the conditional expectation $\mathbb{E}(X/Y)$, we have to evaluate online the correction gain $\frac{cov(X,Y)}{cov(Y,Y)}$, and this is the purpose of the Kalman filter or of the deterministic variational filter. Later, in the case of an Ensemble Kalman Filter (EnKF), we will see that the linear regression is not performed correctly and produces wrong estimation for a non-linear dynamics.

Back to classical Kalman filter equations filter, we assume that (X_q, Y_q) is a Markov chain such that for any $q \geq 1$:

$$\begin{cases} X_q = \boldsymbol{M}_q.X_{q-1} + \sqrt{\boldsymbol{Q}_q}\,W_q \\ Y_q = \boldsymbol{H}_q.X_q + \sqrt{\boldsymbol{R}_q}\,V_q \\ Y_0 = y_0 \end{cases} \qquad (41)$$

where $(\boldsymbol{M}_q, \boldsymbol{Q}_q, \boldsymbol{H}_q, \boldsymbol{R}_q)$ are deterministic matrices and the noises W_q and V_q are independent centered Gaussian random variables with identity covariance matrices. The initial state X_0 is independent of the Gaussian noises with mean $\overline{X}_0 = \mathbb{E}(X_0)$ and covariance $\overline{P}_0 = \mathbb{E}((X_0 - \mathbb{E}(X_0))(X_0 - \mathbb{E}(X_0))^T)$.

The predictor law is $\mathcal{N}(\overline{X}_q, \overline{P}_q)$ and the update law $\mathcal{N}(\hat{X}_q, \hat{P}_q)$ where $\overline{X}_q = \mathbb{E}(X_q/Y_{q-1})$ and $\hat{X}_q = \mathbb{E}(X_q/Y_q)$. The covariance matrices are $\overline{P}_q = \mathbb{E}([X_q - \hat{X}_q][X_q - \hat{X}_q]^T/Y_{q-1})$ and $\hat{P}_q = \mathbb{E}([X_q - \overline{X}_q][X_q - \overline{X}_q]^T/Y_q)$.

The tour-de-force of R.E. Kalman is to establish that these different terms can be computed recursively as the observations are available:

$$\mathcal{N}(\overline{X}_q, \overline{P}_q) \xrightarrow{\text{Update}} \mathcal{N}(\hat{X}_q, \hat{P}_q) \xrightarrow{\text{Prediction}} \mathcal{N}(\overline{X}_{q+1}, \overline{P}_{q+1}) \qquad (42)$$

$\overline{X}_q$ is only the conditional expectation given Y_{q-1} of the dynamical equation, so $\overline{X}_q = \boldsymbol{M}_q \hat{X}_{q-1}$. To prove this first equality, we have to use mainly independence hypotheses. The covariance $\overline{P}_q$ is also a standard computation of $\mathbb{E}[(\boldsymbol{M}_q(X_{q-1} - \hat{X}_{q-1}) + \sqrt{\boldsymbol{Q}_q}W_q)(\boldsymbol{M}_q(X_{q-1} - \hat{X}_{q-1}) + \sqrt{\boldsymbol{Q}_q}W_q)^T]$ and gives $\overline{P}_q = \boldsymbol{M}_q \hat{P}_{q-1} \boldsymbol{M}_q^T + \sqrt{\boldsymbol{Q}_q}\sqrt{\boldsymbol{Q}_q}^T$.

With $\overline{X}_q$ and $\overline{P}_q$, we have calculated the two parameters of the predictor law. Now, for the update step, we want to compute $\hat{X}_q = \mathbb{E}(X_q/Y_q)$ and $\hat{P}_q = \mathbb{E}([X_q - \overline{X}_q][X_q - \overline{X}_q]^T/Y_q)$. The filter estimation $\hat{X}_q$ is a conditional expectation and employing the comment starting this section we can express it as a linear regression. The Kalman gain is therefore equal to the slope of the regression $K_q = \dfrac{cov(X_q, Y_q)}{cov(Y_q, Y_q)}$.

Using the observation equation, a direct computation leads to $cov(X_q, Y_q) = \mathbb{E}((X_q - \overline{X}_q)(\boldsymbol{H}_q(X_q - \overline{X}_q) + \sqrt{\boldsymbol{R}_q}V_q)) = \overline{P}_q \boldsymbol{H}_q^T$. By the same way $cov(Y_q, Y_q) = \mathbb{E}[(\boldsymbol{H}_q(X_q - \overline{X}_q) + \sqrt{\boldsymbol{R}_q}V_q)(\boldsymbol{H}_q(X_q - \overline{X}_q) + \sqrt{\boldsymbol{R}_q}V_q)^T] = \boldsymbol{H}_q \overline{P}_q \boldsymbol{H}_q^T + \sqrt{\boldsymbol{R}_q}\sqrt{\boldsymbol{R}_q}^T$. Finally the Kalman gain is $K_q = \overline{P}_q \boldsymbol{H}_q^T (\boldsymbol{H}_q \overline{P}_q \boldsymbol{H}_q^T + \sqrt{\boldsymbol{R}_q}\sqrt{\boldsymbol{R}_q}^T)^{-1}$ and the estimator equation is $\hat{X}_q = \overline{X}_q + K_q(Y_q - \boldsymbol{H}_q \overline{X}_q)$.

The last parameter to compute is the update covariance $\hat{P}_q$. Using the hilbertian decomposition properties, we establish that $\hat{P}_q = \overline{P}_q - \mathbb{E}\big([\mathbb{E}(\hat{X}_q) - \mathbb{E}(\overline{X}_q)]^2\big)$. With the value of the gain K_q and the update estimation $\hat{X}_q$ it yields $\hat{P}_q = \overline{P}_q - K_q H_q \overline{P}_q$. We have refound all the equations of Kalman.

This way to sense the Kalman filter dynamic is very instructive, and also useful for the validation of data assimilation scheme to test the optimality of the system [26].

Now, when the linearity hypothesis does not hold while the dynamical equation is differentiable, it is possible to come back to the classical filter using a linearization and the use of the increments. This is the very popular Extended Kalman Filter (EKF). Another popular algorithm is the variational formulation. These two ways for solving the filtering issue is detailed in the next section.

3. Practical Implementation of Kalman Filter and Variational Extensions

The equation of the Kalman filter are very costly, in particular the time propagation of the covariance matrix Eq.(33). It follows that the technical fails to be useful as it in very large systems. Various strategy have been design to incorporate the pros of the Kalman approach within a feasible implementation. Among them, we consider two class given by the ensemble discretization of the information dynamic and the variational approach. The former has been introduced by G. Evensen in 1994 [30], while the latter has been popularized from works of F.X. Le Dimet, O. Talagrand and Ph. Courtier [27, 19].

3.1. The Ensemble Kalman Filter

In the linear Gaussian case, the Kalman estimator is optimal and very powerful, moreover for stable dynamics the convergence rate in time is exponential like. Out of this simple case, the particle approximation of the Feynman-Kac distributions converges to the optimal estimator, but with a numerical cost depending on the sufficient number of particle to represent the behavior of the dynamical system. This sufficient number is highly related to the degree of freedom number, the chaotically of the system, the type of the random event **and** the type of the filter selection rule. A natural optimization of the particle approximation of the Feynman-Kac laws is to use the Kalman technology and its performance. For linear dynamics and non-Gaussian perturbation, Kalman filters in interaction can be a solution (see [59]). Another way of thinking is the EnKF. The estimator comes to the ensemble forecast ideas and looks like a particle estimator but with a corrective update instead of a state selection. This correction is a linear weighting of the innovation process learned by an empirical estimation of averages and variances. Consequently the empirical correction introduced is an approximation of the linear regression. Another motivation of the EnKF with respect to the classical Kalman filter is actually the dynamical learning of the prediction error covariance matrices P_q. In variational filter we have to model this matrix (see sections 3.4. and 5.) while in the EnKF it is given by the ensemble itself.

In this section, we present the algorithm of the EnKF and we discuss about the convergence of the approximation. Then from this convergence, some comments and discussion will come up.

3.1.1. The Ensemble Kalman Filter

In the section 2.2. we have establish the Kalman filter equation. Two concern the prediction step, three the update correction. The EnKF take back the same set of equation but using a particle set also named ensemble elements. At any time $q > 0$, we have a nonlinear dynamical model of the system $M_q(X_{q-1})$ where X_{q-1} is the state vector and a temporal series of observations $Y_{0:q}$. If the dynamical model may be non-linear, we suppose that the observation is linear with a transfer matrix H and perturbed by a Gaussian noise with covariance R_q. As done in the classical Kalman, we denote $\overline{X}_q$ a predicted state, and $\hat{X}_q$ a corrected one. To seed the ensemble of N elements, we give a variance matrix $P_0 > 0$ and randomize an ensemble of points $\left(\overline{X}_0^{k,N}\right)_{k \in [1,N]}$ independent and identically distributed with the centered Gaussian law of covariance matrix $\overline{P}_0$, $\mathcal{N}(0, \overline{P}_0)$.

Using the ensemble $\left(\overline{X}_q^{k,N}\right)$, we can compute at any time, $\overline{m}_q^N$ an empirical average of the elements and $\overline{P}_q^N$ an unbiased empirical variance. To prevent possible ensemble collapse i-e only one disctinct element in the ensemble, we perturb the dynamical equations with a Brownian motion $\sqrt{Q_q}.W_q$ where Q_q is a covariance matrix and $W_q \sim \mathcal{N}(0, I)$. Using again the classical Kalman equation and the empirical estimation, it would be possible to compute the Kalman gain $K_q^N = \overline{P}_q^N H_q^T (H_q \overline{P}_q^N H_q^T + R_q)^{-1}$ and then apply for each element $\overline{X}_q^k$ its correction to have $\hat{X}_q^{k,N} = \overline{X}_q^{k,N} + K_q^N \times I_q^{k,N}(Y_q, \overline{X}_q^{k,N})$ where $I_q^{k,N}$ is an innovation process associated to the element $\overline{X}_q^{k,N}$.

We have to modify the formulation of the innovation $I_q^{k,N}$. Indeed, if the Kalman correction is true for a linear Gaussian system, it is also the correction of a mean state. Here we want to apply the same correction not to the average but to one element $\overline{X}_q^{k,N}$. Doing this, we introduce an error that we have to balance it with an additive compensation random process $\xi_q^{k,N}$. Now we have to express what $\xi_q^{k,N}$ can be and this expression will give the formula of the special innovation process $I_q^{k,N}$.

$\xi_q^{k,N}$ is a centered process to retrieve in mean the Kalman correction. For a linear Gaussian system, we can compute its variance using the calculation of $\mathbb{E}([\hat{X}_q^{k,N} - \hat{m}_q^N]^2)$ where $\hat{m}_q^N$ is the empirical of the element $\hat{X}_q^{k,N}$.

Straightly he have $\hat{X}_q^{k,N} - \hat{m}_q^N = (\overline{X}_q^{k,N} - m_q^N)[1 - K_q^N H_q] + \xi_q^{k,N}$ and then $\mathbb{E}((\xi_q^{k,N})^2) = [1 - K_q^N H_q] \overline{P}_q^N H_q^T (K_q^N)^T$. Back to the definition of the Kalman gain K_q^N, we can simplify the variance of the compensator as $[1 - K_q^N H_q] \overline{P}_q^N H_q^T (K_q^N)^T = K_q^N R_q (K_q^N)^T$. In the linear Gaussian case, $\xi_q^{k,N}$ is a Gaussian centered random variable.

This compensator can be integrated in the classical innovation to have $I_q^{k,N}(Y_q, \overline{X}_q^{k,N}) = Y_q - H_q \overline{X}_q^{k,N} + \sqrt{R_q} V_q^{k,N}$ where $V_q^{k,N}$ is distributed according to a reduced centered Gaussian law $\mathcal{N}(0, 1)$. Finally the correction process of the ensemble Kalman filter is $\hat{X}_q^{k,N} = \overline{X}_q^{k,N} + K_q^N [Y_q - H_q \overline{X}_q^{k,N} + \sqrt{R_q} V_q^{k,N}]$.

Once the equations are written, the Ensemble Kalman filter holds in 5 lines:

A nonlinear prediction step :

$$\overline{X}_q^{k,N} = M_q(\hat{X}_{q-1}^{k,N}) + \sqrt{Q_q}.W_q^{k,N}$$

since empirical estimations :

$$m_q^N = \frac{1}{N}\sum_{k=1}^{N} \overline{X}_q^{k,N}$$

$$P_q^N = \frac{1}{N-1}\sum_{k=1}^{N}(\overline{X}_q^{k,N} - m_q^N)(\overline{X}_q^{k,N} - m_q^N)^T$$

$$K_q^N = P_q^N.\boldsymbol{H}_q^T.(\boldsymbol{H}_q.P_q^N.\boldsymbol{H}_q^T + R_q)^{-1}$$

then a correction step :

$$\hat{X}_q^{k,N} = \overline{X}_q^{k,N} + K_q^N\left[Y_q - \boldsymbol{H}_q\overline{X}_q^{k,N} + \sqrt{R_q}V_q^{k,N}\right]$$

We do not describe here details on practical implementation of the EnKF, and redirect interested readers toward specialized articles see [43, 44, 42]

If the construction of the EnKF is very simple, the difficulty is really in the existence of a limit process for this particle estimation and the definition of this process. This is the topic of the next section.

3.1.2. Convergence of the EnKF and Related Problems

The question of the convergence of the EnKF has been tackle recently by François Le Gland et al. [39], and we present here the results with explanations. To have a particle estimation implies some questions. First is the existence of a process limit as the number of element goes to infinity. Then the rate of convergence to this limit process is the second question. In the case of the EnKF there is also the question of the meaning of this limit itself.

We can express for any time and particle $\hat{X}_q^{k,N}$ its evolution in term of $\hat{X}_{q-1}^{k,N}$ and have

$$\hat{X}_q^{k,N} = M_q(\hat{X}_{q-1}^{k,N}) + K_q^N.\left[Y_q - \boldsymbol{H}_q.M_q(\hat{X}_{q-1}^{k,N}) + \sqrt{R_q}.V_q^{k,N}\right] \qquad (43)$$

If we denote $\hat{\pi}_{q-1}^N$ the empirical probability law of $\hat{X}_{q-1}^{k,N}$ the correction gain K_q^N is a function of $\hat{\pi}_{q-1}^N$ and should be denoted $K_q^N(\hat{\pi}_{q-1}^N)$. In that sense, the process $\hat{X}_q^{k,N}$ is called a mean-field process. So we have to know in the general case as the number of particles tends to infinity if the law $\hat{\pi}_q^N$ converges to a probability law $\hat{\pi}_q$ and if $\hat{\pi}_q$ is equal to the Bayesian filter law $\hat{\eta}_q = Law(X_q \,/\, Y_{0:q})$.

The question of the empirical probability laws convergence is relative to the class of functions used for the estimations [83]. LeGland et al show for locally Lipschitz with polynomials growth functions the L^p convergence of the EnKF, i-e the estimation error of $\hat{\pi}_q$ by $\hat{\pi}_q^N$ on this class $\mathcal{C}$ of functions for the L^p norm is controlled by the ratio $\frac{C_p(q)}{\sqrt{N}}$

where $C_p(q) \geq 0$ is a finite constant. The proof of this result requires the control of the empirical Kalman gain (the Kalman gain have to belongs to the class $\mathcal{C}$) the control of the empirical covariance matrices and the contiguity of the empirical elements $\hat{X}_q^{k,N}$ to their exact process $\hat{X}_q^{k,N}$.

The existence of the limit process for the EnKF is an important result. Its convergence rate looks like the other Monte-Carlo methods in $\mathcal{O}(\frac{1}{\sqrt{N}})$. Now we have to compare the limit law $\hat{\pi}_q$ with the Bayesian filter $\hat{\eta}_q$. To detail that, we can back to the equation 43. If the model M_q is linear (a product by a matrix), the dynamical and observational noises are Gaussian, by an average we find the evolution of the Kalman filter process. So in linear Gaussian case, the EnKF is a particle approximation of the Kalman filter. In this case, the EnKF is more costly than the classical KF and brings errors. For a nonlinear model M_q with all the noises i.i.d according Gaussian laws, the EnKF tends to a BLUE. In other case the EnKF is the particle approximation of a mean-field process Z_q distributed with $\hat{\pi}_q$ which is not the state vector X_q and have the evolution equation

$$Z_q = M_q(Z_{q-1}) + K_q(\hat{\pi}_q)[Y_q - H_q.M_q(Z_{q-1}) + \sqrt{R_q}.V_q] \qquad (44)$$

To fully understand the reason that the EnKF is an approximation of the process Z_q in the general case and not of the state vector X_q, we can back to the linear regression. Indeed we have seen section 2.4. that the Kalman filter is a projection onto the subspace generated by the observations and is a linear regression. Using the same correction step, the EnKF would be also a linear regression, but the coefficient K_q is not linear and depend on a probability laws. So the regression fails and projects to another subspace than those of the observations. Unfortunately, there is no specific sense to the process Z_q, except that it is not the filter process. In short, EnKF is a convergent ensemble estimator, and if the dynamic is not too far to a linear model and the noises not too different to Gaussian processes, the EnKF looks like a filter.

3.1.3. Illustration of the EnKF within a Simple Example

To illustrate the EnKF onto a simple framework we consider again the academic model described in section 2.1.5. . Since the assumption of linearity and of Gaussianity are not necessary, the model has not to be approximated compared to the Kalman illustration applied on this example. The figure 6 reproduced the filtering process associated to the EnKF algorithm for same ensemble size than experiment Fig.3.

The estimation converges toward the signal faster than with the Kalman filter (see Fig. 4) when Q is small (see eg Fig. 4, (a)) but the solution is closed to the one obtained with a large Q especially when comparing panel (d) of the figures 4 and 6. However, as described in the previous section, the non-linear filtering process is provided by the particle filter approximation (PF) and not by the EnKF. Thus, differences between the two solutions illustrated on panels (d) of the figures 4 and 6 provide some clues onto the subtle gap between the processes. One in particular is the deviations of the uncertainty tube as observed in the range from iteration 30 to 55: the EnKF tube has more oscillation than the PF one. It appears that the EnKF is much more sensitive to the observations than the PF. As the PF can be seen as the reference, this lead to conclude that this sensitivity is a defect. Another defect is the exponentionalyl like convergence of the EnKF toward the solution at

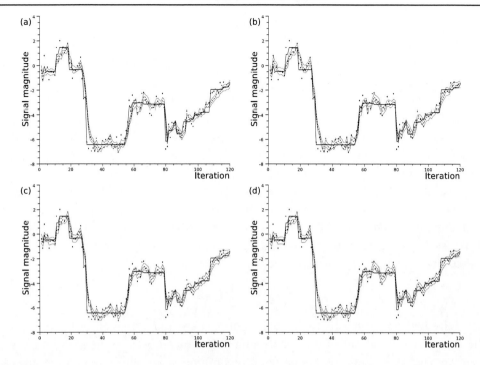

Figure 6. Filtering by using an EnKF for various ensemble size: (a) $Ne = 25$, (b) $Ne = 100$, (c) $Ne = 400$ and (d) $Ne = 1600$.

fast transition (see eg the step at iteration 30). This appears as a residual effect due to the Kalman filter approach.

These comments have to be seen here as food for thought. Actually, the comparison of the both methods is not complete and a full understanding of the differences between EnKF and PF is still an active field of research.

3.2. The Variational Approach: 3D-Var and 4D-Var

It has been seen that the Kalman filter equation can be deduced from the dynamic of the Feynman-Kac measures, where the log-likelihood takes the form of Eq.(28) then Eq.(29). The Kalman filter estimator is obtained as the minimum of the cost function Eq.(28). However, in the very large system as those encountered in numerical weather forecasting with degrees of freedom n of the order of $\mathcal{O}(10^8)$ (at the date of this book), the optimum can not be directly computed since it imply to invert matrices of size $\mathcal{O}(10^{16})$ coefficients. A tricky way to reach the solution has be to compute the variational solution associated to the optimization of the cost function Eq.(28).

In that framework, the classical notation must be introduced following [45]. In the variational framework, x_b is the prior information also called the background, it corresponds to the prior process denoted before $\overline{X}_q$. y_o denotes the observations. The posterior information is x_a and called the analysis, it corresponds to the posterior process denoted before $\widehat{X}_q$. Among important notation, one can site the errors compared to the truth. It is common to notation x_t for the (unknown in general) truth, previously written X_q. Then, the background error is defined as $\varepsilon_b = x_b - x_t$, while the observational error is $\varepsilon_o = y - \boldsymbol{H} x_t$. With these definition, the background error covariance matrix denoted by $\boldsymbol{B} = \mathbb{E}(\varepsilon_b \varepsilon_b^T)$ is the equivalent of matrix $\overline{\boldsymbol{P}}_q$, and the observational covariance matrix is $\boldsymbol{R} = \mathbb{E}(\varepsilon_o \varepsilon_o^T)$. Being familiarized with these notations is essential since data assimilation formalism is quite rigid on it, so readers interested by these developments will faster understand advances articles on this field of research.

The cost function takes thus the form

$$\mathcal{J}(x) = \frac{1}{2}||x - x_b||^2_{\boldsymbol{B}^{-1}} + \frac{1}{2}||y_o - \boldsymbol{H} x||^2_{\boldsymbol{R}^{-1}}, \tag{45}$$

where again $||\cdot||_{\boldsymbol{F}}$ stands for the norm associated to the Mahalanobis distance, *i.e.* $||x||^2_{\boldsymbol{F}} = x^T \boldsymbol{F} x$. The Kalman equation for the correction is written

$$x_a = x_b + \boldsymbol{K} d, \tag{46}$$

where $\boldsymbol{K} = \boldsymbol{B} \boldsymbol{H}^T (\boldsymbol{H} \boldsymbol{B} \boldsymbol{H}^T + \boldsymbol{R})^{-1}$ is the gain matrix and $d = y_o - \boldsymbol{H} x_b$ is the innovation vector. x_a is obtained thanks to an iterated process by using classical minimizer algorithm for instance a conjugated gradient [34] or a newton method. These algorithm require to provide the gradient of $\mathcal{J}$

$$\nabla \mathcal{J}(x) = \boldsymbol{B}^{-1}(x - x_b) - \boldsymbol{H}^T \boldsymbol{R}^{-1}(y_o - \boldsymbol{H} x). \tag{47}$$

With this formalism, the analysis covariance matrix, denoted $\boldsymbol{A}$, is the equivalent of the matrix $\widehat{\boldsymbol{P}}_q$.

This way to solve the analysis Eq.(46) is called the 3D-Var method. It is classical, with this algorithm, to bring together observations at different instants as being valid at the analysis time. Of course, this can not be true, but it has been seen as a necessary assumption to provide an initial state to numerical weather prediction model at time to provide new forecast before the real time of nature.

With progress of computer a better solution has been introduced taking into account the temporal aspects: the 4D-Var. In this framework, the cost function Eq.(45) is replaced the following

$$\mathcal{J}(x) = \frac{1}{2}||x - x_b||^2_{\boldsymbol{B}^{-1}} + \sum_{k=0}^{q} \frac{1}{2}||y_o^k - \boldsymbol{H}_k x^k||^2_{\boldsymbol{R}_k^{-1}}, \tag{48}$$

where $x^k = \boldsymbol{M}_k(x^0)$ is the temporal avolution of the state $x^0 = x$ evolving with the propagator of the model $\boldsymbol{M}_k$ (assumed linear for the sake of simplicity). This cost function can be minimized by classical algorithms providing the gradient

$$\nabla \mathcal{J}(x) = \boldsymbol{B}^{-1}(x - x_b) - \sum_{k=0}^{q} \boldsymbol{M}_k^T \boldsymbol{H}_k^T \boldsymbol{R}_k^{-1}(y_o^k - \boldsymbol{H}_k x^k), \tag{49}$$

where $\boldsymbol{M}_k^T$ is the adjoint of the model. In the non-linear framework $\boldsymbol{M}_k^T$ is the adjoint code of the linear code. The 4D-Var method own to the class of smoother more than filter since it doesn't provide a state analysis but a trajectory analysis. Readers interested by the variational approach are redirect toward some classical article about this subject [79, 17, 18, 19].

Before to continue the description of the variational algorithm. We would like to introduce a little deflection about how it is possible to generalize the 4D-Var within the non-linear filter framework.

3.3. Non-linear Smoother Extension for the 4D-Var: the 4D-PF

The 4D-Var approach seems quite constraining since its relies on the Gaussian assumption of the background dispersion. This can be relaxing by using mathematic tools introduced before.

In section 2.1., the non-linear filtering has been introduced trough the Feynman-Kac measures, and approximated by the particle discretization. This framework can be completed by taking into account the temporal extension of solutions. From this point of view, we recognize the soul of 4D-Var. But the limitation of the 4D-Var are the Gaussianity of the background error distribution and the fact that operational incremental implementation is based on the linear and the adjoint transport of increment. These constraints can be sidestepped by considering a trajectorial extension of the particle filter technical. Hence, the idea is to provide a particle smoother over a given window. From this, the background error statistic is no more necessary Gaussian and there is no need of linear/adjoint model. Because this methods merges the 4D-Var idea with trajectory considerations, and the particle filter approach, it can be called 4D-PF. The formulation of the 4D-PF relates on the particle filter algorithm, *i.e.* it is devised on a selection kernel discretized with particles as described onto the diagram Eq.(22). The difference holds on the definition of the potential function where G_q is replaced by a potential $\mathcal{G}_q$ computed over the qth time window $[1, \tau]$

$$\mathcal{G}_q = G_\tau \wedge \cdots \wedge G_2 \wedge G_1. \tag{50}$$

This computation is equivalent in terms of density to

$$\mathcal{G}(X_{1:\tau}^q) = \prod_{k=1}^{\tau} G_k(X_k^q), \tag{51}$$

where $X_{1:\tau}^q$ denotes the random variable associated to a qth trajectory. This solution is of particular interest when the observational network is heterogeneous in time, space and quality [70].

Thus, it has been shown that the background Gaussian assumption under the 4D-Var formalism can be relaxed. Now this is not a classical approach, and most of time a simple Gaussian assumption is introduced so that it reduces the numerical cost. At the difference from the Kalman filter equations, the covariance matrix B is assumed to be known from side, and it doesn't evolve with the Kalman filter propagation of the covariances. Until recently, B was considered as mean covariance matrix obtained from climatology. From few years, another approach to estimate this covariance matrix is born, and presented now.

3.4. Analysis Ensemble and Estimation of B

In its present form, the background error covariance matrix B is not made evolving with time. An a climatological estimation is used, computed over a large period (more than one

Kalman Filters Family in Geoscience and Beyond

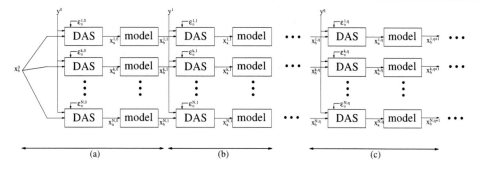

Figure 7. Scheme of the ensemble of perturbed analysis and forecasts: the data assimilation system (DAS) is applied to assimilate perturbed observations from a perturbed background providing the perturbed analysis, that is the initial state of a the forecast step (model). The ensemble is initialized from the assimilation of perturbed observation with the same background field (a), then cycling the analysis and the forecast steps (b-c).

month). Different technicals exist, but the most promising is the ensemble of perturbed analysis by perturbation of observations [31, 7]. This is the approach developed now.

The figure 7, illustrates the philosophy of this state of art. The idea is to benefit of the data assimilation system (DAS) implemented in routine so to mimic the information dynamic.

The attention is first focused onto the subsystem (c) corresponding to the recipe for evolving the ensemble. Assuming a collection of perturbed background $\{x_b^{k,q}\}_{k \in [1,N]}$ at time q and an observations set y^q. Then the ensemble of analysis $\{x_a^{k,q}\}_{k \in [1,N]}$ is obtained by assimilating each member of the ensemble of background $x_b^{k,q}$ with a perturbed observation set $y^q + \varepsilon_o^{k,q}$ where $\varepsilon_o^{k,q}$ is a sample of a Gaussian random vector of covariance R_q and null mean. The observational error is generated as $\varepsilon_o^{k,q} = R_q^{1/2} \zeta^{k,q}$, where $R_q^{1/2}$ is a square-root of the matrix R_q (that is $R_q = R_q^{1/2} R_q^{T/2}$) and $\zeta^{k,q}$ is a sample of a Gaussian random vector with zero mean and covariance matrix the identity I. Hence, each analysis state takes the form $x_a^{k,q} = x_b^{k,q} + K^q(y^q + \varepsilon_o^{k,q} - H_q x_b^{k,q})$, where K^q is the gain matrix associated to the data assimilation system used (DAS). Here, the DAS is an abstract system that can be for instance a 3D-Var, a 4D-Var or any other formulations. At this stage, the analysis covariance matrix A_q can be estimated as

$$A^q = \frac{1}{N} \sum_{k=1}^{N} \left(x_a^{k,q} - \overline{x_a^{k,q}} \right) \left(x_a^{k,q} - \overline{x_a^{k,q}} \right)^T, \qquad (52)$$

where $\overline{x_a^{k,q}} = \frac{1}{N} \sum_{k=1}^{N} x_a^{k,q}$ is the ensemble mean. Then the ensemble of perturbed analysis is made evolving according to the dynamic (model on Fig.7) so to provide a new ensemble of perturbed background for the next cycle. The background error covariance matrix at time $q+1$ is given by

$$B^{q+1} = \frac{1}{N} \sum_{k=1}^{N} \left(x_b^{k,q+1} - \overline{x_b^{k,q+1}} \right) \left(x_b^{k,q+1} - \overline{x_b^{k,q+1}} \right)^T. \qquad (53)$$

The climatological background covariance matrix is as the time averaged covariance matrix

$$B = \frac{1}{N_q} \sum_{i=1}^{q} B^i. \quad (54)$$

Of course, since it is not possible to store any of these matrix, the core of the covariance representation is to model B from the ensemble of estimated background error $\varepsilon_b^{k,q} = x_b^{k,q} - \overline{x_b^{k,q}}$ for $q \in [q_s, q_s + N_q]$ where q_s is the time from which the statistic are stabilized and can be incorporated into the estimation of B. This concept of stationarity of statistics has now to be explained. It is related to the initialization of the perturbed ensemble.

At the ignition (a), a single background is distributed toward each independent DAS, that assimilate the observation y^0 perturbed with their own observational perturbation $\varepsilon_o^{k,0}$. This leads to a first ensemble of perturbed analysis $x_a^{k,0} = x_b^0 + K^0(y^0 + \varepsilon_o^{k,0} - H_0 x_b^0)$. Then each analysis is time integrated by the model to provide a set of perturbed background $\{x_b^{k,1}\}_{k\in[1,N]}$. The ensemble of perturbed background is then assimilated (b) according the cycle described in the previous paragraph, similarly to (c). However, the ensemble of perturbed background can not be used as it is. Actually, the dispersion of the ensemble is not characteristic of the background magnitude. Hence we have to wait some cycling before to used it in the background covariance matrix estimation. Over this relaxation time, the variance of the ensemble of perturbed background $\{x_b^{k,1}\}_{k\in[1,N]}$ increases until saturation. The saturation time can be denoted q_s.

Beyond the stationary component of the background error covariance matrice, this recipe also provides the dispersion at a given time. If the ensemble is large enough, then B^q could be employed instead of B. In that case, the resulting algorithm would be very similar than the EnKF method detailed in section 3.1. It corresponds to a flow dependent formulation of B. To be useful, such an approach requires large ensembles or appropriate covariance model to extract, from the approximated background error $\varepsilon_b^{k,q} = x_b^{k,q} - \overline{x_b^{k,q}}$, a relevant information onto the covariance functions. It leads to sum up what contains a covariance matrix of $\mathcal{O}(10^{16})$ coefficients, from particular diagnosis of the covariance matrix as detailed now.

4. Diagnosis of B

This section is devoted to the diagnosis of the background error covariance matrix B introduced in the formalism of the variational formulation. The first component of interest is the standard-deviation field that represent the characteristic magnitude of the error at a given location. The second attribute is the characteristic length of correlation that feature the local spread of a correlation function. Of course, these diagnoses can by applied to whatever covariance matrix. Lets start with the standard-deviation.

4.1. Standard-Deviation and Illustration on the Circle

The standard-deviation field is the easiest information one can extract from the covariance matrix B, since it is defined as the square-root of the diagonal of B. The variance field is just the diagonal. Despite simplicity, this is a crucial component of the assimilation since it quantify the skill of the background (the prior state) and thus the way observations are incorporate into the analysis (the posterior state).

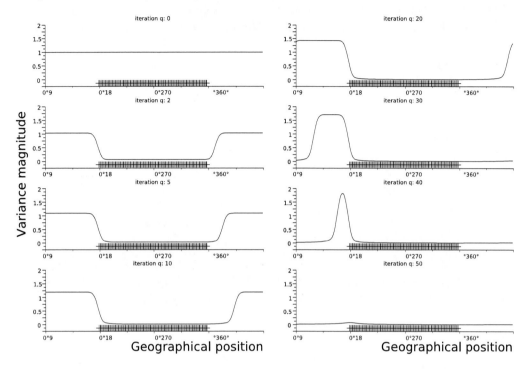

Figure 8. Example of the variance dynamic illustrated in the simple 1D circular framework.

To tackle the question of the variance dynamic, we consider again the academic advection proposed in section 2.3.. The figure 8 discloses the variance field of the covariance evolution. This variance field is the diagonal of the covariance P_q, while the standard-deviation is the square-root of this diagonal. At start, and focussing onto the covariance functions between 0° and 90°, the variance is 1, then it increases toward 1.5 at $q = 20$, and 2 at $q = 40$. However, this growth of the variance is balanced by introduction of observational information over the range 90° and 270°. This lead to a fast decrease of uncertainty since after only two iterations (see $q = 2$), the variance is less than 0.25. Since the advection, the reduction of uncertainty is transported while inflation still occurs from 270° to 90°, along the circle. It develops a front of uncertainty reduction near 270° at $q = 2$, and moving with the velocity U.

There is few to add about the standard-deviation except that an active field of research is devoted to its estimation from a small ensemble, see eg [6, 68, 69]. We do not discuss these aspects here and proposed to detail another important component of covariance matrix diagnosis: the length-scale.

4.2. Length-Scale

One of the main issues for a data assimilation system is to better specify the background error covariance matrix $B = \mathbb{E}(\varepsilon_b \varepsilon_b^T)$, where ε_b is the forecast error assumed unbiased. In order to characterize the curvature of the correlation functions near their origin, length-scale diagnosis is often introduced.

The differential length-scale is defined in data assimilation following Daley (1991, p110). The definition is similar to the turbulent microscale. In this section, the Daley length-scale is reviewed, and formulae are derived to approximate it.

The decomposition of covariances into standard deviations and correlations is common e.g. in variational schemes. This is appropriate if standard deviations and correlations do not vary much on scales smaller than the correlation length-scale.

4.2.1. Definition of the Length-Scale

For a smooth and isotropic correlation function ρ at the origin, the Daley length-scale is given by

$$L_D = \sqrt{-\frac{1}{\nabla^2 \rho(0)}}, \tag{55}$$

in one dimension and $L_D = \sqrt{-\frac{2}{\nabla^2 \rho(0)}}$ in two dimensions. This length-scale is proportional to the turbulent (or Taylor) microscale which is similarly defined. This formula is obtained from a Taylor expansion of the correlation at the origin $\rho(0)$:

$$\rho(\delta x) \approx \rho(0) + \frac{\delta x^2}{2} \frac{d^2 \rho}{dx^2}(0) = 1 - \frac{\delta x^2}{2 L_D^2}. \tag{56}$$

The isotropic assumption is required in order to ensure the continuity of the second order derivative at 0, i.e. $\frac{d^2 \rho}{dx^2}(0^-) = \frac{d^2 \rho}{dx^2}(0^+)$. A geometrical interpretation of this definition of length-scale is given as the scale for which the tangential parabola at the origin is equal to 0.5. This is illustrated in the top panel of Fig. 9, where a correlation function (solid line) and its tangential parabola at the origin (dashed line) are represented. The length-scale deduced from the above geometrical interpretation is $L_D = 250\,km$, for this particular correlation function. The length-scale is also related to the curvature of the correlation function, at the origin. The radius of curvature of the correlation function at the distance r is defined by $R(r) = \frac{\left(1+\left(\frac{d\rho}{dx}(r)\right)^2\right)^{3/2}}{\frac{d^2\rho}{dx^2}(r)}$. At the origin, $\frac{d\rho}{dx}(0) = 0$ leading to $R(0) = \frac{1}{\frac{d^2\rho}{dx^2}(0)} = -L_D^2$.

Note that the Daley length-scale does not give information about the correlation anisotropy. Moreover, it requires the knowledge of the second order derivative of the correlation function. The calculation of this second order derivative can be rather costly, as ideally it should involve the calculation of the whole correlation function.

4.2.2. Approximations and Computation of the Length-Scale

There exists various ways to compute the length-scale, and there are based on direct application of the Daley's expression or some approximations. Among methods only few are presented here.

Belo Pereira and Berre [5] (hereafter noted B&B) have proposed a relatively costless formula for the computation of length-scale. Under local differentiability and local homogeneity assumptions, the variance of the spatial derivative of the forecast error can be

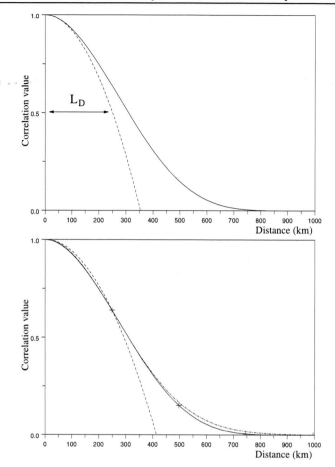

Figure 9. Top panel : Gaspari and Cohn ([37], Eq. 4.10) correlation function (solid line) and its tangential parabola (dashed line). Bottom panel : Parabolic (dashed line) and Gaussian (dash-dotted line) approximations at the origin of the correlation function (solid line), determined by its value for $\delta x = 248 km$, on a regular grid (crosses).

approximated by $(\sigma(\partial_x \varepsilon_b(x)))^2 = (\partial_x \sigma(\varepsilon_b(x)))^2 - (\sigma(\varepsilon_b(x)))^2 \partial_x^2 \rho(0)$, where $\partial_x = \frac{\partial}{\partial x}$ is the derivative along the coordinate. From the Daley length-scale definition, it follows:

$$L_{B\&B} = \sqrt{\frac{(\sigma(\varepsilon_b(x)))^2}{(\sigma(\partial_x \varepsilon_b(x)))^2 - (\partial_x \sigma(\varepsilon_b(x)))^2}}, \qquad (57)$$

where $\sigma(\varepsilon_b(x))$ is the standard deviation of $\varepsilon_b(x)$. This formula requires the computation of forecast error standard deviation, its gradient and also the standard deviation of the gradient of forecast error. In the case of a periodic domain, the computation of the gradient can be done either in grid-point space or in spectral space.

Another approximations can be suggested by equation (56), a direct discretization of the Laplacian appearing in Eq. (55) leads to a simple expression of the length-scale

$$L_{Pb} = \frac{\delta x}{\sqrt{2(1 - \rho(\delta x))}}. \qquad (58)$$

This length-scale is called hereafter the parabola-based length-scale (Pb). It is based on the approximation of the correlation function by a parabolic function, as represented in Fig. 9. As suggested by the example shown in the bottom panel of Fig. 9, for some separation distances (those smaller thant the chosen distance δx), the parabolic function may decrease less quickly (from the origin to the chosen distance δx) than the true correlation function. This suggests that the quality of the parabolic length-scale approximation may depend on the quality of the correlation function approximation and on the considered separation distance δx. Experiments indicate that the sensitivity to the choice of δx is relatively small, and that using a small value for δx provides a somewhat more accurate estimate of the length-scale. In this paper, δx corresponds to the resolution of the grid (*i.e.* the smallest possible δx).

In order to study this sensitivity to the correlation shape approximation, it is thus interesting to consider another analytical model of the correlation function ρ near the origin. By approximating the correlation at the origin with a Gaussian, the following equation is obtained: $\rho(\delta x) = exp(-\frac{\delta x^2}{2 L_D^2})$. Inverting this equation to extract the length-scale formulation, associated to correlation at distance δx, brings

$$L_{Gb} = \frac{\delta x}{\sqrt{-2 \ln \rho(\delta x)}}. \tag{59}$$

This length-scale is called hereafter the Gaussian-based length-scale (Gb). This approximation of the length-scale computation is easy to implement in real applications and costless. The bottom panel of Fig. 9 illustrates the Gaussian approximation at the origin of the discretized correlation function. Note that when the correlation is close to one, then both Parabola-based and Gaussian-based length-scales are equal. Let $\eta = 1 - \rho$, then a Taylor expansion leads to $L_{Pb} = L_{Gb} = \frac{\delta x}{\sqrt{2\eta}}$.

The previous definition of the length-scale are not able to diagnose the anisotropy of a correlation function. This can be introduced as follow now. Formulae (58) and (59) can be defined along an arbitrary direction as follows. Let $\boldsymbol{\delta x}$ be the displacement in a direction $\boldsymbol{u} = \frac{\boldsymbol{\delta x}}{|\boldsymbol{\delta x}|}$ of the domain (circle, plane, 2D-sphere, 3D-sphere,...). Then the vectorial parabola-based and Gaussian-based length-scale are thus defined by replacing δx by $\boldsymbol{\delta x}$ in equation (58) and (59). Thus it offers a characterization of the correlation for different directions. Similarly, formula (55) and (57) can be defined directionally for an anisotropic correlation function. For equation (55), it consists in replacing $\Delta \rho(0)$ by $\frac{\partial^2 \rho}{\partial u^2}(0^+) = \lim_{t \to 0^+} 2 \{\rho(t\,\boldsymbol{u}) - 1\} t^{-2}$, which is the second order derivative, calculated in the oriented direction $\boldsymbol{u}$, of the anisotropic correlation function. For equation (57), the directional length-scale is obtained by calculating the gradients $\partial_{\boldsymbol{u}} \varepsilon_b$ and $\partial_{\boldsymbol{u}} \sigma(\varepsilon_b)$, where $\partial_{\boldsymbol{u}}$ is the derivation along $\boldsymbol{u}$. It should be noted that these length-scales can be calculated whether the domain is bounded or not. Thus such formulations are suitable in oceanography or for a limited area model, as well as for a global meteorological model.

In the particular case of a 1D domain, one can define a directional parabola-based left length-scale as $L_{Pb}(-\delta x)$ and a right length-scale as $L_{Pb}(+\delta x)$. A similar definition is given for the directional Gaussian-based length-scale. Thereafter, the left directional length-scale is designed by a superscript $^-$ and the right one by the superscript $^+$. Note that the ratio $\frac{L^+}{L^-}$ is an indicator of anisotropy.

To conclude this section, dedicated to the length-scale diagnosis, we note that the length-scale can be approximated in other ways, by considering various analytical expressions for correlation

$$\rho(\delta x) = f(\delta x, L_D). \tag{60}$$

The main constraint on the method is the calculation of the reciprocal function. Thus the length-scale can be deduced from the correlation value, associated to a particular grid, with $L_D = f^{-1}(\delta x, \rho(\delta x))$. The choice of a particular relation between correlation and length-scale may arise from estimated correlation functions. It might depend on the physical field, or on the model used to represent the correlation function in the system. For instance, if a SOAR function (Daley, 1991, p117) is a good model to approximate the correlation function, then one has to invert Eq. (60), with $f(\delta x, L) = (1 + \frac{\delta x}{L})e^{-\frac{\delta x}{L}}$. This inversion can be achieved by using a Newton algorithm to resolve $F(L) = 0$, with $F(L) = \rho(\delta x) - f(\delta x, L)$ where δx and $\rho(\delta x)$ are given. Moreover, it can be noticed that such development may be applied on more complex diagnosis in 1D, 2D and 3D. For instance, a Gaussian-based approximation can be lead by assuming local homogeneous Gaussian-type correlation functions. This is then related to the local diffusion tensor that is introduced latter in the chapter, where anisotropy is shown to correspond to the principal axis of the diffusion tensor.

In the following, the 1D circle and the 2D sphere will be considered in order to illustrate the theory.

4.2.3. Simple Illustration on the Circle

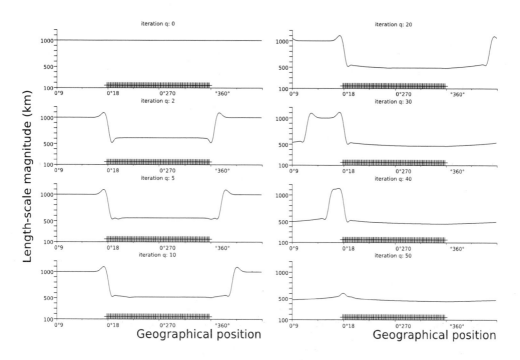

Figure 10. Example of the length-scale dynamic illustrated in the simple 1D circular framework.

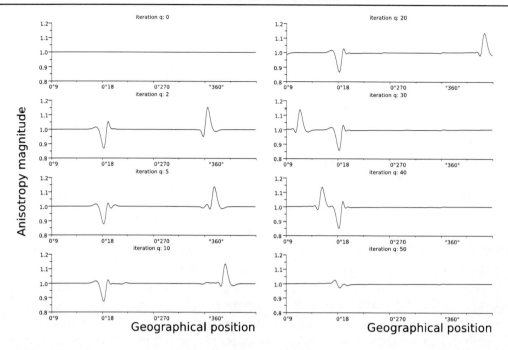

Figure 11. Example of the anisotropy dynamic illustrated in the simple 1D circular framework.

We now apply the computation of the length-scale onto the simple testbed introduced in section 2.3.. The computation of the length-scale is reported onto the figure 10, with the Gaussian-based approximation, defined here as the average of $L_{Pb}(+\delta x)$ and $L_{Pb}(-\delta x)$. The dynamic of length-scale is related to the physical dynamics but also to the network density [10, 11]. These two results are well represented on the different panel. Firstly, the network influence is clearly visible since the length-scale magnitude is significantly reduced from the iteration $q = 0$ to $q = 2$, with a reduction localized over the observational network. Initially at $1000\,km$ the length-scale decreases to saturate at thereafter $500\,km$ There are also some oscillation related to some boundary effects at the border of the observational network. Secondly, the East displacement of the length-scale front of law values is the fingerprint of the advection transport. The front is localized around $270°$ at $q = 2$ and around $360°$ at $q = 20$.

Another item can be deduced from the length-scale computation: the anisotropy $\gamma = L^+/L^-$, reported on the figure 11. If this diagnosis is of poor interest within a simple 1D model, its equivalent in 2D corresponds to an active field of research since anisotropy is a quiet difficult property to represent in covariance modeling. What is interesting here is to understand that length-scale variation (heterogeneity of the length-scale field) over a domain is necessarily related to anisotropy. An homogeneous (heterogeneous) length-scale field is featured by an anisotropy $\gamma = 1$ ($\gamma \neq 1$). If the attention is focused at the vinicity of $270°$ panel $q = 2$, the main variation of γ is a positive lobe corresponding to a positively skewed correlation function matching the transition (Fig.10) from a low length-scale value ($500\,km$) to a larger one ($1000\,km$). Of course the reverse is visible near $90°$ where the correlation function are negatively skewed. The advection of information is clearly visible by observing the displacement of the positive lobe from $270°$ on panel $q = 2$ toward $360°$ on panel $q = 20$.

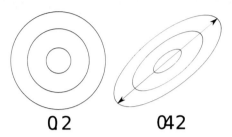

Figure 12. Illustration of isotropic (a) and of anisotropic (b) Gaussian-type correlation function where solid lines represent the contour values for levels each 0.25. The anisotropy in panel (b) is featured by its principal axes (solid line with arrows).

The length-scale and the anisotropy diagnoses can be extended in 2D and 3D by considering the local metric tensor under a local homogeneous Gaussian assumption, as illustrated on the figure 12. This will more detailed at the section (5.3.2.).

With the variance and the length-scale diagnosis, it becomes easy to feature the main component of a background covariance matrix in terms of uncertainty magnitude and of characteristic length of correlation. But these tools are restricted to analyze auto-correlations function of a particular quantity *e.g.* the auto-correlation of pressure in geophysics. Other ingredients complete the diagnosis of correlation function, the diagnostic of balances in multivariate fields.

4.3. Multivariate Components

Here we just present the idea about what balances are and how they influence correlations. Balances lie on physical link among several quantity. For instance, in geophysical science, the effect of the rotation make relevant to link the wind v to the pressure P according to the geostrophic balance equation (the geostrophic wind)

$$v = \frac{1}{\rho f} k \times \nabla P, \qquad (61)$$

where f is the Coriolis parameter related to the Earth rotation, ρ is the density, and k is the local unit vector orthogonal to the sphere. It appears that a correction on the pressure will necessarily imply a correction of the wind. This is quiet important for multivariate modeling. We won't further detail the balances diagnosis and redirect interested readers toward specialized articles [20, 23, 31].

Now we are armed to face the difficult issue of covariance modeling, that is a crucial aspect of variational data assimilation.

5. The B Matrix Modeling

In this section we review some formulations of the background error covariance matrix. The two first relates on the diagonal assumption in a given representation for signal. It can be a basis like the Fourier basis (1D & 2D) or the spherical harmonics on the sphere, and more generally one can consider a frame representation similar to the duality [21]. Another modeling is through a grid point, the modeling based on the diffusion equation. Here, these three formulations are presented with advanced tools. Note that there exits other formulations, and interested readers will be interested by the recursive filter approach see *e.g.* [66, 65]. We start now by the diagonal assumption in spectral space.

5.1. Diagonal Assumption in Spectral Space

The diagonal assumption in spectral space applied to the background error covariance matrix has been introduced and detailed in [15, 67]. Thereafter, the background error covariance matrix $\boldsymbol{B} = \mathbb{E}(\varepsilon_b \varepsilon_b^T)$ is assumed to be a correlation matrix (the variance is homogeneously equal to the unit).

5.1.1. Expansion of a Stationnary Process

The basic idea for the spectral approach is to benefit of the Wiener–Khinchin theorem stating that the energy spectrum of a stationary random process is the Fourier transform of the auto-covariance function. The stationarity assumption means that the correlation between two position x and y only depends on their algebraic distance $r = x - y$. The Fourier transform of a correlation function ρ, defined over a circle of radius 1 (the 1D case), consists in the expansion of ρ into a trigonometric serie

$$\rho(r) = \sum_{j=-T_g}^{T_g} \rho_j e_j(r), \qquad (62)$$

where T_g is the truncations (T_g could be infinite but only at a theoretical level), $e_k(r) = e^{ijr}$ ($i^2 = -1$), and $\rho_j = \langle e_j | \rho \rangle$ with the dot product $\langle f | g \rangle = \frac{1}{2\pi} \int_{[0,2\pi]} f^*(u) g(u) du$, * denotes the complex conjugate operator. Real correlation functions verify that $\rho_j^* = \rho_{-j}$. The theorem of Wiener–Khinchin implies that $\rho_j = \sigma_j^2 = \mathbb{E}(|\varepsilon_{b,j}|^2)$, where $\varepsilon_{b,j}$ is the jth Fourier coefficient of ε_b. Hence, ρ_j is a positive real number, moreover the constraint $\rho_k^* = \rho_{-k}$ imply that $\rho_j = \sigma_{|j|}^2$ do not depends on the sign of j. It follows that $\rho(r)$ can be expanded under a cosine sum (or series)

$$\rho(r) = \sigma_0^2 + \sum_{j=1}^{T_g} \sigma_j^2 \cos(jr). \qquad (63)$$

This means that $\rho(r)$ is an even function ($\rho(-r) = \rho(r)$), or more simply, that ρ is only a function of the distance $|r|$. In this framework, a background error field ε_b appears to be a random function expanded according a Karhunen–Loève decomposition

$$\varepsilon_b(x) = \sum_{j=-T_g}^{T_g} \sigma_{|j|} \zeta_j \, e_j(x), \qquad (64)$$

where for each j, ζ_j is a reduced center Gaussian variable. In terme of vector (imagine the signal is discretized over $2T_g + 1$ regularly distributed over the circle), the background error covariance matrix $\boldsymbol{B} = \mathbb{E}(\varepsilon_b \varepsilon_b^T)$ is thus

$$\boldsymbol{B} = \sum_j \sigma_j^2 \, \boldsymbol{e}_j \boldsymbol{e}_j^T, \tag{65}$$

where $\boldsymbol{e}_j$ is the discretized version of the circular exponential $e_j(x)$.

We recognize, with this kind of expansion, the classical spectral theorem stating that a real symmetric (complex normal) matrix can be diagonalized in an orthonormal basis. An essential property used in statistics *e.g.* the principal component analysis (PCA), also called empirical orthogonal function (EOF). It means that, for the covariance $\boldsymbol{B}$ there exists an orthogonal basis $\boldsymbol{u}_j$ so that $\boldsymbol{B}$ can be expanded under the sum of orthonormal projector

$$\boldsymbol{B} = \sum_j \sigma_j^2 \, \boldsymbol{u}_j \boldsymbol{u}_j^T. \tag{66}$$

Now, the idea is to apply this material in order to construct covariance matrix.

5.1.2. Diagonal Assumption Formulation

The diagonal assumption consists to assume that a given orthonormal basis $\boldsymbol{u}_j$ is the eigen-basis of $\boldsymbol{B}$, even if it is not true of course. In the spectral approach, $\boldsymbol{u}_j$ denotes the Fourier basis (or its equivalent, depending on the geometry). Thus, $\boldsymbol{B}$ is assumed to be well approximated by a matrix $\tilde{\boldsymbol{B}}$ defined by

$$\boldsymbol{B} \approx \tilde{\boldsymbol{B}} = \sum_j \sigma_j^2 \, \boldsymbol{u}_j \boldsymbol{u}_j^T, \tag{67}$$

where σ_j^2 is the variance of the background error liying onto the direction $\boldsymbol{u}_j$. Hence, if $\varepsilon_{b,j} = \langle \boldsymbol{u}_j, \varepsilon_b \rangle$, then the variance σ_j^2 is equal to

$$\sigma_j^2 = \mathbb{E}(|\varepsilon_{b,j}|^2). \tag{68}$$

This leads to a simple algorithm to model the covariance matrix from an ensemble of perturbed forecast $\varepsilon_b^{k,q}$ (see section 3.4.). The estimation of the approximated covariance matrix $\tilde{\boldsymbol{B}}$ reduces to the estimation of the variances as

$$\sigma_j^2 = \frac{1}{N_q N} \sum_q \sum_k |\varepsilon_{b,j}^{k,q}|^2. \tag{69}$$

Hence, this approach replaces the store of a $\mathcal{O}(10^{16})$ coefficient into only a state vector. In term of matrix representation, $\boldsymbol{B}$ is approximated by the expansion

$$\boldsymbol{B} \approx \boldsymbol{S}^{-1} \boldsymbol{D} \boldsymbol{S}^{-T}, \tag{70}$$

where $\boldsymbol{S}$ is the grid-point to spectral transform operator, and $\boldsymbol{D}$ is the diagonal of variances in spectral space.

5.1.3. General Comments on Homogeneity and Isotropy

It results that the correlation tensor B associated to such a 1D stationary process is homogeneous and isotropic. Homogeneity means that the correlation value of the background error is invariant by translation, that is $cor[\varepsilon_b(x), \varepsilon_b(y)] = cor[\varepsilon_b(z+x), \varepsilon_b(z+y)]$ whatever the triplet of points (x, y, z), and thus $cor[\varepsilon_b(x), \varepsilon_b(y)] = \rho(x-y)$. While the isotropy means that the correlation only depends on the distance separating two points (x, y), or $cor[\varepsilon_b(x), \varepsilon_b(y)] = \rho(|x-y|)$. The equivalent to the Fourier transform over the circle is the 2D-Fourier transform over the plan or the spherical harmonics for the sphere. These two geometries are very different. Let see why.

The diagonal assumption in bi-Fourier (torus) corresponds to expand the homogeneous correlation ρ as

$$\rho(\boldsymbol{r}) = \sum_{\boldsymbol{j}} \sigma_{\boldsymbol{j}}^2 e_{\boldsymbol{j}}(\boldsymbol{r}), \tag{71}$$

where $\boldsymbol{j}$ is the 2D wave vector and so that $\sigma_{-\boldsymbol{j}}^2 = \sigma_{\boldsymbol{j}}^2$ for real correlation functions. This formulation is homogeneous but not isotropic since without additional assumption, $\sigma_{\boldsymbol{j}}^2$ depends on the direction of $\boldsymbol{j}$. Homogeneous and isotropic correlation functions are obtained when $\sigma_{\boldsymbol{j}}^2$ doesn't depend on the direction but only the magnitude of $\boldsymbol{j}$, that is on $j = |\boldsymbol{j}|$.

On the sphere $\mathcal{S}^2$, the homogeneity property is equivalent to the isotropy property, and the correlation has the form

$$\rho(\delta) = \sum_{n=0}^{T_g} \sqrt{2n+1} \sigma_n^2 P_n^0(\cos \delta), \tag{72}$$

where δ is the angular distance between two points of the sphere, $Y_n^m(\theta, \phi) = P_n^m(\sin \phi) e^{im\theta}$ is the spherical harmonic basis with θ the longitude and ϕ the latitude, and $\sigma_n^2 = \mathbb{E}(|\varepsilon_{bn}^m|^2)$. By denoting $\eta = (\theta, \phi)$, we have $\varepsilon_b(\eta) = \sum_{n=0}^{T_g} \sum_{m=-n}^{n} \varepsilon_{bn}^m Y_n^m(\eta)$ and $\varepsilon_{bn}^m = \frac{1}{4\pi} \int_{\mathcal{S}^2} Y_n^m(\eta)^* \varepsilon_b(\eta) d\omega(\eta)$, with the metric $d\omega(\eta) = \sin \phi d\theta d\phi$. The question of the equivalence between the homogeneity and anisotropy on the sphere can be understood under a simple thought experiment. Let consider an anisotropic Gaussian-type correlation function on the sphere similar to the one presented on panel Fig.12.(b), localized at, and oriented along, the equator says latitude 0°, longitude 0°. This correlation is then displaced toward the North pole, but according to two different ways. The former is a direct translation of the anisotropic function along the meridian 0°. The latter starts with a 90° translation along the equator, then a direct translation toward the North pole along the meridian 90°. It results that, at the North pole, the principal axes of the two correlation functions are orthogonal: we do not obtain the same correlation function. It results that anisotropic correlations are not invariant by translation on the sphere while of course isotropic function are. Said differently, we have obtained that homogeneous correlations on the sphere are necessarily isotropic [37].

Now able to construct isotropic correlations, it would be interesting to go further in order to represent more realistic correlation function. A simple way to achieve that is to consider other representation of informations: the wavelets.

5.2. Diagonal Assumption in Wavelet Space

The advances afforded by the diagonal assumption in spectral space is to provide a simple way to construct non trivial correlation function. However, in counterpart it is not possible to build heterogeneous formulations with anisotropic correlation. A first improvement of this first formulation is made by considering the diagonal assumption in other bases localized in both grid-point and spectral space. For instance, by using wavelets. In this part, two spherical wavelet decompositions are detailed and then applied within the covariance modeling framework.

5.2.1. Spectrally Based Wavelets

The basic ingredients of a wavelet transform on the real line are the dilatation and the translation of a locally supported function. The former analyzes a signal in physical space while the latter analyzes the signal in spectral space. A wavelet transformation splits a signal into different levels of details. Thus, it can be considered as a set of band-pass filters and it is possible to formulate the wavelet transformation in term of convolution. On the sphere S^2 (also on the circle), it is difficult to define a good dilatation operator since a dilatation at a point along a great circle will lead to a contraction at the antipode [40, 41]. In despite of these difficulties, different wavelet transformations on the sphere exist [72, 1].

In order to cope with the dilatation problem, [36] (see also [35]) have proposed a wavelet transformation based on the Legendre spectrum. It consists of defining an appropriate set of band-pass filters ψ^j, given by their Legendre spectrum, and then convolute it with the spherical function to decompose. This tool has been introduced in data assimilation for covariance modeling by Fisher (see e.g. [33]), who has made this choice because such a spectrally based wavelet (SBW) formulation is invariant with spherical rotation. This last point is a strong constraint for the isotropic representation of dynamics (e.g. with quasi-spectral model and triangular truncation), but it is less true in a grid point model as those encountered in ocean modeling, with a heterogeneous grid and resolution.

The direct and the inverse SBW transforms are defined as follows. For an arbitrary field ε, the wavelet coefficients at scale j, denoted by ε_j, are computed as the convolution by the band-pass filter ψ^j: $\varepsilon_w^j = \psi^j \otimes \varepsilon$, where $\otimes$ denotes the convolution product. The field ε can be obtained again as the sum of the details ε_w^j with $\varepsilon = \sum_j \psi^j \otimes \varepsilon_w^j$. In general, spherical convolution is defined according to the group of rotations of $\mathbb{R}^3$, denoted $SO(3)$. This group is also the group of 3x3 orthogonal matrices with determinant one. But in this particular case, thanks to the Addition Theorem for spherical harmonics with appropriate normalization [60], the convolution reduces to a spectrum product as $(\psi^j \otimes \varepsilon)_q^m = \frac{1}{\sqrt{2n+1}} \psi_q^j \varepsilon_q^m$, where ψ_q^j is the Legendre spectrum of ψ^j and ε_q^m is the spherical harmonic spectrum of ε. This is similar to the convolution product in the Fourier representation in a one dimensional circle.

The formulation of [31] is designed with band-limited ψ^j functions so that their spectrum representation is as follows. For a set of arbitrary chosen integers, $(N_j)_{j \in [0, J]}$, with $N_j < N_{j+1}$, the spectral coefficients of function ψ^j are given by, for $j \neq 0$ (n being the total wave number in the spherical case):

$$\frac{1}{\sqrt{2n+1}}\psi_q^j = \begin{cases} \sqrt{\frac{n-N_{j-1}}{N_j-N_{j-1}}} & \text{for } N_{j-1} \leq n < N_j, \\ \sqrt{\frac{N_{j+1}-n}{N_{j+1}-N_j}} & \text{for } N_j \leq n < N_j + 1, \\ 0 & \text{otherwise.} \end{cases}$$

For $j = 0$, the definition is the same, except that the range $N_{j-1} \leq n < N_j$ is replaced by $0 \leq n < N_0$, for which $(2n+1)^{-1/2}\psi_q^j = 1$.

Note that these SBW are not orthogonal and do not form a basis, but rather a frame [32, 21, 55]. In finite expansions, if $\boldsymbol{W}$ represents the linear operator associated with this wavelet transformation, then $\boldsymbol{W}$ is not invertible (in particular, $\boldsymbol{W}$ is not square but rectangular). Of course, $\boldsymbol{W}$ has a left inverse (the Moore-Penrose pseudo-inverse) $\boldsymbol{W}^{-1}$ so that $\boldsymbol{W}^{-1}\boldsymbol{W} = \boldsymbol{I}$ but $\boldsymbol{W}\boldsymbol{W}^{-1} \neq \boldsymbol{I}$, where $\boldsymbol{I}$ represents the identity operator. Moreover, this particular kind of wavelets forms a tight frame, implying that $\boldsymbol{W}^{-1} = \boldsymbol{W}^T$.

Figure 13 illustrates the band-pass filters for the particular set $\{N_j\} = \{$ 0, 1, 2, 3, 4, 5, 7, 10, 15, 21, 30, 63, 130$\}$. As ψ^j are band limited, the wavelet coefficients ε_w^j can be stored on a coarse grid in physical space, associated with an adapted triangular truncation. By convention, the smaller the truncation, the smaller the index j. Note that the truncation for the two last scales is the truncation of the full resolution grid. Thus, the number of wavelet coefficients is at least two times greater than the degree of freedom of the grid point representation (which is why $\boldsymbol{W}$ is rectangular). The wavelet function associated with a given position x on the sphere is $\psi_x^j(x') = (\psi^j \otimes \delta_x)(x')$ where δ_x is the Dirac distribution at point x. Wavelet functions are such that the larger the spectral band, the more localized in physical space it is, constrained by Heisenberg's uncertainty principle. Figure 14 represents an example of SBW expansion of a spherical field (a). There is one resolution for each scale, with a low resolution for the large scales (b) and a higher resolution for the small scales (d).

The choice of $\{N_j\}$ is a degree of freedom for the formulation and can be optimized in order to match geographical variations of local correlation from a given set of statistics. For instance, it is possible to specify $\{N_j\}$ so that the modeled correlation length-scale corresponds to the diagnosed length-scale under least-square criterion. This discrete optimization problem can be solved by using meta-heuristic algorithms such as simulated annealing. Note that for orthogonal wavelets, some algorithms are available for selecting the best basis among a dictionary of basis (in particular dictionary of wavelet packets, see [14]). In this framework, it is also possible to select the best approximation of the Karhunen-Loève basis (an orthonormal basis for which a covariance matrix is represented by a diagonal) ([56]; appendix E in [61]).

However, the choice of $\{N_j\}$ constraints, in correlation modeling, the resolution of the variation of vertical correlations with horizontal scale (spectrally defined): the larger bands $[N_j, N_{j+1}]$ are lower in the spectral resolution of the vertical-horizontal non-separability [33]. Because of some atmospheric balances, (*e.g.* shallow water behavior in the large scales and deep water behavior in the small scales with a fast transition in the low wavenumbers) it appears better to construct a fine (coarse) representation of low (high) wavenumbers *i.e.* the spectral support of the band-pass filters is larger in the small scales than in the large scales.

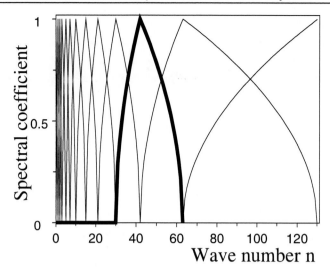

Figure 13. Spectral coefficients $(2n+1)^{-1/2}\psi_n^j$ of the wavelet band-pass filters ψ^j. The spectrum associated to the particular ψ^{10} function is shown in bold.

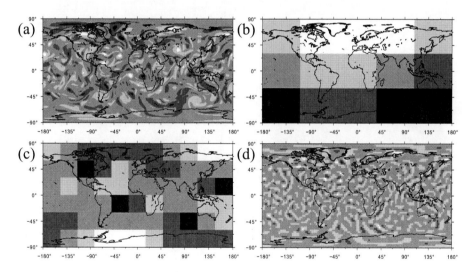

Figure 14. Illustration of a wavelet expansion of a field (a). Some wavelet coefficients are represented for only few scales: the large scale $j = 1$ (b), the medium scale $j = 4$ (c) and the small scale $j = 10$ (d). On these last panels, each gray bloc is a wavelet coefficient.

5.2.2. Grid Point Based Wavelets

Subdividing the sphere and splitting

A hierarchical subdivision of the sphere is first described. It relies on recursive partitioning of the sphere into geodesic spherical triangles. Starting from an icosahedron (a convex polyhedron composed of twenty triangular faces with twelve vertices) at each iteration, four triangles are generated from each triangle of the previous subdivision. Figure 15 represents the initial grid (a) and the first three subdivisions (b)-(d). The degree of a point is the number of connection it has in the mesh. The point A, on panel (a) and (b), is of degree 5, while the point B on panel (b) is of degree 6. Thus the grid is not homogeneous. It is worth noting that this kind of mesh on the sphere has been already used by [73] and also by [86] for numerical modeling. Recently the use of this kind of mesh has been extended *e.g.* with an adaptive wavelet collocation method [58].

The set of all vertices p_k^j, after j subdivisions is denoted $S^j = \{p_k^j\}_{k \in K^j}$, where K^j is an index set. S^0 corresponds to the original icosahedron. Since $S^j \subset S^{j+1}$, we so let $K^j \subset K^{j+1}$. The point p_k^{j+1} for $k \in K^j$ is defined by $p_k^{j+1} = p_k^j$. The cardinal of K^j is $Card(K^j) = 10 \cdot 4^j + 2$. The index set of the vertices added when going from level j to level $j+1$ is denoted $M^j = K^{j+1} - K^j$.

Thus at a given resolution $j+1$ (high resolution) the set K^{j+1} can be split into two sets K^j and M^j so that $K^{j+1} = K^j + M^j$. If a field ε is represented by its discretization on a grid S^{j+1} as $\varepsilon^{j+1} = (\varepsilon_k^{j+1})_{k \in K^{j+1}}$, then its low resolution version is defined as $\varepsilon^j = (\varepsilon_{k'}^j)_{k' \in K^j}$ so that

$$\forall k \in K^j, \varepsilon_k^j = \varepsilon_k^{j+1}. \tag{73}$$

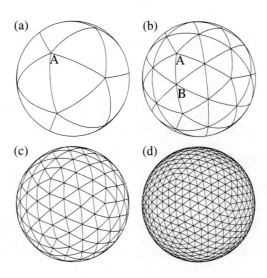

Figure 15. Representation of grid subdivisions. The initial grid, formed from the projection on sphere of the icosahedron (a), is subdivided along a recursive process after one (b), two (c) and three (d) iterations.

Wavelet coefficients as corrections to a prediction

Now, the idea consists in predicting the higher resolution field ε^{j+1} from the low resolution field ε^j. Following Eq. (73), only the value of points in M^j remains to be found. [74, 75] have used a butterfly scheme, a technique inherited from Computer Aided Geometric Design [28], where what is wanted is to construct a smooth C^1 surface out of a control polyhedron. For each m in M^j, the butterfly basis uses a stencil of 8 neighboring points $\mathcal{K}_m \subset K^j$ as illustrated in Fig. 16. The prediction is $\mathcal{P}(\varepsilon^j) = \sum_{k \in \mathcal{K}_m} w_{k,m}^j \varepsilon_k^j$, with the particular weight (independent of j): $w_{v_1} = w_{v_2} = 1/2$, $w_{f_1} = w_{f_2} = 1/8$ and

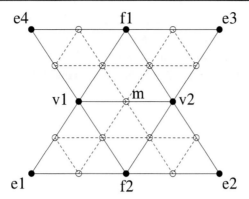

Figure 16. The members of the index sets used in the prediction $\mathcal{P}$ based on the butterfly scheme, with m in M^j and with $\mathcal{K}_m = \{v_1, v_2, f_1, f_2, e_1, e_2, e_3, e_4\}$ included in K^j. Circles and dashed lines represent the next finer level of resolution $j+1$, while black points and solid lines represent the grid at level resolution j.

$w_{e_1} = w_{e_2} = w_{e_3} = w_{e_4} = -1/16$. The wavelet coefficients $\varepsilon_w^j = \left(\varepsilon_{wm}^j\right)_{m \in M^j}$ at scale j are thus defined as the misfit between the true values of ε^{j+1} and the predicted ones

$$\varepsilon_{wm}^j = \varepsilon_m^{j+1} - \sum_{k \in \mathcal{K}_m} w_{k,m}^j \varepsilon_k^j. \tag{74}$$

This expansion could be sufficient, but an additional step called lifting is added, after the computation of ε_w^j, in order to ensure that wavelet functions are of zero mean, which is realized as follows: For a given j and a given k in K^j, the scaling function ϕ_k^j is defined as the inverse of the wavelet coefficients ϕ_w so that $\phi_w{}_{k'}^{j'} = \delta_{j,j'}\delta_{k,k'}$, where $\delta_{i,j}$ stands for the Kronecker symbol. A wavelet function ψ_m^j is defined from these ϕ_k^j functions by

$$\psi_m^j = \phi_m^{j+1} - \sum_{k \in \mathcal{K}_m = \{v_1, v_2\}} s_{k,m}^j \phi_k^j, \tag{75}$$

where $s_{k,m}^j = I_m^{j+1}/2I_k^j$ with $I_k^j = \int_{S^2} \phi_k^j d\omega$ the integral over the sphere S^2 with its classical measure $d\omega$ (see appendix A for the computation of $I_{j,k}$). Finally, the wavelet coefficient is defined as

$$\forall m \in M^j : \forall k \in \mathcal{K}_m = \{v_1, v_2\}, \varepsilon_k^j = \varepsilon_k^j + s_{k,m}^j \varepsilon_{wm}^j. \tag{76}$$

Thus the direct wavelet transformation consists in computing recursively the wavelet coefficients $\varepsilon_w{}^j$. In practice this recursion is initialized by considering that the discrete signal samples ε belong to S^J, and is stopped when $j = 0$. Each level is divided into two steps, with a first step corresponding to the subsampling, described by Eq.(73) and Eq.(74), and a second step, corresponding to the lifting, described by Eq.(76). Note that these wavelets are not radial (they do not have an axis of symmetry). This is related to the butterfly scheme, and is also related to the directional definition of the wavelet functions

ψ^j in Eq. (75). However, this anisotropy is not very strong: the wavelet functions of large scales are quasi-isotropic (not shown here).

The inverse wavelet transform is also computed under a recursive form but in the reverse, from scale $j = 0$ to $j = J$. Each level is divided into two steps with a first step described by the opposite of Eq.(76):

$$\forall m \in M^j : \forall k \in \mathcal{K}_m = \{v_1, v_2\}, \varepsilon_k^j = \varepsilon_k^j - s_{k,m}^j \varepsilon_{w_m}^j, \qquad (77)$$

and then the upsampling described by the opposite of Eq.(74)

$$\varepsilon_m^{j+1} = \varepsilon_{w_m}^j + \sum_{k \in \mathcal{K}_m} w_{k,m}^j \varepsilon_k^j, \qquad (78)$$

and by the reverse of Eq.(73)

$$\forall k \in K^j, \varepsilon_k^{j+1} = \varepsilon_k^j. \qquad (79)$$

The linear operator associated to the direct (the inverse) wavelet transformation is denoted by $\boldsymbol{W}$ ($\boldsymbol{W}^{-1}$).

The mathematical background of this wavelet decomposition is given next. In terms of functional analysis, the space V^{j+1}, engendered by scaling functions ϕ^{j+1} (which forms a biorthogonal basis, or a Riesz basis, of V^{j+1}), is included in the space of square-integrable functions on the sphere $\mathcal{L}^2(S)$, while the wavelet functions ψ^j engender a subspace $W^j \subset V^{j+1}$ so that $V^{j+1} = V^j \oplus W^j$ (ψ^j forms a Riesz basis of W^j). Thus, V^J is decomposed along the recursive process into $V^J = V^0 \oplus W^0 \oplus W^1 \oplus \cdots \oplus W^{J-1}$. Moreover, the closure of $\cup_j V^j$ is $\mathcal{L}^2(S)$. This kind of construction is called a Multi Resolution Analysis and is important for orthogonal wavelets. In this case, Riesz bases are replaced by orthogonal bases, with additional constraints [21, 55].

5.2.3. Formalism of the Diagonal Assumption in Wavelet Space

Similarly to the diagonal formulation in spectral space [15], the wavelet equivalent is that the background correlation tensor is represented by a diagonal in wavelet space. Thus $\boldsymbol{B}$ is expanded as the product of $\boldsymbol{B} = \boldsymbol{\Sigma}_g \boldsymbol{C} \boldsymbol{\Sigma}_g^T$ where $\boldsymbol{\Sigma}_g^2$ is the diagonal matrix of $\boldsymbol{B}$ and $\boldsymbol{C}$ is the correlation matrix. With the previous notation, the wavelet formulation of the matrix $\boldsymbol{B}$ is specified following [31] and [22] as $\boldsymbol{B}_{dw} = \boldsymbol{\Sigma}_g \boldsymbol{C}_{dw} \boldsymbol{\Sigma}_g^T$ with

$$\boldsymbol{C}_{dw} = \boldsymbol{\Lambda} \boldsymbol{\Sigma}_s \boldsymbol{W}^{-1} \boldsymbol{D}_w \boldsymbol{W}^{-T} \boldsymbol{\Sigma}_s^T \boldsymbol{\Lambda}^T, \qquad (80)$$

where $\boldsymbol{\Sigma}_s$ is a diagonal normalization by spectral standard deviation, $\boldsymbol{D}_w^{1/2}$ is a diagonal of wavelet standard deviations, and $\boldsymbol{\Lambda}$ is a diagonal normalization by standard deviations of $\boldsymbol{\Sigma}_s \boldsymbol{W}^{-1} \boldsymbol{D}_w \boldsymbol{W}^{-T} \boldsymbol{\Sigma}_s^T$. Thereafter, this normalization is not recalled, but is applied. Note that a non-separable formulation can be constructed following [33] (see also [31]). In that correlation model, the matrix $\boldsymbol{D}_w$ is a block diagonal matrix whose blocks $\boldsymbol{D}_{w,\alpha}$ are the vertical covariances associated with wavelet index α, where $\alpha = (j, \boldsymbol{x})$ is a couple of scale and position. Similarly to the spectral formulation, the dependence on frequency is ensured by the dependence on the scale j. But unlike the spectral formulation, the dependence on position $\boldsymbol{x}$ involves a geographical variation of the vertical correlation by scale.

One of the consequences of the diagonal assumption in a frame is that the diagonal matrix C_d is related to the initial matrix C following a relation of the form

$$C_d(x', x) = \sum_{z,z'} C(z', z) \Phi_{(x',x)}(z', z),$$

where the weight matrix $\Phi_{(x',x)}$ is function of the frame [62]. This means that C_d is a spatial average of C. In the spectral basis, the weight only depends on the distance between x' and x. Thus, the diagonal assumption in spectral space C_{ds} can be seen as a global spatial average of C, so that the resulting tensor is homogeneous (and isotropic according to [37]). In the wavelet representation, the weight $\Phi_{(x',x)}$ is localized around the position (x', x), because the wavelets are localized in both space and spectrum. C_{dw} can be seen as a local spatial average of C. Then, this formulation is able to represent geographical variations of the local correlation functions. However, as the wavelets ψ_x^j are radial (i.e. isotropic) functions and thus not polarized (i.e. anisotropic or directional, see [1]), the local average at each scale tends to damped anisotropy.

The result of this averaging is that the geographical variations of the local correlation functions, modeled with the wavelet diagonal assumption, are smooth, associated with a filtering of sampling noise [62]. Figure 17 mimics the estimation of wavelet statistics at different scales and for two points A and B. For the large scales, the spatial average is done over a large area (horizontally hashed area for A and vertically hashed area for B). This eliminates small scale contribution and thus damps a part of the sampling noise. At the opposite extreme, for small scales, the averaging is done over a local area near the point and they are more sensitive to sampling noise. The statistics computed at two different points vary slowly for the large scale: areas around A and B overlap. Then the statistics vary more rapidly in the medium scales: areas around A and B overlap slightly. Finally, they are independent in the small scales: areas around A and B are disjointed.

In order to illustrate the diagonal assumption in wavelet space and to give another point of view of why sampling noise is eliminated in this formulation, a one dimensional example is now introduced.

5.3. Formulation Based on the Diffusion Equation

5.3.1. Formalism of the Correlation Modeling with a Diffusion Operator

Another way to construct a heterogeneous correlation matrix has been introduced by [84] and is based on the generalized diffusion equation. It is noticeable that on a plane, a particular solution of the diffusion equation

$$\partial_t \eta = \nabla \cdot (\nu \nabla \eta), \tag{81}$$

with constant diffusion tensor ν (corresponding to the homogeneous diffusion equation) is the Gaussian function

$$G^{\nu,t}(x, x') = \frac{1}{2\pi |\Gamma|^{1/2}} exp\left(-\frac{1}{2}(x - x')^T \Gamma^{-1}(x - x')\right),$$

where

$$\Gamma = 2t\nu, \tag{82}$$

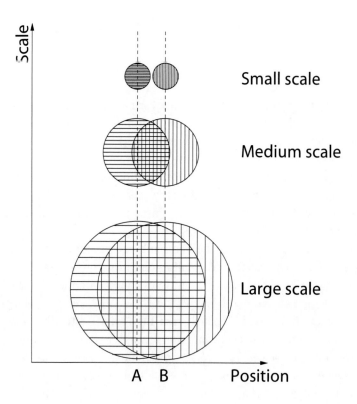

Figure 17. Sketch of the local spatial average at different scales with the diagonal wavelet assumption (see text for details).

with t the time of integration, and $|\Gamma|$ is the determinant of Γ. The expansion of Γ^{-1} in the 2D case according to

$$\Gamma^{-1} = \begin{pmatrix} \frac{1}{L_x^2} & \frac{1}{L_{xy}} \\ \frac{1}{L_{xy}} & \frac{1}{L_y^2} \end{pmatrix},$$

gives rise to the length-scale L_x (L_y) along axis x (y). This length-scale corresponds to the differential length-scale ([20], p110), and it can be approximated under assumption of correlation behaviour in the vicinity of small separation distance [63].

The idea is then to construct Gaussian-like correlation functions as the result of the diffusion of a Dirac distribution. By using this approach, it is feasible to generate heterogeneous correlation functions when the local diffusion tensor $\nu(x)$ varies with the position $x = (x, y)$. The correlation tensor is then modeled as

$$C = \Lambda L^{1/2} E^{-1} L^{T/2} \Lambda^T,$$

where $L^{1/2}$ is an half time integration (or the propagator) from initial time 0 and final time $t/2$, E is a local metric tensor, and (similarly to the wavelet formulation) Λ is a normalization so that C is a correlation matrix. Note that in the homogeneous sphere of triangular truncation T, the inverse metric E^{-1} is expressed by $E^{-1} = (T+1)^2/4\pi a^2 I$, where a is the radius of the sphere and I denotes the identity operator.

For practical applications, a numerical time integration of the diffusion equation (81) is achieved in a quasi-spectral approach (triangular truncation T), with an Euler time scheme of time step defined by $dt \leq (dx)^2/(CFL\, \nu_{max})$ where dx is the regular step associated to truncation T and the Courant Friedrichs Levy condition of stability is set to $CFL = 4$ [84]. ν_{max} corresponds to the maximum of the spectrum of all the local diffusion tensor. The convention is that local diffusion tensors are defined for time $t = 1$ in Eq. (82). Thus, $L^{1/2}$ corresponds to a time integration over the interval $[0, 1/2]$.

5.3.2. Estimation of the Local Diffusion Tensor

The question of how to estimate this local diffusion tensor remains. A first possibility is to estimate the local tensor from the dynamical aspects of the geophysical flow considered (Weaver, personal correspondence). The underling assumption is that the spread of the local correlation function is only related to the dynamics. But this is not entirely satisfactory as it is known that background error correlation functions are also influenced by the data network [11].

Another possibility is to estimate the local diffusion tensor from ensemble statistics [64]. This estimation relies on the estimation of the local anisotropic tensor, under the assumption of a local homogeneous diffusion. This means that the local correlation function $\rho(x, \cdot)$ is considered as being locally Gaussian in the limit of small separation distance, *i.e.*

$$\rho(x, x + \delta x) \equiv exp\left(-\frac{1}{2}\delta x^T \Gamma^{-1} \delta x\right),$$

where $\delta x = (\delta x, \delta y)$ is a small displacement. Γ^{-1} can be numerically estimated as the least-square solution from an ensemble of $K \geq 3$ position $\delta x_k = (\delta x_k, \delta y_k)$ associated with numerical correlation values $\rho_k = cor(x, x + \delta x_k) > 0$. Thus, local length-scale and anisotropy play an important role in this approach (contrary to the wavelet diagonal assumption where the length-scale is not necessary to estimate the formulation). Note that the question of how to estimate parameters of the generalized diffusion equation is not yet answered.

5.4. Illustration with a Toy Ensemble of Perturbed Forecasts

This paper focuses on the representation of horizontal heterogeneity of the background error covariances. In real applications, the true background error correlation matrix is not known. This matrix is generally estimated from a small ensemble and thus its estimation is affected by sampling noise [44, 52]. Thus it is not possible to compare the correlation model with the true correlation matrix, as done for an analyticaly known example that corresponds to an infinite ensemble framework. To overcome these difficulties, a simple toy model on the sphere has been implemented to generate a large ensemble of perturbed forecasts, providing a correlation matrix of reference. The ensemble is large enough to be considered as infinite. In other words, this correlation matrix corresponds to the truth unknown in practice. The heterogeneous correlation functions produced with this approach are sufficiently complex to be considered as a testbed for the formulation presented in the previous sections. The model is first described and the construction and analysis of the ensemble are also detailed. The analysis of previous formulations is then presented.

5.4.1. Description of the Model and the Ensemble

The spherical model considered is the non-rotating non-divergent barotropic vorticity equation

$$\partial_t \zeta + J(\Delta^{-1}\zeta, \zeta) = -\nu_2 \Delta^2 \zeta,$$

where ζ is the vorticity of the horizontal wind, t is time, J is the Jacobian operator, Δ is the horizontal laplacian, Δ^2 is the horizontal bi-laplacian, and $\nu_2 = 5x\,10^{14}$ a superviscosity coefficient [87]. The numerical integration is based on a quasi-spectral approach in a triangular truncation $T = 130$, with a classical 3/2-rules for de-aliasing procedure [12]. The time integration is done by a leap-frog scheme [49] with Asselin's filter to damp numerical modes (Asselin's parameter is set to 1%) [2].

An initial state of vorticity $\zeta^r(t=0)$ has been randomly generated as follows: From an energy density spectrum $E(n) = n(1 + n/n_c)^{-7}$ (n denotes the wave number) and normalized so that the total energy $E_{tot} = \sum_{n=0}^{T} E(n)$ is $E_{tot} = 300\,m^2.s^{-2}$, and the enstrophy spectrum G_q has been computed according to $G_n = n(n+1)/a^2\,E_q$, with the earth radius $a = 6400$ km [8, 9]. The cut off wavenumber n_c is set to 20. Then, the spherical harmonic spectrum ζ_n^m of the vorticity field has been randomly generated with $\zeta_n^m = \sqrt{2G_n/(2n+1)}\xi_n^m$ where ξ_n^m is a sample of a complex Gaussian variable of mean 0 and standard deviation 1. The initial state $\zeta^r(t=0)$ of reference is represented Fig. 18-(a). It corresponds to a heterogeneous field with patterns randomly distributed. The 2D-characteristic length of patterns is given by the differential length-scale ([20], p110) $l = \sqrt{2a^2 \sum_n G_n / \sum_n n(n+1)G_n} = 216$ km. The coast lines are represented to facilitate the discussion on localized structures, but no orography is present in the model.

Similarly, an ensemble of $N_e = 800$ independent perturbations $\delta\zeta_k(t=0)$ (k denotes the number of the member) has been randomly generated with energy spectrum $E'_n = n(1+n/n_c)^{-4}$ normalized so that the total energy of each perturbation is $E'_{tot} = 30\,m^2.s^{-2}$. An example of perturbation, $\delta\zeta_1(0)$ is represented in Fig. 18-(d). Similarly to the initial field, the 2D-characteristic length of the patterns is $l' = 113$ km. The patterns of the perturbation are of smaller scales than in the initial field, and the magnitude of the perturbation is 26% of the magnitude of the reference $\zeta^r(0)$.

The reference $\zeta^r(t = 0)$ (respectively each member of the ensemble of perturbed initial state $\zeta_k(0) = \zeta^r(0) + \delta\zeta_k(0)$) has been integrated over $\tau = 54$ hours, to obtain $\zeta^r(\tau)$ (respectively $\zeta_k(\tau) = \zeta^r(\tau) + \delta\zeta_k(\tau)$). The field $\zeta^r(\tau)$ is represented in Fig. 18-(b). The interaction of coherent structures have emerged during the integration from the initial patterns [57, 87], with formation of long filaments of vorticity similar to some front like-structure. Any perturbed forecasts exhibit a similar time evolution. Slight differences, resurgences of the initial independent perturbations, still exist however. The perturbation $\delta\zeta_1(\tau)$ is represented in Fig. 18-(e). Several dipoles appear as near the point at latitude $-45°$ and longitude

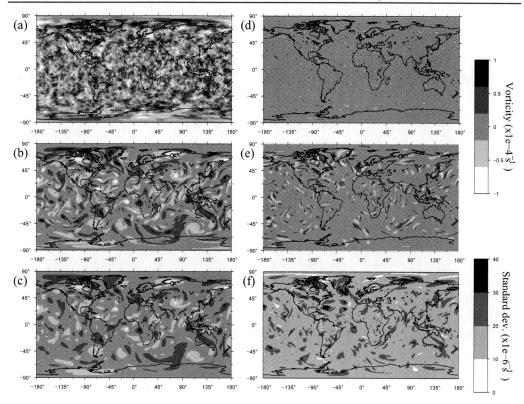

Figure 18. Representation of initial vorticity field (in $10^{-4}\ s^{-1}$) (a) and the final field (b) after time integration of the non-divergent barotropic model. An example of initial perturbation is represented in panel (d) while its non-linear time evolution is represented in panel (e). The mean of the ensemble of perturbed forecasts is represented in panel (c). The standard deviation (in $10^{-6}\ s^{-1}$) of this ensemble is represented in panel (f). Coast lines are represented to facilitate the presentation but the model does not include any orography.

$0°$, or near the point at latitude $-45°$ and longitude $-180°$. Each dipole corresponds to a phase shift error. Moreover, the error magnitude is now of the same order as the reference magnitude $\zeta^r(\tau)$. Subsequently, statistics are computed at time τ.

Some simple diagnostics of the ensemble can be computed, such as the ensemble mean $\overline{\zeta_k}$ represented in Fig. 18-(c). The field is clearly of larger scales than the reference forecast, this is the fingerprint of slight nonlinearities. The standard deviation field $\sigma(\zeta_k)$ is reported in Fig. 18-(f). Large values of standard deviation are concentrated on coherent structures and on filaments, this is consistent with the fact that coherent structures are the dynamical part while they inhibit the nonlinear instability of the background flow, *i.e.* the incoherent part of the flow [50].

Other diagnostics of correlation functions can be computed as the the local anisotropy (not shown here) or the local length-scale. Figure 19 represents the zonal length-scale L_x on the panel (a) and the meridional length-scale L_y on the panel (b). The length-scale field is computed according to the method described section (5.3.2.). In that case, zonal and meridional displacement are the finest scale associated to the homogeneous resolution at truncation T. Both length-scale fields are very organized and fingerprints of coherent structures are visible. Areas of strong anisotropy are visible *e.g.* over the North-East of the Greenland, where L_x are almost 100 km while L_y are nearly 300 km. Both length-scale fields appear to be intermittent, which is a difficulty for covariance modeling. Now this diagnosis is used to compare the different formulations introduced in previous sections.

5.4.2. Comparison of the Different Correlation Models

The three correlation models presented in the preliminary sections are now compared. Each of them has been estimated from the previous ensemble of $Ne = 800$ perturbed forecasts $\zeta_k = \zeta^r + \delta\zeta_k$. Then an ensemble of $Ne = 800$ simulated background errors has been generating as $\varepsilon_b = \boldsymbol{B}_{model}^{1/2}\boldsymbol{\xi}$ where $\boldsymbol{\xi} \sim \mathcal{N}(0, \boldsymbol{I})$ is a sample of a Gaussian random vector with zero mean and the identity as covariance matrix [34]. In this section, only qualitative results are presented in order to illustrate the feasibility of each method and their behaviour. Quantitative results can be found in previous studies about the wavelets formulation and the diffusion operator (see *e.g.* [62, 64, 61]).

The diagnosed length-scale associated with the SBW formulation is reported in Fig. 19 on panel (c) for L_x and (d) for L_y. The heterogeneity is accurately represented and the main structures present on the reference (a) and (b) are retrieved. For instance, one can notice the long length-scale pattern over the Himalayas (compare panels (a) and (c)) and the short length-scale area in West Africa (compare panels (b) and (d)). But in this experiment, the SBW formulation misses the extremes in the length-scale (a & b). Moreover, the range of magnitude variation of the modeled length-scale is smaller than the original reference. On the reference, the order of the minimum length-scale represented is 100 km while this minimum is now up to 150 km for the wavelets. The height values are also reduced as they are nearby 300 km on the reference and 250 km with the model. This corresponds to the spatial average discussed in section (5.2.3.). Even if this spatial average is an advantage for ensemble of small size [62], in this simulated example, the SBW formulations seems to fail. This should be further investigated. A first guess may be that the SBW formulation is less appropriate to represent filament structures, which are particularly strong in this toy experiment. Similar to the bad representation of the anisotropy, this could be related to the isotropy of the SBW functions.

Now the SGW diagonal assumption is analyzed. The modeled length-scale fields of this formulation are reported on panel (e) for L_x and on panel (f) for L_y. The geographical variations of the length-scale are present with a good range of length-scale value. Nevertheless, the most visible defect is that the length-scale fields are dependent on the grid. This might be attenuated by using a spectral normalization [22]. This solution seems difficult to achieve as the hierarchical mesh is not adapted for spectral computation, results should depend on the interpolation strategy. Moreover, it appears that anisotropy is not well represented, even if there exists a part anisotropy represented such as over Himalaya.

Finally, the length-scales of the model based on the diffusion equation are represented on panel (g) for L_x and on panel (h) for L_y. With this approach, the geographical variations of the local correlation functions are better represented. Both length-scale maps are well correlated with their associated reference (a) for L_x or (b) for L_y. Representation of

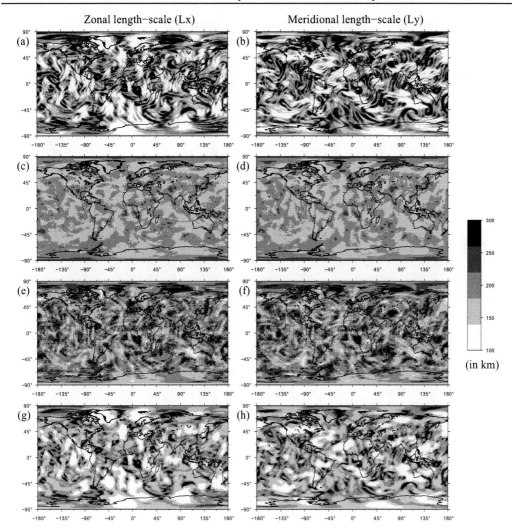

Figure 19. Diagnosis of the length-scale estimated from the raw ensemble or from a randomization: raw zonal length-scale L_x (a) and raw meridional length-scale L_y (b); the length-scales modeled with SBW L_x (c) and L_y (d); the length-scales modeled with SGW L_x (e) and L_y (f); and the length-scale modeled with the diffusion equation L_x (g) and L_y (h). (length-scale values low to 100km (up to 300 km) are in white (black))

anisotropy is better than the two previous models *e.g.* West Africa. Some slight known defects occur like the difficulty to catch fast oscillations [64] *e.g.* at the south of the South Africa.

This simulated experiment has shown the various abilities of the these formulations to represent geographical variations. In this framework, the correlation model based on the diffusion has been shown better than the SBW and the SGW formulations. Of course the length-scale analysis solely is not sufficient, but it gives a preliminary view of the heterogeneity of each formulation.

When the ensemble is small, the above results are affected by the sampling noise. As reminded in section (5.2.3.), the wavelet formulation reduces the sampling noise. However, the correlation model with the diffusion is estimated from the computation of the local correlation, which is sensitive to the ensemble size. Under a local ergodicity assumption, the knowledge of the correlation sampling distribution, versus the ensemble size, is a possible way to improve the estimation and to reduce the sampling effects [63].

6. Conclusion

All along the chapter, we have introduced theoretical tools and practical algorithms that are encountered in data assimilation. These various technicals are derived from idea of the Kalman filter dynamic of information viewed within a larger filtering process. The non-linear filtering, that relies on the Bayes' rule, is the natural way to merge information together. This fusion of informations has been introduced by the use of the conjunction binary operator applied on probability distributions. The formalism corresponds to the Feynman-Kac measures.

The Feynman-Kac represent the filtering process, that is the reference in terms of information dynamic. However this algorithms is not usable in practice since it would require the use of a huge amount of particle of the discretization. Despite on its limitations, the Kalman filter equations are very popular providing a low cost estimation from observations and a robust algorithm.

We have discussed classical algorithms like the EnKF, and shown it was possible to use it out of the Kalman assumptions of Gaussianity and linearity of the dynamic. However, compared with the filtering process as design by the particle filter, the EnKF differs from the non-linear filtering. This new result has to be better understood, and research studies on this aspect are in expansion to understand what imply this limit.

Another formulation to solve the Kalman equations has been remind the 4D-Var. In particular the attention has been focused on the diagnosis of the background error covariance matrix. Then a large developments has been detailed about several background error covariance matrix model. This as lead to present the wavelets and their use within this framework.

The various algorithms have been illustrated within appropriate example to focuses the attention onto particular aspect of interest. The limit of the Kalman filter when applied on a linear but non-Gaussian dynamic. with a comparison to the particle filter and the EnKF. It has been shown that in that case the EnKF and the Kalman filter provide the same results when the model error is correctly specified within the Kalman filter equations. Then, the uncertainty tube provided by the EnKF present an unappropriated jigsaw profile compared to the filtering process. Moreover, the uncertainty diagnosed with the EnKF appeared to underestimate the true magnitude.

Dynamics of covariance, illustrated on a simple 1D advection over a circular framework has shown, the impact of both the dynamic and the observation network on the time evolution of the variance and the length-scale. This has been also illustrated over the sphere by the mean of a large ensemble experiment offering a known and complex covariance matrix. This was the ideal testbed to compare some covariance model. For instance, the formulations based on the wavelets decomposition. Among the various covariance model, the length-scale diagnosis has appeared as a way to design a background error covariance matrix thanks to the diffusion operator. Some defects of the the covariance model appearing from particular wanted properties have not been details *e.g.* the non-separability issue for the atmosphere. These points can be now addressed from specialized articles.

Beyond these materials, the reader is now able to tackle the question of previsibility that is now crucial for a right use of weather forecast and their societal impact. This also relies on the Feynman-kac measures. But requires new tool like validation scores of ensemble spread (Brier score, rank histogram,...). The applications are of crucial importance since we can reduce and document the risk (human, economical activities,...). This is another landscape to discover.

References

[1] J.P. Antoine, L. Demanet, L. Jacques, and P. Vandergheynst. Wavelets on the sphere: Implementation and approximation. *Appl. Comput. Harmon. Anal.*, **13**:177–200, 2002.

[2] R. Asselin. Frequency filter for time integrations. *Mon. Wea. Rev.*, **100**:487–490, 1972.

[3] L. Auger and A. Tangborn. A wavelet-based reduced rank kalman filter for assimilation of stratospheric chemical tracer observations. *Mon. Wea. Rev.*, **132**:1220–1237, 2004.

[4] C. Baehr and O. Pannekoucke. *chapter 4 : Some Issues and results on the EnKF and particule filters for meteorological models*. World Scientific, 2010.

[5] M. Belo Pereira and L. Berre. The use of an ensemble approach to study the background error covariances in a global nwp model. *Mon. Wea. Rev.*, **134**:2466–2489, 2006.

[6] L. Berre, O. Pannekoucke, G. Desroziers, S. Stefanescu, B. Chapnik, and L. Raynaud. A variational assimilation ensemble and the spatial filtering of its error covariances: increase of sample size by local spatial averaging. In ECMWF, editor, *Proceeding of workshop on flow-dependent aspects of data assimilation*, pages 151–168, Reading, UK, 11-13 June 2007.

[7] L. Berre, S.E. Stefanescu, and M. Belo Pereira. The representation of the analysis effect in the three error simulation techniques. *Tellus A*, **58**:196–209, 2006.

[8] G. Boer. Homogeneous and isotropic turbulence on the sphere. *J. Atmos. Sci.*, **40**:154–163, 1983.

[9] G. Boer and T. Shepherd. Large-scale two-dimensional turbulence in the atmosphere. *J. Atmos. Sci.*, **40**:164–184, 1983.

[10] F. Bouttier. The dynamics of error covariances in a barotropic model. *Tellus A*, **45**:408–423, 1993.

[11] F. Bouttier. A dynamical estimation of error covariances in an assimilation system. *Mon. Wea. Rev.*, **122**:2376–2390, 1994.

[12] J.P. Boyd. *Chebyshev and Fourier Spectral Methods*. Dover Publications, second edition, 2001.

[13] M. Buehner and M. Charron. Spectral and spatial localization of background error correlations for data assimilation. *Q.J.R. Meteorol. Soc.*, **133**:615–630, 2007.

[14] R. Coifman and W. Wickerhauser. Entropy-based algorithms for best-basis selection. *IEEE Trans. Inform. Theory*, **38**:713–718, 1992.

[15] P. Courtier, E. Andersson, W. Heckley ans J. Pailleux, D. Vasiljević, M. Hamrud, A. Hollingsworth, F. Rabier, and M. Fisher. The ecmwf implementation of three-dimensional variational assimilation (3d-var). i: Formulation. *Q.J.R. Meteorol. Soc.*, **124**:1783–1807, 1998.

[16] P. Courtier and J.F. Geleyn. A global numerical weather prediction model with variable resolution : Application to the shallow-water equations. *Q.J.R. Meteorol. Soc.*, **114**:1321–1346, 1988.

[17] P. Courtier and O. Talagrand. Variational assimilation of meteorological observations with adjoint vorticity equation. ii: Numerical results. *Q.J.R. Meteorol. Soc.*, **113**:1329–1347, 1987.

[18] P. Courtier and O. Talagrand. Variational assimilation of meteorological observations with the direct and the adjoint shallow-water equations. *Tellus A*, **42**:531–549, 1990.

[19] P. Courtier, J.-N. Thepaut, and A. Hollingsworth. A strategy for operational implementation of 4d-var, using an incremental approach. *Q.J.R. Meteorol. Soc.*, **120**:1367–1387, 1994.

[20] R. Daley. *Atmospheric Data Analysis*. Cambridge University Press, 1991.

[21] I. Daubechies. *Ten lectures on wavelets*. Society for Industrial and Applied Mathematics, 1992.

[22] A. Deckmyn and L. Berre. A wavelet approach to representing background error covariances in a lam. *Mon. Wea. Rev.*, **133**:1279–1294, 2005.

[23] J. Derber and F. Bouttier. A reformulation of the background error covariance in the ecmwf global data assimilation system. *Tellus A*, **51**:195–221, 1999.

[24] J. Derber and A. Rosati. A global oceanic data assimilation system. *J. Phys. Oceanogr.*, **19**:1333–1347, 1989.

[25] G. Desroziers. A coordinate change for data assimilation in spherical geometry of frontal structures. *Mon. Wea. Rev.*, **125**:3030–3038, 1997.

[26] G. Desroziers, L. Berre, B. Chapnik, and P. Poli. Diagnosis of observation, background and analysis error statistics in observation space. *Quart. Jour. Roy. Meteor. Soc.*, **131**:3385–3396, 2006.

[27] F.X. Le Dimet and O. Talagrand. Variational algorithms for analysis and assimilation of meteorological observations: theoretical aspects. *Tellus*, **38**:97–110, 1986.

[28] N. Dyn, D. Levin, and J. Andgregory. Butterfly subdivision scheme for surface interpolation with tension control. *Transactions on Graphics* **9**, 2:160–169, 1990.

[29] M. Ehrendorfer. The liouville equation and its potential usefulness for the prediction of forecast skill. part i: Theory. *Mon. Wea. Rev.*, **122**:703–713, 1994.

[30] G. Evensen. Sequential data assimilation with a nonlinear quasi-geostrophic model using monte carlo methods to forecast error. *J. Geophys. Res.*, **99**:10 143–10 162, 1994.

[31] M. Fisher. Background error covariance modelling. In ECMWF, editor, *Proc. ECMWF Seminar on "Recent developments in data assimilation for atmosphere and ocean"*, pages 45–63, Reading, UK, 8–12 September 2003.

[32] M. Fisher. Generalized frames on the sphere, with application to background error covariance modelling. In ECMWF, editor, *Proc. ECMWF Seminar on "Recent developments in numerical methods for atmospheric and ocean modelling"*, pages 87–102, Reading, UK, 6-10 September 2004.

[33] M. Fisher and E. Andersson. Developments in 4d-var and kalman filtering. *Technical Report* **347**, ECMWF Technical Memorandum, 2001.

[34] M. Fisher and P. Courtier. Estimating the covariance matrices of analysis and forecast error in variational data assimilation. Technical Report 220, ECMWF Technical Memorandum, 1995.

[35] W. Freeden and M. Schreiner. Orthogonal and non-orthogonal multiresolution analysis, scale discrete and exact fully discrete wavelet transform on the sphere. *Constructive approximation*, **14**:493–515, 1998.

[36] W. Freeden and U. Windheuser. Spherical wavelet transform and its discretization. *Adv. Comput. Math.*, **5**:51–94, 1996.

[37] G. Gaspari and S. Cohn. Construction of correlation functions in two and three dimensions. *Q.J.R. Meteorol. Soc.*, **125**:723–757, 1999.

[38] MGG Foreman G.D. Egbert, A.F. Bennett. Topex/poseidon tides estimated using a global inverse model. *J. Geophys. Res.*, **99**:24821–24852, 1994.

[39] F. Le Gland, V. Monbet, and V.-D. Tran. *Large sample asymptotics for the ensemble Kalman filter*. Oxford University Press, 2010.

[40] M. Holshneider. Wavelet analysis on the circle. *J. Math. Phys.*, **31**:39–44, 1990.

[41] M. Holshneider. Continuous wavelet transforms on the sphere. *J. Math. Phys.*, **37**:4156–4165, 1996.

[42] P. L. Houtekamer and Herschel L. Mitchell. A sequential ensemble kalman filter for atmospheric data assimilation. *Monthly Weather Review*, **129**(1):123–137, 2001.

[43] P.L. Houtekamer, L. Lefaivre, J. Derome, H. Ritchie, and H.L. Mitchell. A system simulation approach to ensemble prediction. *Mon. Wea. Rev.*, **124**:1225–1242, 1996.

[44] P.L. Houtekamer and H. Mitchell. Data assimilation using an ensemble kalman filter technique. *Mon. Wea. Rev.*, **126**:796–811, 1998.

[45] K. Ide, P. Courtier, M. Ghil, and A. Lorenc. Unified notation for data assimilation: Operational, sequential and variational. *J. Meteor. Soc. Japan*, **75**:181–189, 1997.

[46] B. Ingleby. The statistical structure of forecast errors and its representation in the met. office global. *Q.J.R. Meteorol. Soc.*, **124**:1783–1807, 2001.

[47] E. T. Jaynes. Information theory and statistical mechanics. *Physical Review Series*, **106**:620630, 1957.

[48] A.H. Jazwinski. *Stochastic processes and filtering theory*. Academic Press, 1970.

[49] E. Kalnay. *Atmospheric modeling, data assimilation and predictability*. Cambridge University Press, 2002.

[50] N. Kevlahan and M. Farge. Vorticity filaments in two-dimensional turbulence: creation, stability and effect. *Journal of Fluid Mechanics*, **346**:49–76, 1997.

[51] P. Lönnberg. Developpements in the ecmwf analysis scheme. In ECMWF, editor, *Proc. ECMWF Seminar on "Data assimilation and the use of satellite data"*, pages 75–120, Reading, UK, 5-9 September 1988.

[52] A. Lorenc. The potential of ensemble kalman filter for nwp – a comparison with 4d-var. *Q.J.R. Meteorol. Soc.*, **129**:3183–3203, 2003.

[53] E.N. Lorenz. Deterministic nonperiodic flow. *Journal of the Atmospheric Sciences*, **20**(2):130–141, 1963.

[54] Lei M., Del Moral P., and Baehr C. Analysis of approximated pcrlbs for nonlinear dynamics using different moments of state estimate. In IEEE transaction on Control and Automationv Error analysis of approximated, editors, *Proceedings of the 8th IEEE International Conference on Control and Automation (ICCA2010)*, Xiamen, China, 9-11 June 2009.

[55] S. Mallat. *A Wavelet Tour of Signal Processing*. Academic Press, 1999.

[56] S. Mallat, G. Papanicolaou, and Z. Zhang. Adaptive covariance estimation of locally stationary processes. *Annals of Statistic*, **26**:1–47, 1998.

[57] J.C. McWilliams. The emergence of isolated coherent vorticies in turbulent flow. *J. Fluid. Mech.*, **146**:21–43, 1984.

[58] M. Mehra and N. Kevlahan. An adaptive wavelet collocation method for the solution of partial differential equations on the sphere. *J. Comp. Phys.*, **227**:5610–5632, 2008.

[59] P. Del Moral. *Feynman-Kac Formulae: Genealogical and Interacting Particle Systems with Applications*. Springer, 2004.

[60] C. Müller. *Spherical harmonics*. Springer, 1966.

[61] O. Pannekoucke. *Modélisation des structures locales de covariance des erreurs de prévision l'aide des ondelettes*. PhD thesis, Univertisté de Toulouse, France, 2008.

[62] O. Pannekoucke, L. Berre, and G. Desroziers. Filtering properties of wavelets for local background-error correlations. *Q.J.R. Meteorol. Soc.*, **133**:363–37, 2007.

[63] O. Pannekoucke, L. Berre, and G. Desroziers. Background error correlation lengthscale estimates and their sampling statistics. *Q.J.R. Meteorol. Soc.*, **134**:497508, 2008.

[64] O. Pannekoucke and S. Massart. Estimation of the local diffusion tensor and normalization for heterogeneous correlation modelling using a diffusion equation. *Q.J.R. Meteorol. Soc.*, **134**:1425–1438, 2008.

[65] R.J. Purser, W.-S. Wu, D.Parrish, and N. Roberts. Numerical aspects of the application of recursive filters to variational statistical analysis. part ii: Spatially inhomogeneous and anisotropic general covariances. *Mon. Wea. Rev.*, **131**:1536–1548, 2003.

[66] R.J. Purser, W.-S. Wu, D. Parrish, and N. Roberts. Numerical aspects of the application of recursive filters to variational statistical analysis. part i: Spatially homogeneous and isotropic gaussian covariances. *Mon. Wea. Rev.*, **131**:1524–1535, 2003.

[67] F. Rabier, A. McNally, E. Andersson, P. Courtier, P. Undén, J. Eyre, A. Hollingsworth, and F. Bouttier. The ecmwf implementation of three-dimensional variational assimilation (3d-var). ii: Structure functions. *Q.J.R. Meteorol. Soc.*, **124**:1809–1829, 1998.

[68] L. Raynaud, L. Berre, and G. Desroziers. Spatial averaging of ensemble-based background-error variances. *Q.J.R. Meteorol. Soc.*, **134**:1003–1014, 2008.

[69] L. Raynaud, L. Berre, and G. Desroziers. Objective filtering of ensemble-based background-error variances. *Q.J.R. Meteorol. Soc.*, **135**:1177–1199, 2009.

[70] S. Remy, O. Pannekoucke, T. Bergot, and C. Baehr. Adaptation of a particle filtering method for data assimilation in a 1d numerical model used for fog forecasting. in revision. *Q.J.R. Meteorol. Soc.*, 2011.

[71] C. Rodgers. *Inverse methods for atmospheric sounding*. World Scientific, 2008.

[72] D. Rosca. Locally supported rational spline wavelets on the sphere. *Math. Comp*, **74**:1803–1829, 2005.

[73] R. Sadourny, A. Arakawa, and Y. Mintz. Integration of the nondivergent barotropic vorticity equation with an icosahedral-hexagonal grid for the sphere. *Mon. Wea. Rev.*, **96**:351–356, 1968.

[74] P. Schröder and W. Sweldens. Spherical wavelets: Efficiently representing functions on the sphere. In *Computer Graphics (SIGGRAPH '95 Proceedings)*, pages 161–172, 1995.

[75] P. Schröder and W. Sweldens. Spherical wavelets: Texture processing. Technical Report 4, Industrial Mathematics Initiative, Department of Mathematics, University of South California, 1995.

[76] W. Sweldens. The lifting scheme: A new philosophy in biorthogonal wavelet constructions. In *Wavelet Applications in Signal and Image Processing III*, pages 68–79, 1995.

[77] W. Sweldens. The lifting scheme: A construction of second generation wavelets. *SIAM journal on mathematical analysis*, **29**:511–546, 1998.

[78] A.S. Sznitman. Topics in propagation of chaos. In *Ecole dt de Probabilits de St Flour XIX - 1989*, pages 165–251. Springer-Verlag, Berlin., 1991.

[79] O. Talagrand and P. Courtier. Variational assimilation of meteorological observations with adjoint vorticity equation. i: Theory. *Q.J.R. Meteorol. Soc.*, **113**:1311–1328, 1987.

[80] A. Tangborn. Wavelet approximation of error covariance propagation in data assimilation. *Tellus A*, **56**:16–28, 2004.

[81] A. Tarantola. *Inverse problem theory (and methods for model parameter estimation)*. Society for Industrial and Applied Mathematics, 2005.

[82] A. Tarantola and B. Vallet. Inverse problem = quest for information. *Journal of Geophysics*, **50**:159–170, 1982.

[83] van der Vaart A.W. and Wellner J.A. *Stochastic Convergence and Empirical Processes*. Springer-Verlag, New York., 1996.

[84] A. Weaver and P. Courtier. Correlation modelling on the sphere using a generalized diffusion equation. *Q. J. R. Meteor. Soc.*, **127**:1815–1846, 2001.

[85] A. Weaver and S. Ricci. Constructing a background-error correlation model using generalized diffusion operators. In ECMWF, editor, *Proc. ECMWF Seminar on "Recent developments in data assimilation for atmosphere and ocean"*, pages 327–340, Reading, UK, 8-12 September 2003.

[86] D.L. Williamson. Integration of the barotropic vorticity equation on a spherical geodesic grid. *Tellus*, **20**:642–653, 1968.

[87] S. Yoden and M. Yamada. A numerical experiment on two-dimensional decaying turbulence on. a rotating sphere. *J.Atmos.Sci.*, **50**:631–643, 1993.

INDEX

A

accuracy, vii, 1, 2, 3, 4, 14, 15, 17, 18, 20, 23, 24, 38, 40, 41, 52, 113, 120, 121, 122, 123, 126, 127, 128, 129, 130, 136, 161, 162, 191, 198, 217, 218, 222
actuators, 170
adaptation, vii, 1, 103
ADC, 60, 68
aerospace, 162
Africa, 368
age, 97, 98, 99, 108, 257
aggregation, 56, 257
Air Force, 319
air temperature, 139, 141, 149
Alaska, 121, 123
Algeria, 161
alpha activity, 239
amplitude, 50, 52, 53, 55, 57, 58, 61, 67, 70, 71, 244, 271, 322
Amplitudes, 62
anisotropy, 348, 350, 351, 352, 353, 356, 361, 363, 365, 367, 368
annealing, 358
ANOVA, 102
anthropologists, 96
anthropometric characteristics, 97
APA, 283
appraisals, 130
aquifers, 223
Artificial Neural Networks, 253
ASEAN, 302
Asia, 290, 299, 300, 301, 302, 304, 307
Asian countries, 302
assessment, 82, 99, 158, 159, 250
assimilation, viii, 99, 135, 136, 137, 138, 139, 144, 145, 154, 155, 157, 158, 159, 321, 323, 324, 326, 327, 338, 343, 345, 346, 347, 348, 353, 357, 370, 371, 372, 373, 374, 375, 376
asymmetry, 116
asymptotic standard errors, 103
asymptotics, 373
atmosphere, 313, 315, 321, 322, 323, 325, 370, 371, 373, 376
attitude measurement, 162

B

ban, 287
bandwidth, 269, 287, 288
base, 4, 14, 74, 136, 140, 142, 149, 157
benefits, 41, 112, 302
bias, 39, 152, 238, 257
biological processes, 292
biological systems, 297
blood pressure, 291
BMA, 138
body size, 96
Boeing, 121, 123
bounds, 91, 279
brain, 246
Brazil, 256, 261, 262
Brazilian price index, x, 255, 256, 267
Brownian motion, 291, 339

C

cables, 54
calculation process, vii, 1, 2, 3, 5, 18, 19, 29, 31, 40, 41
calibration, 7, 108, 142, 151, 154
CDC, 109
cell size, 3, 4
central bank, 290

certificate, 307, 308, 309
challenges, xi, 3, 158, 269, 321
chaos, 329, 376
chemical, 290, 371
children, viii, 77, 97, 98, 101, 108
China, 269, 300, 302, 306, 310, 311, 374
circulation, 239, 240, 313
class, x, 79, 81, 125
classes, 198
classification, 49, 52, 53, 74, 75, 76, 158
climate, 136, 149, 151, 157, 313, 322
closure, 362
clustering, 157
commercial, 2, 3, 40, 54, 101
communication, x, 17, 54, 144, 198, 215, 269, 271, 276, 279, 286, 288
communication systems, 54, 198, 215
community, x, 137, 269
company stock, 115, 117, 118, 123
comparative analysis, 121, 130
compatibility, 73, 76
compensation, 51, 60, 61, 62, 339
competition, 123, 257, 258
complement, 282
complex numbers, 248
complexity, 3, 63, 70, 99, 101, 245, 322
compliance, 97
composition, 258
computation, ix, 17, 50, 58, 65, 75, 81, 139, 161, 183, 246, 322, 337, 344, 348, 349, 350, 352, 361, 368, 369
computational capacity, 272
computational grid, ix, 135, 139, 149
computer, 49, 53, 54, 73, 74, 107, 163, 170, 262, 343
computing, viii, 17, 50, 60, 70, 77, 90, 107, 135, 142, 158, 233, 272, 317, 361
conditional mean, 114, 117, 118, 336, 337
conditioning, 114, 336
conductivity, 141, 142, 143
configuration, ix, 161, 180, 191, 271
construction, 97, 340, 362, 365, 376
consumer goods, 302
consumption, viii, 2, 4, 16, 78, 101, 102, 103, 106, 108, 109, 247, 264, 265
contiguity, 341
contour, 353
convention, 230, 249, 358, 365
convergence, 119, 142, 145, 270, 277, 283, 286, 325, 328, 329, 330, 333, 338, 339, 340, 341
Coriolis effect, 313
correlation, 41, 42, 85, 91, 93, 97, 99, 103, 104, 106, 107, 116, 125, 151, 213, 229, 233, 234, 236, 238, 262, 310, 346, 348, 349, 350, 351, 352, 353, 354, 356, 357, 358, 362, 363, 364, 365, 367, 368, 369, 373, 375, 376
correlation analysis, 310
correlation coefficient, 151
correlation function, 346, 348, 349, 350, 351, 352, 353, 354, 356, 357, 363, 364, 365, 367, 368, 373
correlations, x, 86, 91, 225, 226, 227, 228, 229, 244, 252, 348, 353, 356, 358, 372, 375
cost, 2, 3, 4, 49, 191, 271, 288, 332, 338, 342, 343, 344, 370
CPU, 247
critical state, 138
Croatia, 76
currency, 255, 316
cycles, 51, 56, 58, 65, 68, 167
cycling, 327, 345, 346
Cyprus, 41
Czech Republic, 77, 102, 107

D

damping, 216, 234
data analysis, viii, 77, 108, 109, 226
data processing, 162, 294
data set, 97, 225, 226, 235, 240, 241, 242, 243, 246, 247, 251
database, 4
datasets, 149
decay, 89
decomposition, viii, 77, 81, 83, 90, 99, 101, 103, 107, 233, 247, 338, 348, 354, 362, 370
deconvolution, 250, 253
defects, 330, 368, 370
deflation, 235
degenerate, 198
Delta Air Lines, 121
demonstrations, 323
dependent variable, 290
depreciation, 258
depth, 139, 140, 141, 142
derivatives, 48, 80, 173, 176, 191, 192
detectable, 283
detection, vii, viii, 43, 49, 50, 51, 52, 53, 64, 65, 66, 68, 69, 70, 72, 74, 75, 76, 149, 158, 198
developed countries, 257
developing countries, 307
deviation, 5, 14, 18, 21, 38, 44, 49, 51, 63, 74, 191, 258, 346, 347, 349, 367
DFT, 55
differential equations, viii, 77, 166, 216, 218, 221, 290, 292, 293, 297, 298, 300, 301, 307, 310, 311, 315, 316

diffusion, 323, 351, 354, 363, 364, 365, 368, 369, 370, 375, 376
directional antennas, 4
discharges, 142
discretization, 239, 325, 329, 338, 344, 349, 360, 370, 373
dispersion, 344, 346
displacement, 335, 350, 352, 353, 365, 367
distortion, vii, 43, 44, 45, 49, 50, 51, 53, 54, 55, 56, 57, 58, 61, 243
distributed modeling, viii, 135
distribution, 6, 7, 13, 17, 28, 30, 34, 35, 44, 45, 50, 51, 53, 54, 55, 56, 63, 66, 67, 68, 72, 73, 74, 75, 76, 78, 80, 107, 116, 119, 123, 124, 125, 126, 127, 128, 129, 138, 145, 146, 226, 239, 241, 243, 244, 246, 272, 273, 323, 324, 325, 326, 327, 328, 329, 344, 358, 364
distribution function, 145
disturbances, vii, 43, 44, 45, 63, 73, 76
DNA, 297
DOI, 157
domestic economy, 257
dominance, 128, 129
drainage, 139
drought, 136
dynamical systems, 247, 252, 253, 291

E

economic crisis, 102
economic growth, 302
economic power, 301
economic theory, 261
eigenvalues, 27
EKG, 291
electric power, vii, 43, 44, 45, 47, 49, 58, 75
electricity, 44, 73
electroencephalogram, 239
emergency, 49
empirical studies, 112
employment, 3, 16, 17, 21, 306
energy, ix, x, 55, 82, 161, 269, 313, 336, 354, 366
energy density, 366
engineering, 119, 198, 294, 323
environment, 2, 4, 6, 14, 18, 41, 158, 258, 260, 290, 299
environmental variables, ix, 135
enzyme, 297
equality, 274, 282, 337
equilibrium, 114, 118
equipment, 44, 53, 54, 56, 63, 76
equity, 112, 116, 123
equity market, 112, 123

error estimation, 181, 183
estimation problems, 163, 198, 288
estimation process, 3, 14, 17, 19, 85, 170, 256
eukaryotic, 297
Europe, 290
evapotranspiration, ix, 135, 139, 140, 141
evidence, 113, 120, 121, 125, 128, 130, 131, 256, 258, 262, 264, 265, 300
evolution, viii, 44, 45, 56, 58, 78, 185, 259, 262, 291, 321, 322, 324, 326, 327, 334, 340, 341, 347, 366, 367, 370
evolutionary computation, 139
exchange rate, viii, x, 77, 82, 88, 89, 112, 113, 255, 256, 257, 258, 260, 261, 262, 263, 264, 265, 267, 268, 291, 307, 308, 316
exchange rate policy, 257
experimental design, 138
exports, 258, 261, 302
extended recursive Wiener fixed-point smoother and filter, ix, 197, 198, 204
external influences, 82, 103
external shocks, 255, 257
extraction, 50, 74, 297, 300, 306, 310

F

factor analysis, 250, 251
fault detection, 49, 74, 75
faults, vii, 43, 44, 49, 50, 51, 63
feature selection, 297
FFT, 76
filament, 368
filters, vii, 43, 49, 51, 52, 56, 60, 66, 67, 68, 74, 75, 137, 162, 198, 217, 224, 270, 272, 284, 286, 287, 288, 294, 338, 357, 358, 359, 371, 375
filtration, vii, viii, 77
financial, xi, 112, 130, 289, 300, 306, 307, 316
financial crisis, 306, 307
fingerprints, 367
fixed-point smoother, x, 197, 198, 199, 202, 222, 223
flexibility, 3, 15, 107
flight, 180
floods, 136
fluctuations, 50, 63, 74, 116, 291
force, 138, 337
Ford, 121
forecast errors, viii, 111, 113, 122, 123, 125, 126, 127, 128, 129, 130, 374
forecasting, viii, 108, 109, 111, 112, 113, 120, 121, 122, 125, 126, 127, 128, 129, 130, 136, 137, 138, 147, 149, 155, 156, 157, 158, 251, 256, 257, 342, 375

formation, 225, 270, 366
formula, 296, 332, 339, 348, 349, 350
Fourier analysis, 44, 51, 55, 56
France, 257, 317, 319, 321, 375
free trade, 302
freedom, 102, 322, 325, 338, 342, 358
frequencies, x, 4, 55, 113
functional analysis, 362
fusion, 324, 370

G

GARCH models, viii, 111, 112, 113, 114, 116, 119, 121, 123, 124, 125, 126, 127, 130, 131
Gaussian random variables, 337
GDP, 261
gene expression, 291, 294, 297, 299, 300
gene regulation, 297
genes, x, 289, 290, 297, 298, 299, 300
geometry, 251, 355, 372
Germany, 157, 225, 257, 310, 311, 319
GNP, 258
GPS, 2, 4, 5, 14, 17, 41
grants, 247
gravity, 163, 164, 180
Greece, 1, 42
growth, viii, 77, 96, 97, 98, 99, 100, 101, 108, 109, 340, 347
growth spurt, 98
GSA, 137

H

harmonics, viii, 43, 49, 50, 51, 53, 54, 55, 56, 58, 60, 67, 73, 75, 76, 354, 356, 357, 375
health, 96, 109
health status, 96
height, 6, 97, 98, 99, 100, 108, 368
heterogeneity, 324, 352, 365, 368
heteroscedasticity, 113, 117
high impedance faults, 50
histogram, 235, 371
historical data, 108, 147, 148
homogeneity, 348, 356
Hong Kong, 108, 300, 301, 302
Hopf equation, 198
House, 41
human, x, 99, 239, 246, 289, 290, 324, 371
human body, 290
human brain, 246
human ignorance, 324
hybrid, 2, 3, 4, 51, 137

hypothesis, 107, 121, 122, 123, 130, 261, 262, 273, 338
hypothesis test, 107, 121, 130
hysteresis, 50

I

identity, 47, 57, 67, 79, 106, 232, 235, 248, 330, 337, 345, 358, 364, 368
impacts, 116
imports, 258, 261
improvements, 137, 138
independence, 85, 98, 102, 103, 105, 228, 232, 244, 246, 329, 333, 337
Independence, 106, 243
Independent Components State Space Model (IC-SSM), x
independent variable, 290
individuals, 98, 99, 100, 102, 103, 105, 107, 142
Indonesia, 290, 302, 307, 309, 310, 311
industries, 257
industry, 101, 120, 258
inequality, 281, 282
inertia, 164, 171, 173, 186
inflation, 256, 257, 258, 261, 290, 334, 335, 347
inflation target, 257, 261
information dynamics, xi, 321, 324
information matrix, 81
information retrieval, viii, 78
information technology, 54, 56, 63
Information Technology, 76
infrastructure, 302
ingredients, 353, 357
initial state, 78, 145, 249, 259, 271, 284, 343, 345, 366
injections, 51
insulation, 102, 103
integration, 169, 218, 221, 239, 306, 310, 333, 334, 364, 365, 366, 367
intercepts, 119
interdependence, 300, 309
interest rates, 307
interference, 54
interharmonics, viii, 43, 53, 55, 76
interruptions, vii, viii, 43, 44, 63
inversion, 81, 351
investment, 2, 112, 130, 131
investment appraisal, 112, 130
investors, 112, 113, 116, 130, 302
Iowa, 73
issues, 2, 49, 52, 56, 137, 154, 257, 347
Italy, 76, 257
iteration, 27, 167, 275, 330, 341, 342, 352, 359

Index

J

Japan, 135, 197, 225, 257, 301, 302, 310, 311, 317, 374

K

Korea, 302

L

landscape, 371
laws, 291, 329, 335, 336, 338, 340, 341
lead, xi, 2, 23, 38, 114, 152, 243, 270, 289, 290, 297, 300, 307, 315, 329, 341, 347, 351, 357, 370
leakage, 55
learning, 233, 252, 253, 299, 302, 304, 338
learning process, 302
Least squares, 287
lifetime, 54, 76, 163
light, 257, 279, 281, 319
linear function, 141, 142, 144
linear model, 81, 98, 106, 109, 113, 177, 227, 294, 341
linear systems, 52, 287, 288, 294
linearity, 79, 99
localization, 137, 372
LTD, 268
Luo, 269, 287

M

machinery, 80
macro economic factors, xi, 289
macroeconomic policy, 255
macroeconomics, 112, 291, 306, 307, 309, 316
magnetic field, 187
magnetic resonance, 250
magnetic resonance imaging, 250
magnetometers, 161, 163
magnitude, viii, 43, 44, 49, 50, 51, 52, 53, 54, 55, 56, 57, 58, 62, 63, 64, 65, 66, 68, 69, 70, 71, 72, 73, 74, 116, 152, 179, 183, 185, 191, 333, 335, 346, 352, 353, 356, 366, 368, 370
Malaysia, 301, 302
mapping, 293, 322, 331
marginal distribution, 326
marginalization, 81
Markov chain, 137, 326
Maryland, 109
mass, 139, 148, 149
matrix algebra, 106
matrixes, 20, 25, 52
maximum likelihood estimate (MLE), viii, 77
measurements, vii, viii, x, 1, 2, 4, 5, 7, 14, 15, 17, 18, 19, 21, 25, 26, 28, 29, 40, 45, 58, 60, 68, 78, 83, 84, 85, 88, 89, 90, 91, 92, 94, 97, 98, 99, 107, 135, 137, 144, 145, 149, 162, 167, 172, 173, 269, 270, 291, 292, 296, 297, 302, 304, 316
media, 158, 159
median, 5
medical, 291
melt, ix, 135, 139, 140, 142, 151, 154
memory, 58, 60, 68
meridian, 356
messages, 272
metaphor, 142
meter, 102, 103, 149
methodology, viii, 75, 77, 108, 118, 121, 138, 149, 152, 156, 157, 290, 291, 292, 293, 295, 316, 317
Metropolis algorithm, 137, 159
microeconomics, 112
microsatellites, vii, ix, 161, 162, 191
Microsoft, 121, 123
military, vii, 269
missions, 290
mixing, x, 225, 226, 228, 229, 234, 235, 239, 241, 243, 245
MMA, 74
mobile communication, 1, 5
model specification, 112
modeling, vii, viii, x, 77, 78, 101, 108, 135, 145, 157, 158, 159
modelling, 49, 52, 114, 118, 157, 198, 290, 292, 293, 294, 297, 299, 300, 301, 303, 311, 315, 316, 317, 373, 375, 376
Moderate Resolution Imaging Spectroradiometer, 138
modernization, 2
modification, 3, 4, 116, 279
MODIS, 137, 138, 139, 154
modulus, 249
moisture, 137, 138, 140, 141, 142, 148, 149, 150, 313
moisture state, 148, 150
molecules, 297
momentum, 163, 164, 171, 173, 186, 187
monetary policy, 257, 261
monitoring, 2, 15, 58, 67, 68, 73, 96
Montana, 158
motivation, 338
multiples, 53, 55, 65
multiplication, 83, 106, 176, 229

N

National Operational Hydrologic Remote Sensing Center's (NOHRSC), ix, 135, 138
National Research Council, 158
natural gas, viii, 78, 108, 109
navigation system, 15
negative relation, 303
neural network, 50, 74
neuroscience, 226
New England, 268
New Zealand, 134, 289, 290, 301, 304, 314, 315, 316
nodes, x, 269, 302
noise, vii, ix, x, 13, 25, 26, 40, 45, 46, 49, 50, 51, 52, 57, 74, 83, 89, 103, 144, 148, 162, 167, 168, 169, 170, 171, 172, 174, 176, 178, 179, 180, 181, 197, 198, 199, 201, 202, 204, 206, 210, 211, 213, 214, 215, 216, 217, 219, 220, 221, 222, 223
nonlinear dynamics, 239, 243
non-linear equations, 47
nonlinear systems, 198, 269
normal distribution, 17, 78, 119, 123, 124, 271
null hypothesis, 122

O

oil, 137, 302
omission, 114
one dimension, 348, 357, 363
open economy, 255, 258
openness, 258, 260, 261
operations, 163, 275
optimal performance, 276
optimization, 3, 40, 41, 81, 90, 95, 97, 99, 103, 137, 157, 158, 159, 167, 246, 247, 270, 271, 276, 277, 279, 338, 342, 358
orbit, vii, ix, 161, 162, 164, 165, 166, 172, 173, 179, 180, 186, 187, 188, 189, 190, 191, 194
orthogonality, 107, 200, 336
oscillation, 341, 352
output gap, 258

P

Pacific, 109, 136, 149, 156, 290, 299, 300, 301, 302, 304
parallel, 17, 99
parameter estimates, 250
parameter estimation, 107, 109, 137, 138, 157, 158, 159, 269, 292, 295, 376
partial differential equations, 374
partition, 230, 324
PCA, 226, 246, 355
PCM, 287
percolation, 139, 140, 142
performance, vii, x, 1, 3, 12, 13, 14, 18, 23, 24, 26, 29, 34, 38, 39, 40, 44, 49, 51, 52, 60, 63, 65, 68, 70, 71, 72, 103, 105, 112, 113, 120, 121, 122, 123, 125, 127, 128, 129, 130, 137, 156, 181
periodicity, 245
personal computers, 54
PET, 141
pitch, 162, 163, 165, 173, 180, 186, 187
platform, 163, 284, 285
plausibility, 260
polar, 313
polarity, 226, 234
policy choice, 267
poor performance, 246
portfolio, 111, 114, 119
POWER, 74, 75, 76
precipitation, 136, 137, 138, 139, 140, 141, 147, 149, 313, 316
predictability, 374
prediction models, 6, 7, 138
present value, 119, 292
price changes, 114
price index, x, 117, 255, 256, 260, 261, 267
principal component analysis, 355
principal component regression, 149
principles, 14, 49, 51
prior knowledge, 146
probability, 5, 108, 145, 152, 225, 228, 232, 235, 270, 271, 273, 276, 291, 323, 324, 325, 326, 327, 329, 330, 335, 336, 340, 341, 370
probability density function, 5, 271
probability distribution, 145, 152, 225, 228, 232, 323, 324, 326, 370
probability theory, 291, 324
project, 41, 109, 149, 296
propagation, 2, 6, 7, 17, 41, 167, 273, 329, 338, 344, 376
protection, 49, 54, 74
protection systems, 54
pruning, ix, 136, 152, 156

Q

quality standards, 44, 58
quantization, 269, 270, 271, 272, 273, 275, 276, 277, 279, 280, 281, 286, 287, 288
Queensland, 319

R

radiation, 82, 107, 109, 141, 164
radio, 2, 6, 41
radius, 13, 14, 23, 316, 333, 348, 354, 364, 366
radon, viii, 77, 82, 88, 89, 90, 91, 92, 94, 95, 96, 107, 108, 109
rainfall, 137, 138, 140, 159
random errors, 259
random walk, 103, 119, 291
rate of return, 120
reading, 101, 102, 103, 104, 105
real time, 58, 75, 163, 337, 343
Received Signal Strength (RSS), vii, 1
reconstruction, 234
recovery, 252, 275, 276, 277, 279
regression, ix, 107, 119, 136, 149, 157, 268, 323, 336, 337, 338, 341
regression model, 268
relative prices, 255, 258
relevance, 246, 249
reliability, 2, 44, 301
reputation, 163, 268
requirements, 16, 17, 95, 101, 105, 190, 191, 272
researchers, 131, 306, 309
residuals, 52, 84, 88, 102, 103, 117, 118, 261, 262, 264
resolution, viii, 55, 78, 90, 98, 101, 139, 149, 154, 350, 357, 358, 360, 361, 367, 372
resources, ix, 96, 142, 161, 163
response, 52, 67, 200, 207, 211, 256, 293
restrictions, 102, 259, 260, 309
Richland, 135
risk, 112, 114, 120, 130, 131, 245, 246, 260, 371
risk management, 131
RNA, 297
root, 8, 44, 63, 89, 117, 124, 138, 142, 152, 304, 309, 345, 346, 347
rotations, 162, 357
routes, 251
RTS, 232, 233, 235, 236, 238, 241, 246
rules, 233, 366
runoff, ix, 135, 136, 137, 138, 139, 159

S

SACE, ix, 135, 149
sampling distribution, 369
satellites, 4, 121
saturation, 186, 346
scaling, viii, 78, 226, 230, 361, 362
science, 137, 290, 353
scope, 19
screening, 96
seasonality, 265
second generation, 376
security, 299
sensing, 137, 139
sensitivity, 51, 52, 54, 63, 90, 94, 137, 143, 159, 215, 341, 350
sensors, x, 149, 161, 162, 163, 168, 171, 172, 180, 269, 285
services, 3, 121
shape, viii, 66, 73, 78, 244, 249, 316, 350
showing, 55, 56, 58, 128, 150, 154, 246, 304, 310
signal processing, vii, 43, 44, 45, 60, 61, 68, 70
signals, x, 4, 44, 50, 51, 56, 74, 75, 144, 229, 252, 269, 291
significance level, 125, 126, 127, 128, 129
simulation, x, 13, 23, 29, 31, 38, 41, 60, 123, 164, 180, 181, 197, 199, 218, 222, 227, 233, 234, 238, 239, 240, 241, 244, 245, 247, 272, 292, 298, 299, 300, 371, 374
simulations, x, 151, 225, 234, 243, 244, 245, 246, 250, 284, 285, 286
Singapore, 194, 269, 301, 302, 306, 319
skewness, 124
Slovakia, 102
smoothing, 109, 137, 199, 200, 201, 202, 203, 204, 205, 208, 210, 213, 216, 217, 218, 219, 220, 221, 222, 224, 259, 293
smoothness, 98
Snow Data Assimilation (SNODAS), ix, 135, 138
social welfare, 257
software, 14, 16, 68, 121
SOI, 270, 277, 280, 284, 286, 287, 288
solution, 10, 11, 17, 21, 22, 27, 35, 37, 39, 40, 52, 167, 168, 270, 276, 283, 290, 322, 326, 332, 334, 335, 336, 338, 341, 342, 343, 344, 363, 365, 368, 374
South Africa, 310, 311, 318, 319
South Korea, 301, 302
Southeast Asia, 302
Spain, 43, 76
specifications, 118, 256, 260
spectral component, 53
Spring, 42
stability, 83, 87, 90, 99, 136, 163, 256, 271, 277, 283, 284, 365, 374
stabilization, 235, 288
standard deviation, 5, 14, 18, 21, 38, 330, 334, 348, 349, 362, 366
standard error, 80, 83, 87, 103, 107, 109
state estimation, 51, 52, 75, 79, 80, 81, 144, 145, 170, 224

states, xi, 79, 82, 109, 144, 145, 148, 150, 151, 154, 158, 162, 229, 232, 262, 265, 289, 290, 292, 293, 295, 316, 324, 329
state-space, vii, viii, 26, 77, 78, 79, 80, 81, 82, 83, 90, 107, 108, 144, 198
statistical modeling, viii, 77, 108
statistical processing, 22
statistics, 103, 108, 109, 124, 125, 145, 146, 251, 346, 355, 358, 363, 365, 366, 372, 375
stimulus, 307
stochastic processes, 119, 243, 250
stock market, xi, 112, 113, 131, 289, 290, 291, 294, 300, 301, 302, 304, 306, 307, 309, 310, 311, 312, 313, 316
stock price, 116, 294, 307
storage, 16, 95, 101, 105, 140, 141, 142, 143, 148, 158
stratification, 101
streamflow forecasting, 136, 137, 138, 147, 149, 155, 158
streams, 149
stress, viii, 77, 97, 107
structure, 60, 61, 81, 98, 102, 106, 112, 116, 119, 124, 130, 136, 137, 138, 227, 229, 230, 232, 233, 235, 238, 240, 243, 244, 245, 246, 252, 253, 256, 259, 290, 291, 366, 374
style, 81, 82, 107
subsidy, 102
subsurface flow, ix, 135, 139
subtraction, 106
surveillance, 101
Swarm Intelligence, ix, 135
Sweden, 158
Switzerland, 73, 76
symmetry, 361
syndrome, 138

T

Taiwan, 301, 302
target, 14, 283, 284, 285, 286, 287, 297
techniques, 1, 2, 3, 4, 5, 14, 15, 17, 18, 19, 73, 76, 113, 136, 137, 145, 157, 162, 246, 323, 371
technologies, 2, 3
technology, vii, 54, 56, 63, 338
telecommunications, 42, 226
telomere, 297
temperature, 101, 108, 139, 140, 141, 142, 147, 149, 291, 313
tempo, 244
temporal window, 326
tension, 373
terminals, 44

test data, 306
test statistic, 123, 126, 127, 128, 129
testing, 111, 123, 132, 261, 301, 302
Thailand, 306, 307
theoretical concepts, 162
tides, 373
time lags, 227
time periods, 5, 101, 131
time resolution, 90, 98, 101
time series, vii, viii, x, 1, 2, 77, 78, 81, 97, 109, 118, 119, 225, 226, 227, 228, 229, 231, 232, 233, 234, 235, 238, 239, 241, 243, 244, 245, 246, 247, 252, 259, 291, 333
topology, 21
torus, 356
total energy, 366
tracks, 62
trade, 3, 256, 257
training, 292, 301, 303, 304, 307
trajectory, 15, 20, 99, 162, 287, 294, 316, 325, 326, 330, 343, 344
transformation, 200, 224, 248, 249, 250, 327, 333, 357, 358, 361, 362
transformation matrix, 249
transformations, 249, 357
translation, 334, 356, 357
transmission, 41, 258, 270, 276
transparency, 257
transport, 61, 101, 323, 327, 328, 333, 344, 352
treatment, 138, 159
triangulation, 4
trucks, 121
turbulence, 322, 371, 374, 376
Turkey, 193, 194, 195, 287

U

U.S. Army Corps of Engineers, ix, 135, 149
United Kingdom (UK), 109, 111, 112, 131, 257, 299, 310, 311, 371, 373, 374, 376
United States (USA), 76, 135, 136, 159, 194, 268, 290, 307, 310, 311
updating, ix, 64, 135, 136, 137, 138, 144, 145, 149, 150, 151, 152, 154, 156, 159, 302, 305
urban, 2, 4, 5, 6, 14, 18, 41
urban areas, 4

V

validation, 338, 370
variables, viii, ix, x, xi, 10, 20, 40, 52, 56, 57, 58, 60, 66, 67, 77, 81, 89, 90, 107, 113, 117, 118, 119,

136, 138, 215, 225, 226, 229, 231, 232, 239, 252, 255, 256, 259, 267, 289, 290, 291, 292, 293, 294, 296, 297, 298, 299, 300, 301, 302, 307, 309, 310, 313, 315, 316, 317, 324, 327, 328, 330, 336
variance-covariance matrix, 118
variations, vii, 44, 256, 257, 358, 363, 368
vegetation, 137, 158
vehicles, 121
velocity, 16, 21, 100, 165, 169, 313, 347
ventilation, 82
vertebrates, 297
violence, 249
vision, 324
volatility, 112, 113, 116, 117, 256, 257, 262, 299
voltage dips, vii, 43, 44, 52, 53, 70, 76
voltage supply, vii, viii, 43, 44, 49, 50, 52, 53, 54, 55, 56, 62, 63, 64, 66, 67, 68, 69, 70, 71, 72, 73, 76

W

Walt Disney, 121, 123
Washington, 76, 158
water, ix, 135, 136, 139, 140, 141, 142, 149, 150, 151, 152, 154, 156, 158, 358, 372
water supplies, 136
watershed, ix, 135, 138, 149, 150, 155
wave number, 358, 366
wave vector, 356
wavelengths, 5
wavelet, 44, 52, 53, 76, 323, 357, 358, 359, 360, 361, 362, 363, 364, 365, 369, 371, 372, 373, 374, 376
wavelet analysis, 44, 52
West Africa, 368
Western Australia, 317
wireless sensor networks, x, 269, 286